TWENTY FIRST CENTURY
science

GCSE Chemistry

Project Directors

Jenifer Burden, Andrew Hunt
John Holman, Robin Millar

Project Officers

Peter Campbell, John Lazonby
Angela Hall, Peter Nicolson

Authors and Editors

Andrew Hunt, Anna Grayson

Contributors

David Brodie, Peter Nicolson
John Holman, Cliff Porter
John Lazonby, Charles Tracy
Allan Mann

OXFORD

Contents

Introduction

Welcome to *Twenty First Century Science*

GCSE Chemistry

This is a course in three parts. The first three Modules look at questions that matter to everyone and which chemistry can help to answer. In these Modules you learn some of the most important chemical explanations. An understanding of chemical reactions helps to show how we are affected by the air we breathe and the food we eat. Chemical explanations allow people to choose more sustainable ways to use materials and grow food. In these Modules you will also learn about how science works to produce reliable knowledge that can be the basis for action and wise decisions.

Modules C4–C6 look more deeply at chemical explanations. Here you see how chemists have discovered that a knowledge of atomic structure can explain the properties of all the elements and their compounds. You meet theories of structure and bonding and develop a greater understanding of the world around you. Also you learn something about the methods that chemists use to make new chemicals.

The final Module (C7) has four topics. The first topic introduces organic chemistry and the importance of carbon compounds. The second topic tackles three challenging questions for chemists: How much?, How fast?, and How far? The third topic is about analytical chemistry, which has a vital part to play in making sure that the food we eat and the air we breathe are safe. The final topic, about green chemistry, shows how modern industry is reinventing the processes it uses to make our use of chemicals more sustainable. Stories of chemists in action in this module show how these chemical ideas are put to use in the twenty-first century.

How to use this book

If you want to find a particular topic, use the **Contents** and **Index** pages. You can also use the **Glossary**. This explains all the key words used in the book.

Each module has two introduction pages, which tell you the main ideas you will study. They look like this:

Why study chemical synthesis?
Why it is useful to know about this topic.

The science
The scientific information you will learn about in this Module.

Chemistry in action
What you will learn from this Module about how science works.

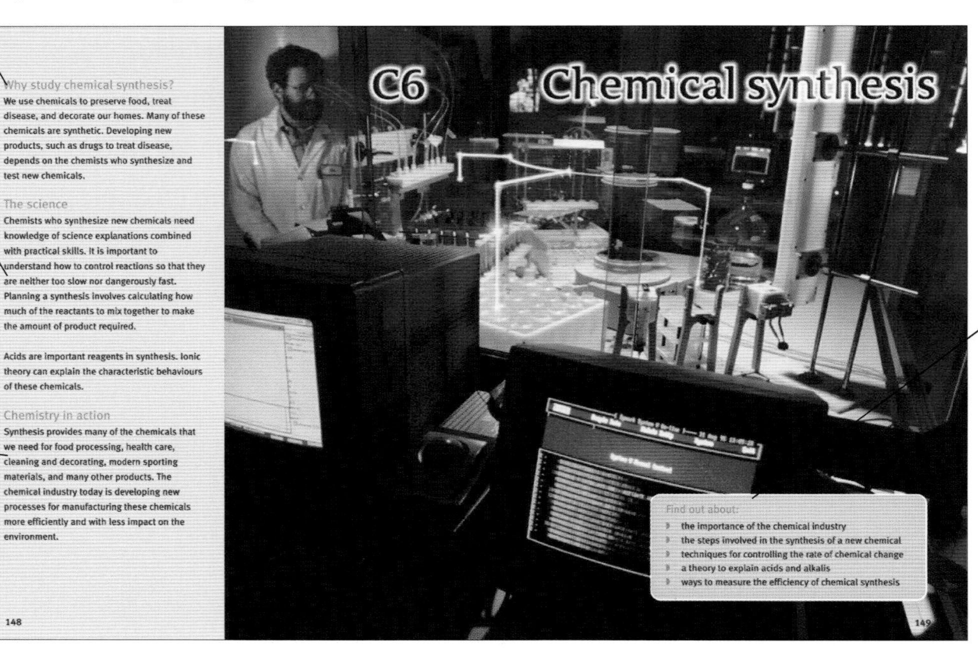

Why study chemical synthesis?

We use chemicals to preserve food, treat disease, and decorate our homes. Many of these chemicals are synthetic. Developing new products, such as drugs to treat disease, depends on the chemists who synthesize and test new chemicals.

The science

Chemists who synthesize new chemicals need knowledge of science explanations combined with practical skills. It is important to understand how to control reactions so that they are neither too slow nor dangerously fast. Planning a synthesis involves calculating how much of the reactants to mix together to make the amount of product required.

Acids are important reagents in synthesis. Ionic theory can explain the characteristic behaviours of these chemicals.

Chemistry in action

Synthesis provides many of the chemicals that we need for food processing, health care, cleaning and decorating, modern sporting materials, and many other products. The chemical industry today is developing new processes for manufacturing these chemicals more efficiently and with less impact on the environment.

148

C6 Chemical synthesis

Find out about:

- the importance of the chemical industry
- the steps involved in the synthesis of a new chemical
- techniques for controlling the rate of chemical change
- a theory to explain acids and alkalis
- ways to measure the efficiency of chemical synthesis

149

Find out about
The main ideas explored in this Module.

Each Module is split into Sections. Pages in a Section look like this:

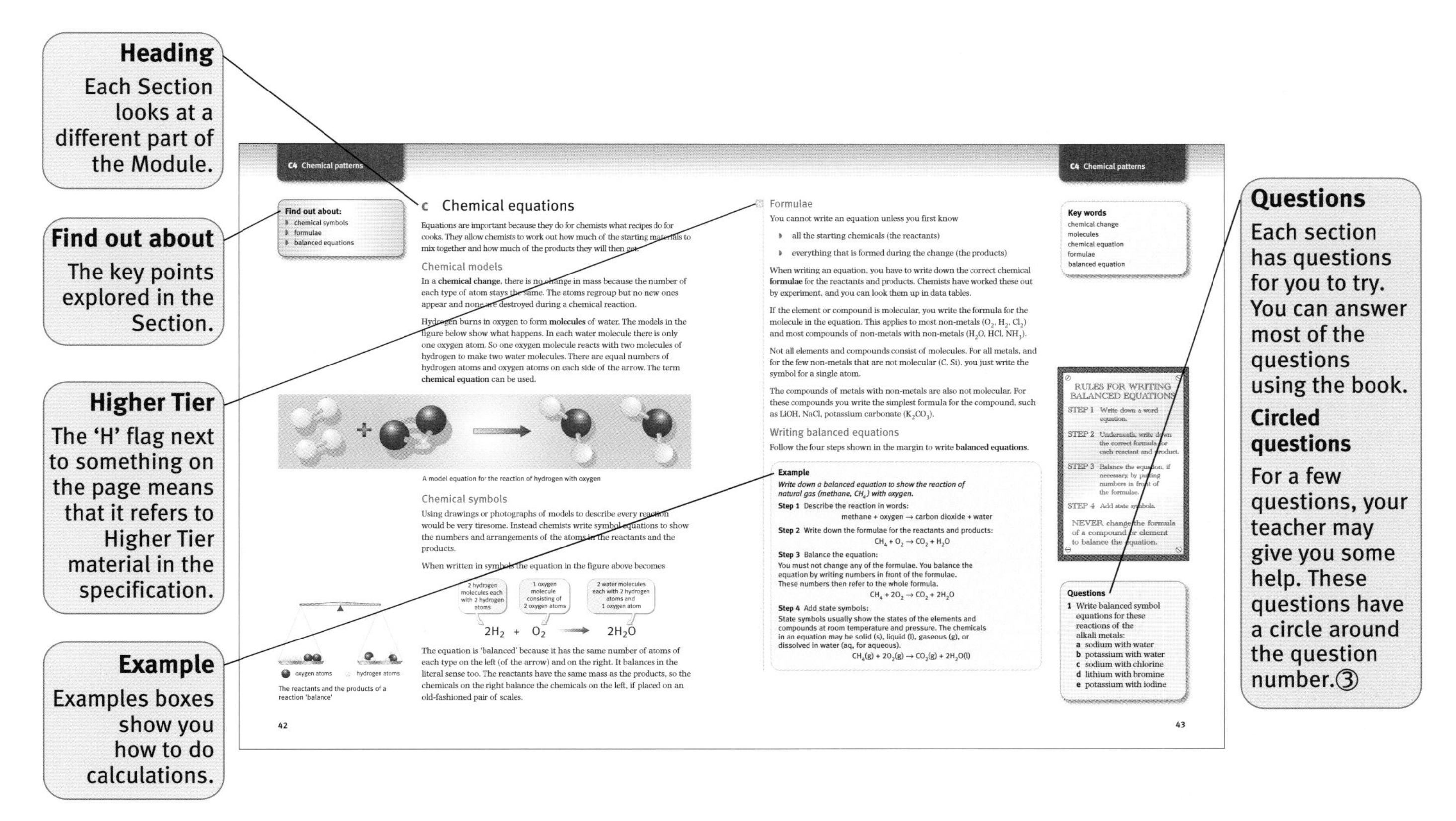

C4 Chemical patterns

Find out about:
- chemical symbols
- formulae
- balanced equations

C Chemical equations

Equations are important because they do for chemists what recipes do for cooks. They allow chemists to work out how much of the starting materials to mix together and how much of the products they will then get.

Chemical models

In a **chemical change**, there is no change in mass because the number of each type of atom stays the same. The atoms regroup but no new ones appear and none are destroyed during a chemical reaction.

Hydrogen burns in oxygen to form **molecules** of water. The models in the figure below show what happens. In each water molecule there is only one oxygen atom. So one oxygen molecule reacts with two molecules of hydrogen to make two water molecules. There are equal numbers of hydrogen atoms and oxygen atoms on each side of the arrow. The term **chemical equation** can be used.

A model equation for the reaction of hydrogen with oxygen

Chemical symbols

Using drawings or photographs of models to describe every reaction would be very tiresome. Instead chemists write symbol equations to show the numbers and arrangements of the atoms in the reactants and the products.

When written in symbols the equation in the figure above becomes

2 hydrogen molecules each with 2 hydrogen atoms | 1 oxygen molecule consisting of 2 oxygen atoms | 2 water molecules each with 2 hydrogen atoms and 1 oxygen atom

$$2H_2 + O_2 \longrightarrow 2H_2O$$

The equation is 'balanced' because it has the same number of atoms of each type on the left (of the arrow) and on the right. It balances in the literal sense too. The reactants have the same mass as the products, so the chemicals on the right balance the chemicals on the left, if placed on an old-fashioned pair of scales.

oxygen atoms hydrogen atoms

The reactants and the products of a reaction 'balance'

42

C4 Chemical patterns

Formulae

You cannot write an equation unless you first know

- all the starting chemicals (the reactants)
- everything that is formed during the change (the products)

When writing an equation, you have to write down the correct chemical **formulae** for the reactants and products. Chemists have worked these out by experiment, and you can look them up in data tables.

If the element or compound is molecular, you write the formula for the molecule in the equation. This applies to most non-metals (O_2, H_2, Cl_2) and most compounds of non-metals with non-metals (H_2O, HCl, NH_3).

Not all elements and compounds consist of molecules. For all metals, and for the few non-metals that are not molecular (C, Si), you just write the symbol for a single atom.

The compounds of metals with non-metals are also not molecular. For these compounds you write the simplest formula for the compound, such as LiOH, NaCl, potassium carbonate (K_2CO_3).

Writing balanced equations

Follow the four steps shown in the margin to write **balanced equations**.

Example

Write down a balanced equation to show the reaction of natural gas (methane, CH_4) with oxygen.

Step 1 Describe the reaction in words:

methane + oxygen → carbon dioxide + water

Step 2 Write down the formulae for the reactants and products:

$$CH_4 + O_2 \rightarrow CO_2 + H_2O$$

Step 3 Balance the equation:
You must not change any of the formulae. You balance the equation by writing numbers in front of the formulae. These numbers then refer to the whole formula.

$$CH_4 + 2O_2 \rightarrow CO_2 + 2H_2O$$

Step 4 Add state symbols:
State symbols usually show the states of the elements and compounds at room temperature and pressure. The chemicals in an equation may be solid (s), liquid (l), gaseous (g), or dissolved in water (aq, for aqueous).

$$CH_4(g) + 2O_2(g) \rightarrow CO_2(g) + 2H_2O(l)$$

Key words
chemical change
molecules
chemical equation
formulae
balanced equation

RULES FOR WRITING BALANCED EQUATIONS
STEP 1 Write down a word equation.
STEP 2 Underneath, write down the correct formula for each reactant and product.
STEP 3 Balance the equation, if necessary, by putting numbers in front of the formulae.
STEP 4 Add state symbols.
NEVER change the formula of a compound or element to balance the equation.

Questions
1 Write balanced symbol equations for these reactions of the alkali metals:
a sodium with water
b potassium with water
c sodium with chlorine
d lithium with bromine
e potassium with iodine

43

Each Module ends with a summary, and some also have questions. Here is an example:

Summary
A checklist of key points explained in this Module.

C4 Chemical Patterns

Summary

You have now met some of the key patterns and theories that chemists use to make sense of the world and to explain how roughly 100 elements can give rise to such a huge variety of chemical compounds.

Atomic structure and the periodic table
- Atoms have a tiny central nucleus surrounded by negative electrons.
- The chemistry of an element is largely determined by the number and arrangement of the electrons in its atoms.
- The number of electrons is equal to the proton number of the atom.

Electrons in atoms
- Electrons in an atom have definite energies.
- The electron shell with the lowest energy fills first until it contains as many electrons as possible, then the next shell starts to fill.

Periodic table
- Arranging the elements in order of their protons numbers gives rise to the periodic table.
- For the first two rows in the table each period corresponds to the filling of an electron shell.
- After calcium the relationship between atomic structure and the periodic table becomes more complex and gives rise to the block of transition elements.

Groups
- Each column in the periodic table consists of a group of related elements.
- The elements in a group have similar chemistries because they have the same number of electrons in the outer shell.
- There are trends in the properties of the elements down a group because of the increasing number of inner full shells.

Atoms into ions
- When metals react with non-metals, the metal atoms lose electrons while the non-metal atoms gain electrons.
- This produces ionic compounds such as sodium chloride, Na+Cl−.
- The properties of an ionic compound are the properties of its ions which behave in a different way from the atoms or molecules in the elements.

Ions into atoms
- Electrolysis turns ions back into atoms.
- At a negative electrode, positive metal ions gain electrons and become metal atoms.
- At the positive electrode, negative ions lose electrons and turn back into atoms.
- These non-metal atoms often join up to make molecules.
- In this way electrolysis splits an ionic compound and turns it back into its elements.

62

Questions

1 Copy this table and extend it to include the first 20 elements.

Element	Proton number	Number of electrons in each shell			
		First shell	Second shell	Third shell	Fourth shell
hydrogen	1	1			
helium	2	2			
lithium	3	2	1		

2 Summarize in outline how the model of atomic structure with electrons in shells can account for:
a the line spectra of elements
b the arrangement of the elements in the periodic table
c the charges on simple ions.

3 This table shows part of the periodic table with only a few symbols included.

period \ group	1	2	3	4	5	6	7	8
1								He
2						O		
3	Na	Mg	Al	Si		S	Cl	
4				Ge			Br	
5								
6	Cs						As	

a Using only the elements in the table, write down the symbols for the following:
i a metal stored in oil
ii two non-metals that are gases at room temperature
iii an element used to kill bacteria
iv a metal that floats on water
v an element with similar properties to silicon (Si)
vi an element that has molecules made up of two atoms
vii the element with the largest number of protons in the nucleus of its atoms
viii the most reactive metal
ix an element with two electrons in the outer shell of its atoms
x an element X that form an ionic chloride with the formula XCl.

b Predict the formula of the compound of sulfur (S) with hydrogen.
c Is astatine (As) a solid, liquid or gas at room temperature? Explain how you decide on your answer.

4 Potassium chloride (KCl) melts at 772 °C and boils at 1407 °C. Crystals of potassium chloride are colourless and shaped like cubes. The compound is soluble in water. A solution of potassium chloride conducts electricity.
a Is potassium chloride a solid, liquid or gas at room temperature?
b Why do crystals of KCl all have the same shape?
c Why does a solution of potassium chloride conduct electricity?
d State two ways that potassium chloride differs from the element potassium.
e State two ways that potassium chloride differs from the element chlorine.
f Why are the properties of potassium chloride so different from its elements?

63

Questions
Some further questions to help bring together what you have learnt in this Module.

Internal assessment

In GCSE Chemistry your internal assessment counts for 33.3% of your total grade. Marks are given for:

- *either* a practical Investigation
- *or* a Case Study and a Data Analysis

Your school or college will decide on the type of internal assessment. You may be given the marking schemes to help you understand how to get the most credit for your work.

Internal assessment (33.3% of total marks)

EITHER: Investigation (33.3%)

Investigations are carried out by scientists to try and find the answers to scientific questions. The skills you learn from this work will help prepare you to study any science course after GCSE.

To succeed with any Investigation you will need to:

- choose a question to explore
- select equipment and use it appropriately and safely
- design ways of making accurate and reliable observations

Your Investigation report will be based on the data you collect from your own experiments. You may also use information from other people's work. This is called secondary data.

Marks will be awarded under five different headings.

Strategy

- Choose the task for your Investigation.
- Decide how much data you need to collect.
- Choose a procedure to give you reliable data.

Collecting data

- Take careful, accurate measurements safely.
- Collect enough data and check its reliability.
- Collect data across a wide enough range.
- Control factors that might affect the results.

Interpreting data

- Present your data to make clear patterns in the results.
- State your conclusions from the results.
- Use chemical knowledge to explain your conclusion.

Evaluation

- Say how you could improve your method.
- Explain how reliable your evidence is.
- Suggest ways to increase the confidence in your conclusions.

Presentation

- Write a full report of your Investigation.
- Lay out your report clearly and logically.
- Describe you apparatus and procedure.
- Show all units correctly.
- Take care with spelling, grammar, and scientific terms.

OR: Case Study and Data Analysis (33.3%)

A **Case Study** is a report which weighs up evidence about a scientific question. You find out what different people have said about the issue. Then you evaluate the information and make your own conclusions.

You choose a topic from one of these categories:

- A question where the scientific knowledge is not certain.
- A question about decision-making using scientific information.
- A question about a personal issue involving science.

Selecting information

- Collect information from a range of sources.
- Decide how reliable each source is.
- Choose relevant information.
- Say where your information came from.

Understanding the question

- Use science knowledge to explain your topic.
- Report on the scientific evidence used by people with views on the issue.

Making your own conclusion

- Compare different evidence and points of view.
- Weigh the risks and benefits of different courses of action.
- Say what you think should be done based on the evidence.

Presenting your study

- Set out your report clearly and logically.
- Use an appropriate style of presentation.
- Illustrate your report.
- Take care with spelling, grammar and scientific terms.

A **Data Analysis** task is based on a practical experiment which you carry out. You may do this alone or work in groups and pool all your data. Then you interpret and evaluate the data.

Interpreting data

- Present your data in tables, charts, or graphs.
- State your conclusions from the data.
- Use chemical knowledge to explain your conclusions.

Evaluation

- Say how you could improve your method.
- Explain how reliable your evidence is.
- Suggest ways to increase the confidence in your conclusions.

Why study air quality?

We breath air every second of our lives. If it contains any pollutants they go into our lungs. If the quality of the air is poor then it can affect people's health.

Chemicals that harm the air quality are called atmospheric pollutants. To improve air quality we need to understand how atmospheric pollutants are made.

The science

Most air pollutants are made by burning fossil fuels. When a fuel burns, the chemicals in the fuel combine with oxygen from the air. They form new chemicals. Some of the new chemicals are air pollutants which escape into the atmosphere. Burning is a chemical reaction. Knowing about chemical reactions helps people understand better what needs to be done to improve air quality.

Ideas about science

Scientists who are trying to improve air quality measure the amounts of pollutants in the air. They use special methods to make sure their data are as accurate as possible. Some scientists use their data to see if they can find a link between air quality and illnesses such as hay fever.

Air quality

Find out about:

- the difference between 'poor', and 'good' quality air
- where the chemicals come from that make the air quality poor
- what can be done to improve air quality
- how scientists collect and use data on air quality
- how scientists investigate links between air quality and certain illnesses

Find out about:

- the gases that make up Earth's atmosphere
- some of the main air pollutants

A The Earth's atmosphere

The Earth's atmosphere provides a protective blanket that supports life. It is a fragile environment that can be damaged easily by pollution.

The table below shows the gases in 'clean' air.

Gas	Percentage by volume
nitrogen (N_2)	78
oxygen (O_2)	21
argon (Ar)	1
carbon dioxide (CO_2)	0.04
water (H_2O)	Variable 0–4

The Earth's atmosphere is just 15 km thick. That sounds a lot but the diameter of the Earth is over 12 000 km. The atmosphere is like a very thin skin around the Earth. The mixture of chemicals it contains is just right to support life. Human activities have altered the balance of these chemicals and this can affect the air quality.

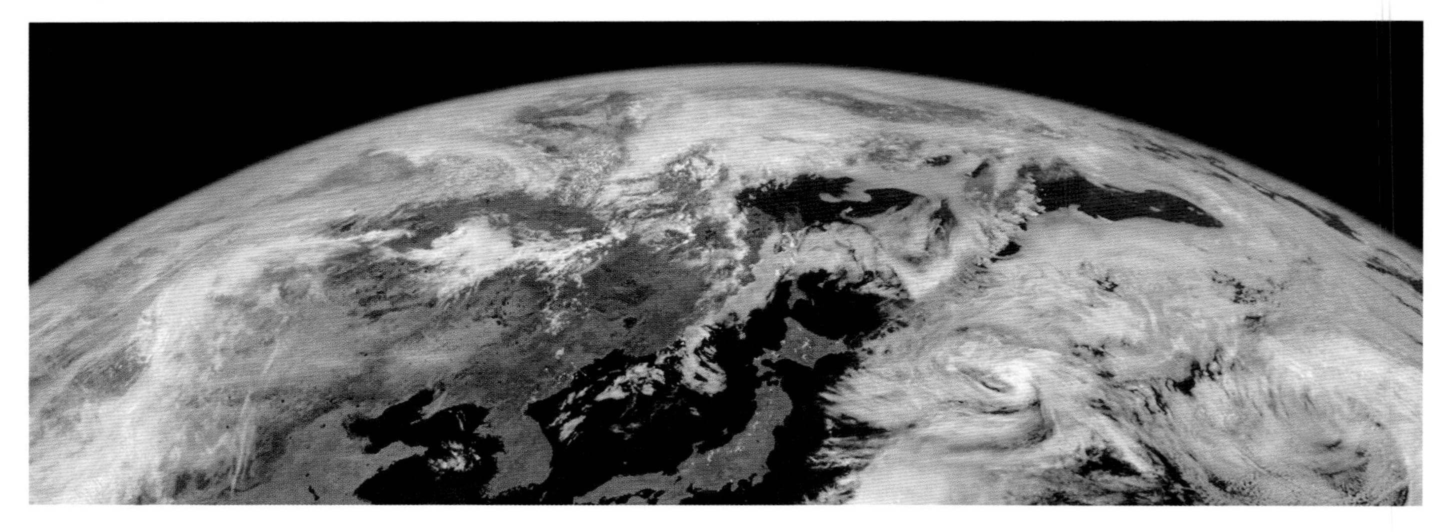

The Earth from space. White clouds of water vapour can be seen in the atmosphere.

Air pollutants

Human activities have been adding **pollutants** to the atmosphere. These include:

- sulfur dioxide (SO_2)
- carbon monoxide (CO)
- nitrogen dioxide (NO_2)
- particulates (microscopic particles of carbon)

A lot of the air pollution comes from burning fossil fuels. The main ones are gas, coal, and oil.

Carbon dioxide and the Earth's temperature

Even though there is only a tiny amount of carbon dioxide in the atmosphere, it helps to keep the Earth warm enough for life.

But the concentration of carbon dioxide in the atmosphere has doubled in the century since humans started burning fossil fuels in huge amounts. Climate scientists are very concerned about possible effects of this increase.

Look at the table below to see the different atmospheric carbon dioxide concentrations on the nearest planets in the solar system.

	Earth	Mars	Venus
Atmosphere	0.04% CO_2	Mostly carbon dioxide but the atmosphere is very thin	97% CO_2 and the atmosphere is very dense
Mean surface temperature	15°C	−53°C	420°C
Effect on life	Just right for life. Water is found as liquid at this temperature	Too cold for life. Any water would be solid ice.	Too hot for life. Any water would boil away.

Questions

1 Oxygen, carbon dioxide, and water vapour are three of the gases in the atmosphere.
- **a** Why do you think oxygen is important?
- **b** Why do you think carbon dioxide is important?
- **c** Why do you think water vapour is important?

2 Draw a bar graph showing the composition of the gases in the Earth's atmosphere. Use data from the table on page 10.

3 List three gases that are air pollutants.

(4) Write a note to a friend suggesting what factors, in addition to the concentration of CO_2, contribute to the differences in the surface temperatures on the planets.

Key words

pollutants

Find out about:

- the most important air pollutants
- the problems pollutants cause
- what can influence the air quality in different locations

B What are the main air pollutants?

Power station cooling towers can look as though they are giving off a lot of pollution. But the white clouds are just condensed water vapour.

Most of the harmful pollutants are usually invisible.

Smoke is a pollutant that can easily be seen. This is because it contains billions of tiny bits of solid. These float in the air. They are called 'particulates'. Smoke makes things dirty. It can give you health problems if you breathe it in.

The table lists air pollutants that scientists are most concerned about.

The clouds coming from the cooling towers are just harmless water vapour. There may be invisible pollutants coming out of the tall chimney.

Smoke contains microscopic particles of carbon. Some of these are just 10 micrometres (10 millionths of a metre) in size. These are called PM10 particles. Although they are very small, they are very much bigger than atoms or molecules. Each particle contains billions of carbon atoms.

Name	What problem does it cause?
sulfur dioxide SO_2 (S, O, O)	Acid rain.
carbon monoxide CO (C, O)	A poisonous gas. It reacts with blood and can kill you.
nitrogen dioxide NO_2 (N, O, O)	Acid rain. Causes breathing problems. Can make asthma worse.
particulates (tiny bits of solid suspended in the air)	Make things dirty. Can be breathed into your lungs. Can make asthma worse. Can make lung infections worse.

How can you find out about air quality?

Some people suffer from asthma or hay fever. They may be able to feel when the air quality is poor. But most people do not know whether the air quality is good or bad.

Some newspapers print a report which gives the day's air quality as a number. Or they may describe it as low, medium, or high quality.

Newspaper reports are quite general. It is often helpful to have more detail. To get this, scientists monitor the air quality. This means that they measure the concentrations of particular pollutants.

You can get more detail about air quality by looking at one of the Government's air quality websites. These give details about the concentrations of individual pollutants.

Some people may have particular reasons for knowing about particular air pollutants.

Name	Concerns/problems
Sulfur dioxide	Some people care a lot about wildlife. SO_2 harms wildlife. It causes acid rain. There is more about acid rain on pages 24–25.
Nitrogen dioxide	Some people suffer from asthma. NO_2 can make asthma worse.
Carbon monoxide	Some people have things wrong with their heart. CO changes the amount of oxygen in the blood. This can make people's heart conditions worse.
Pollen	Some people suffer from hay fever. High levels of pollen can trigger their hay fever.

Measuring the concentration of a pollutant

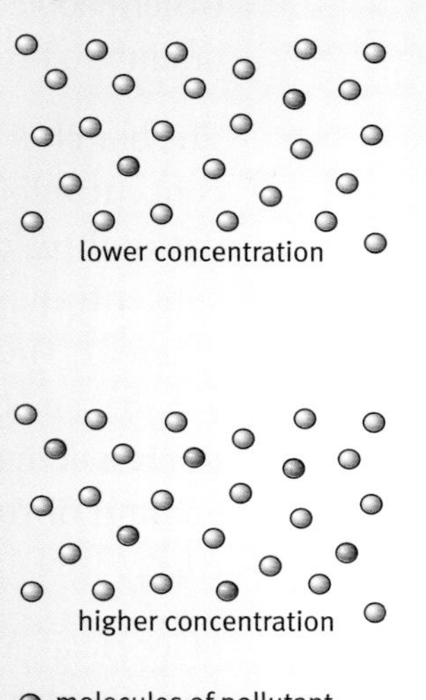

A low concentration of pollutants. There are very few pollutant molecules in a certain volume of air. This is an indication of good air quality

A high concentration of pollutants. There is a large number of pollutant molecules in a certain volume of air. This shows that the air quality is poor.

Concentration is the amount of pollutant in a certain volume of air.

Note: the air molecules are normally much more spread out than shown in the diagrams.

Key words

concentration

Questions

1 Write down one problem that can be caused by each of these air pollutants:

a SO_2

b NO_2

c particulates

2 A newspaper article on air quality included a photograph of white clouds coming out of power station cooling towers. Write a note to the paper explaining why the clouds are not polluting the atmosphere.

Does it matter where you live?

Is the air quality the same all over the country? Some people live in cities. Other people live in the countryside. Will they all have air of the same quality to breathe?

The bar chart shows the concentration of NO_2 on the same day at three different places. The concentrations are clearly different. The concentration of NO_2 depends a lot on the level of human activity in the area. The amount of road traffic has a big effect.

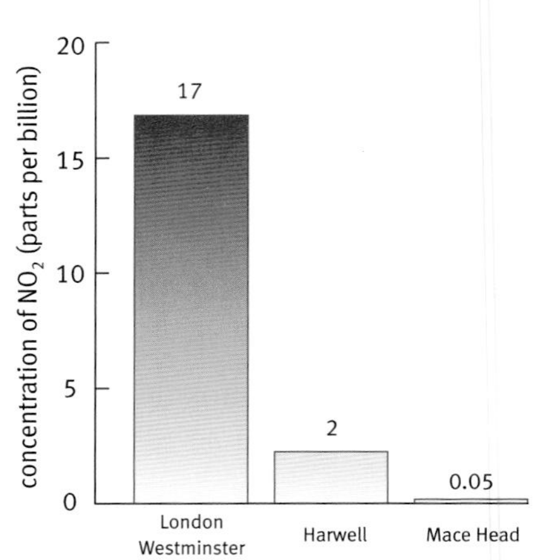

The concentration of NO_2 in three places: London (large city), Harwell (rural part of Britain), and Mace Head (west coast of Ireland).

Mace Head, in Ireland, has very pure air when the wind blows in from across the Atlantic Ocean. Scientists use it as a baseline to see what air would be like without the effects of human activities.

Most of us live in environments where the air quality is much poorer than at Mace Head.

Nitrogen dioxide in London

Nitrogen dioxide levels increase when traffic is heavy. Can you see any patterns in the graph that back this up?

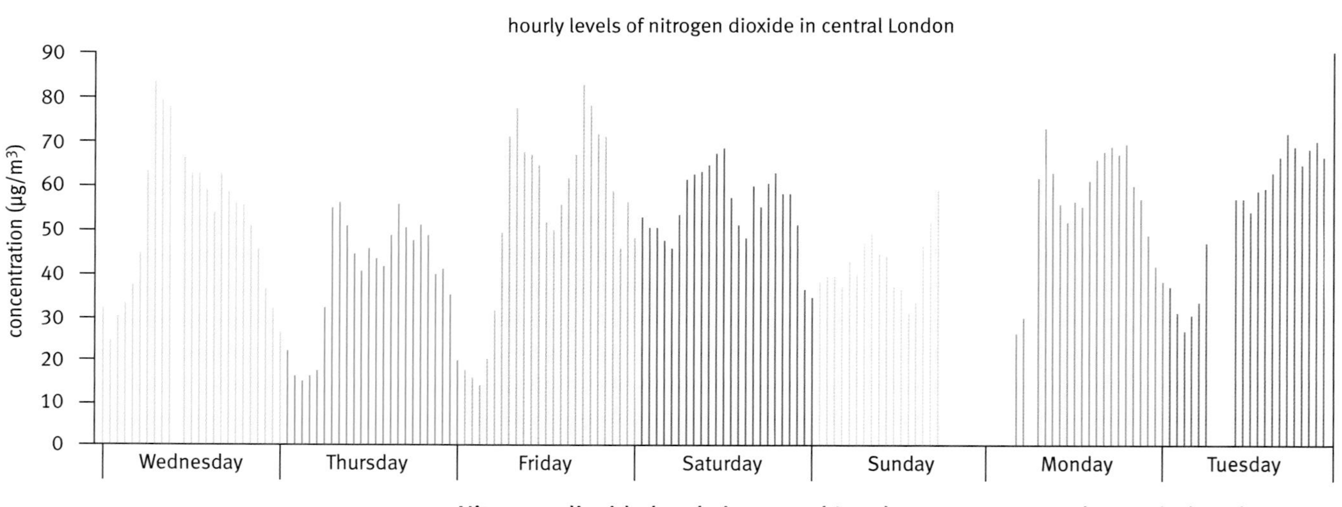

Nitrogen dioxide levels in central London over a seven-day period at the beginning of January 2005.

What influences air quality?

The quality of the air where you live depends mostly on two things.

- **Emissions:** vehicles, power stations, industry, and other sources put pollutants into the air. Homes and factories are called stationary sources. Regulations have greatly reduced pollution from these sources. Most emissions now come from cars and lorries. These are called mobile sources.
- **Weather:** pollutants are mixed up and carried around by the winds. Wind can move pollutants many miles and even carry them from one country to another.

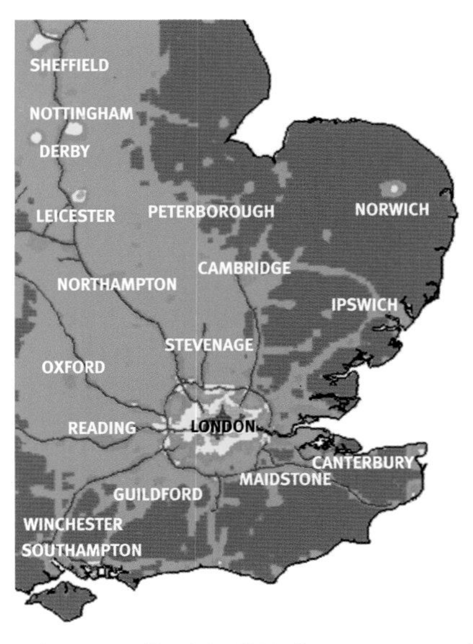

Nitrogen dioxide (NO_2) over parts of England. Red shows the highest concentration of NO_2. Why do you think the high levels follow the motorway routes?

Buildings and air pollution

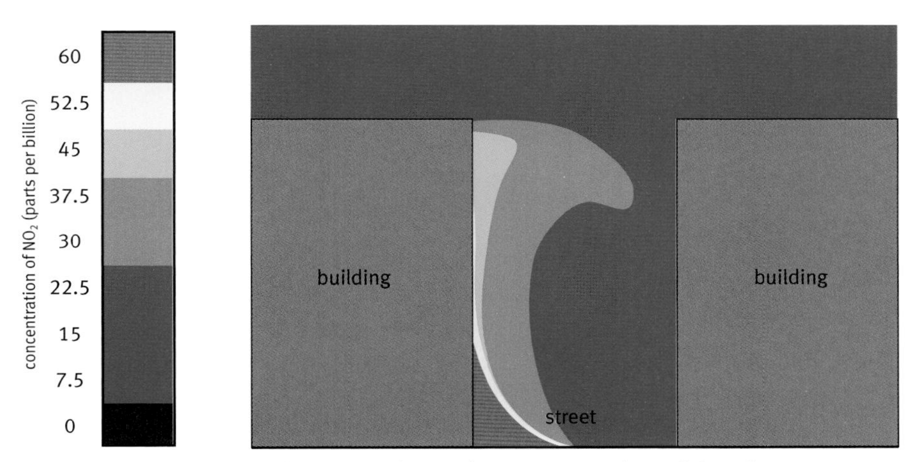

This computer-generated picture shows the concentration of the air pollutant NO_2 in a city street. The red area shows that invisible air currents have channelled the NO_2 onto just one side of the street. The tall buildings on either side make the street into a kind of 'canyon' and trap the NO_2 at street-level.

Ozone and the wind

Some pollutants can react with other gases in the air. This can make new pollutants. Ozone is an example of a pollutant made in this way.

Ozone is formed when the sun shines on polluted air. Polluted air is formed by traffic in towns. It gets carried by the wind into the country. This can happen before the ozone has time to form. This means that ozone concentrations are often higher in the countryside than in the towns.

Key words

emissions
weather

Questions

3. Look at the chart showing levels of NO_2 in London. Suggest reasons for the pattern of readings for Wednesday.
4. The weather moves air pollutants from one place to another. If we reduce air pollution in our own town, we can still get pollution from other areas. Explain why it is still important to try to reduce the air pollution.

Find out about:

- how air quality is measured
- how data are checked and used

C Measuring an air pollutant

If you measure the concentration of NO_2 in a sample of air several times, you will probably get different results. This is because:

- you used the equipment differently
- there were differences in the equipment itself

If you take just one reading, you cannot be sure it is very accurate. So, it is better to take several measurements. Then you can use them to estimate the true value.

The true value is what the measurement should really be. The **accuracy** of a result is how close it is to the true value.

How can you make sure your data are accurate?

The table shows what you should do to get a measurement of the NO_2 level that is as accurate as possible. *Read the table now.*

What you do	Data	Describing what you do
Take several measurements from the same air sample. Not all the measurements will be the same.	Concentration of NO_2 in parts per billion (ppb) 18.8, 19.1, 18.9, 19.4, 19.0, 19.2, 19.1, 19.0, 18.3, 19.3	The measurements (10 in this case) are called the data set.
Plot the results on a graph. This shows that the 18.3 ppb measurement is very different from the others.	19.6, 19.5, 19.4, 19.3, 19.2, 19.1, 19.0, 18.9, 18.8, 18.7, 18.6, 18.5, 18.4, 18.3 this result is an outlier – so leave it out of the mean	A result that is very different from the others is called an **outlier.**
The outlier must have been a mistake so you ignore it. Add the other nine results together. Divide the total by 9. The answer is 19.1 ppb of NO_2.	Total of nine readings = 171.8 $\frac{171.8}{9} = 19.1$ ppb	19.1 is called the **mean value** of the nine measurements.
You can use the mean value rather than any of the nine measurements.	The best estimate for the concentration of NO_2 is 19.1 ppb	The mean value is used as the **best estimate** of the true value.
When you write down the mean value you also record: • the lowest, 18.8 ppb, • and the highest, 19.4 ppb, measurements.	The range is 18.8 ppb – 19.4 ppb	18.8 ppb – 19.4 ppb is called the **range** of the measurements.

The mean value is 19.1 ppb. This is the best estimate of the concentration of NO_2 in the sample of air. You cannot be absolutely sure that it is the true value. But you can be sure that:

- the true value is within the range, 18.8 – 19.4 ppb
- the estimate of the true value is 19.1 ppb

If you had taken only one measurement, you wouldn't have been sure it was accurate. If the range had been more narrow, say 19.0 – 19.3 ppb, you would have been even more confident about your best estimate of the true value.

Comparing NO_2 concentrations

The graph shows the mean and range for the NO_2 concentration in three different places.

- Compare London and York. The means are different but the ranges overlap.
- The range for London overlaps the range for York. So the true value for London could be the same as the true value for York. You cannot be confident that their NO_2 concentrations are different.
- Compare London and Harwell. The means are different and the ranges do not overlap.
- You can be very confident that there is a real difference between the NO_2 concentrations in London and Harwell.

When you compare data, do not just look at the means. To make sure that there is a **real difference**, check that the ranges do not overlap.

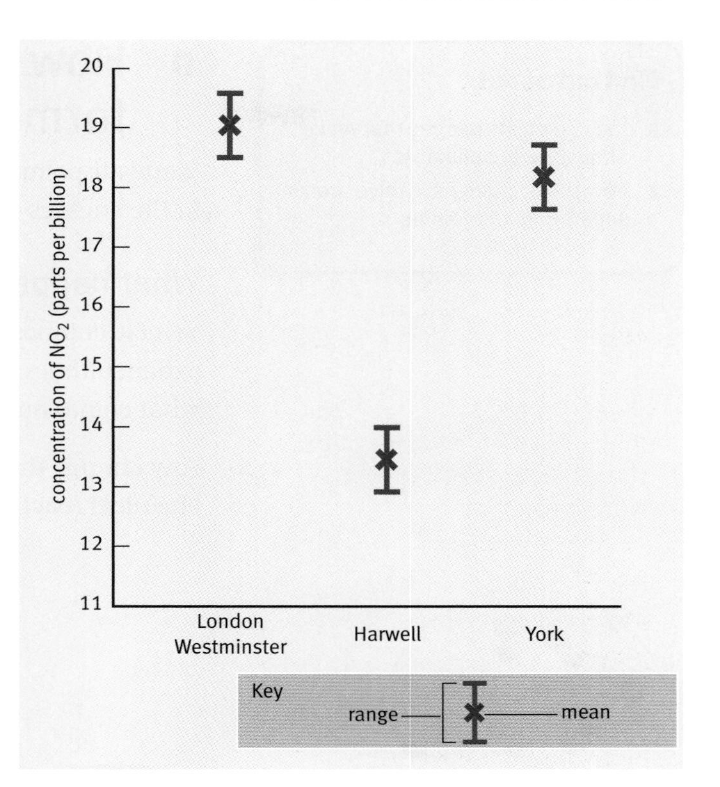

NO_2 concentrations in air from three places in England. All the measurements were made at the same time of day.

Key words

accuracy	mean value	range
outlier	best estimate	real difference

Questions

1 Jess measured the NO_2 concentration in the middle of a town. She took six readings: 22 ppb, 20 ppb, 16 ppb, 24 ppb, 21 ppb, 23 ppb.

a Explain which one of these readings she should decide is an outlier.

b Calculate the mean value of the remaining five measurements.

c Write down the best estimate and the range for the NO_2 concentration in this sample of air.

(2) Look at the graph above. Does it show that there is a real difference in NO_2 levels between Harwell and York? Explain your answer.

3 Repeat measurements on an air sample produced these results for the NO_2 concentration:

Reading 1 – 39.4 ppb Reading 2 – 45.8 ppb
Reading 3 – 42.3 ppb Reading 4 – 38.7 ppb
Reading 5 – 39.7 ppb Reading 6 – 32.7 ppb

a Draw a graph to show the range for these results.

b Work out the mean NO_2 concentration and range for this sample.

(c) Another sample was taken from a second place in the same town. The mean NO_2 concentration for this sample was found to be 44.1 ppb. Can you say with confidence that the second location had a higher NO_2 concentration than the first? Explain your answer.

4 A scientist took one measurement of NO_2 in an air sample. Explain why this would not give you much confidence in the accuracy of the result.

Find out about:

- the chemical changes that make atmospheric pollutants
- how these changes involve atoms separating and joining

D How are atmospheric pollutants formed?

Many air pollutants are made by the burning of fossil fuels. This happens in the engines of vehicles and in power stations.

What happens when fuel burns in a car engine?

Vehicle engines burn petrol or diesel. Fuel and air go into the engine and exhaust fumes come out. Use the diagram to compare what goes in with what comes out.

Any change that forms a new chemical is called a **chemical change** or a **chemical reaction**.

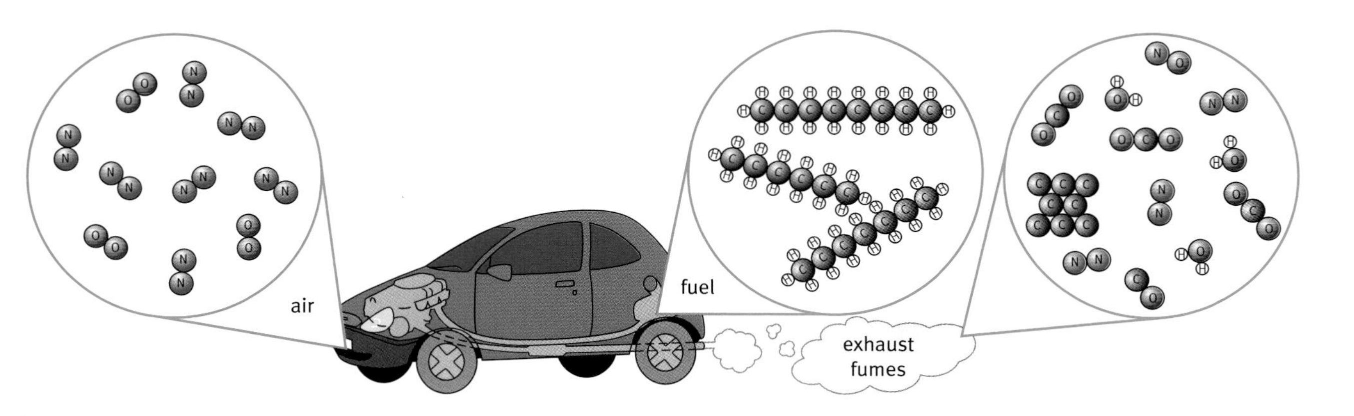

The chemicals going into and coming out of a car engine.

The overall change can be summarized as:

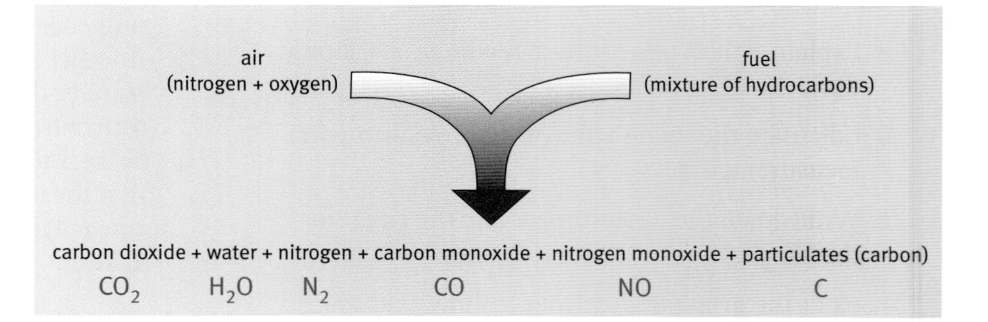

Check that you know which of the pictures in the three circles represents each of the chemicals mentioned in the summary. You can use these pictures to work out the chemical changes happening in the engine: for example, nitrogen monoxide (NO) is one of the new chemicals in the exhaust emissions. It must have been formed from nitrogen (N_2) and oxygen (O_2) in the air. These must have first split apart into **atoms** and then reformed to make nitrogen monoxide (NO).

What happens when fuel burns in a power station furnace?

Most fossil-fuelled power stations burn:

- either coal which is mainly carbon (C)
- or natural gas which is mainly methane (CH_4)

You can compare what goes into a power station with what comes out. Then you can work out some of the chemical changes that take place inside the furnace.

The main product that comes out of the chimney at a coal power station is CO_2. It must have been formed by oxygen atoms in O_2 separating and then combining with carbon atoms.

The main products from the burning of natural gas are CO_2 and H_2O.

These must have been formed by:

- carbon atoms and hydrogen atoms in CH_4 separating
- then carbon atoms combining with oxygen atoms to form CO_2
- and hydrogen atoms combining with oxygen atoms to form H_2O

Burning coal and gas can also produce smaller amounts of these air pollutants:

- particulates – small pieces of unburned carbon
- carbon monoxide (CO) – formed when carbon burns in a limited supply of oxygen
- nitrogen monoxide (NO) – formed when some of the nitrogen in the air reacts with oxygen at the high temperatures in the furnace
- sulfur dioxide (SO_2) – forms if the fuel contains some sulfur atoms

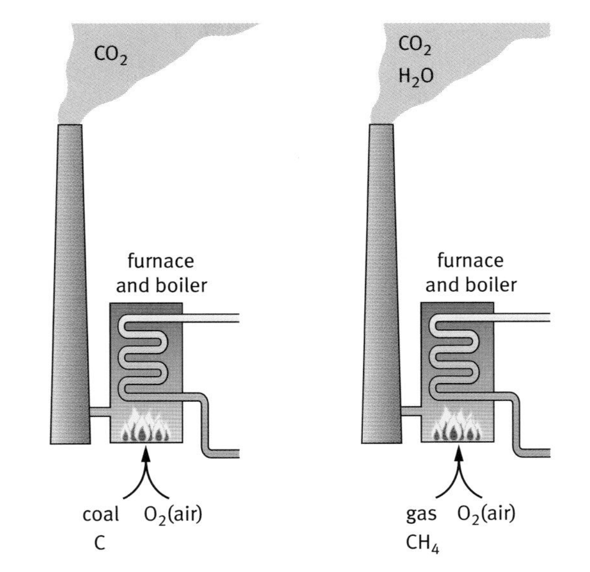

The chemicals going into and coming out of power station furnaces.

Key words

chemical change/reaction
atoms

Questions

1 List the air pollutants released from a car engine when it burns fuel.

2 List the air pollutants that can be released from a coal-burning power station.

3 Use ideas about atoms separating and then joining together in different ways to explain how:

a H_2O forms when methane gas (CH_4) burns in a power station

b CO forms when coal (C) burns in a power station

c CO_2 forms when petrol burns in a car

Find out about:

- the chemical changes involved in combustion
- different ways of representing chemical changes

E What happens during combustion reactions?

Some chemicals can react rapidly with oxygen to release energy and possibly light. This type of reaction is called **combustion** or burning.

A controlled combustion reaction between natural gas (methane) and oxygen occurs when you use a gas cooker.

Fuel has escaped during this racing car crash. An uncontrolled combustion reaction is happening. The fuel and air mixture has been heated by either a spark or the hot engine.

Burning charcoal

Burning charcoal on a barbeque is one of the simplest combustion reactions.

Charcoal is almost pure carbon. You can picture the surface of a piece of charcoal as a layer of carbon atoms tightly packed together.

Oxygen is a gas. All the atoms of this gas are joined together in pairs (O_2). These are called **molecules** of oxygen.

It will help you to understand this reaction if you can picture what happens to the atoms and molecules involved.

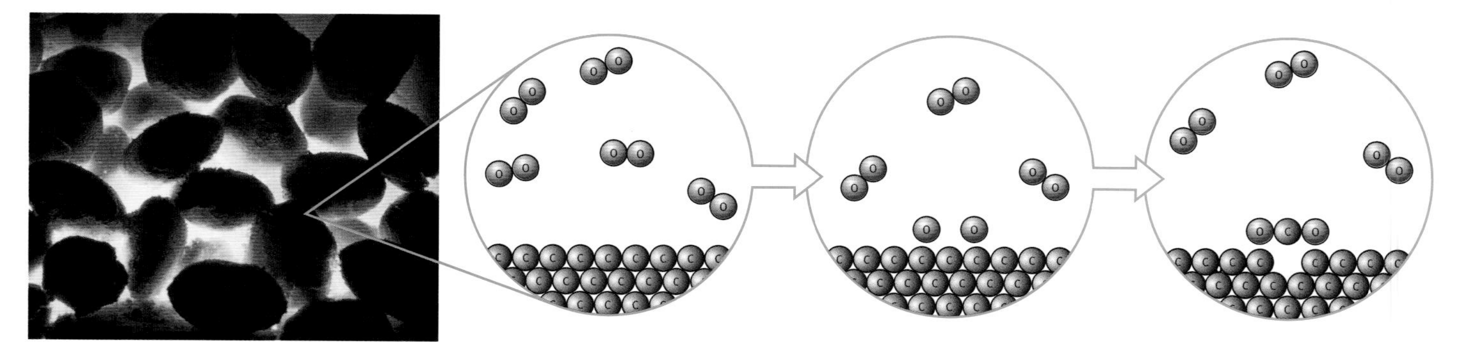

Air contains oxygen gas. One molecule of oxygen is two oxygen atoms joined together (O)(O). Oxygen molecules split and react with carbon atoms in the charcoal. This forms carbon dioxide gas (O)(C)(O).

Describing combustion reactions

You can use pictures to describe the chemical change that happens when carbon dioxide burns.

The chemicals before the arrow are the ones that react together. We call them **reactants**.

The chemicals after the arrow are the new chemicals that are made. We call them **products**.

It would be time consuming if you always had to draw pictures to describe chemical reactions. So scientists use equations to summarize the pictures.

The combustion of charcoal can be summarized in this **word equation**:

carbon + oxygen → carbon dioxide

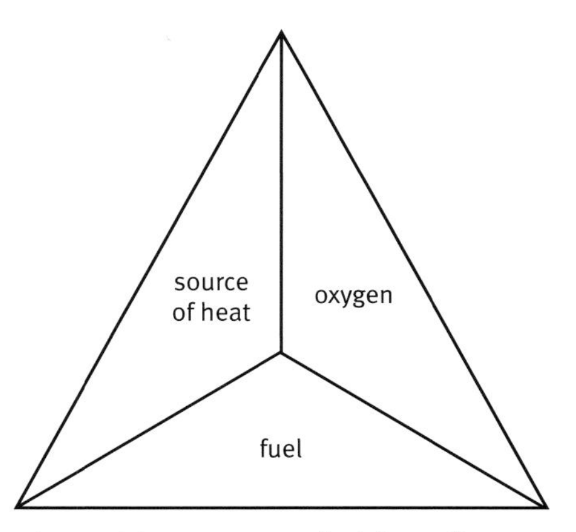

Three things are needed for a fire, or combustion reaction:
- a fuel mixed with
- oxygen (air) and a
- source of heat to raise the temperature of the mixture

If you want more detail, you can write the **chemical equation** to show the atoms that make up each of the chemicals involved. This uses symbols for each chemical. These are called **chemical formulae.**

This is the chemical equation for the combustion of charcoal.

$C + O_2 \rightarrow CO_2$

The chemical equation is a more useful description of the reaction than the word equation. It tells you how many atoms and molecules are involved and what happens to each atom.

Questions

1 What are the reactants and what are the products in each of the following chemical changes:
  a carbon combines with oxygen to form carbon dioxide
  (b) a hydrocarbon in petrol burns in oxygen to form carbon dioxide and water

(2) Draw pictures to represent these chemical changes:
  a hydrogen burning in oxygen to form water
  b methane burning in oxygen to form water and carbon dioxide

(3) You and your cousin are having a barbecue. Your cousin asks you what happens to the charcoal when it burns. Write down what you would say. Include the words: atom, molecule, combustion, reactants, products, chemical change.

Key words

combustion
molecules
reactants
products
word equation
chemical equation
chemical formula

Find out about:

- what happens to atoms during chemical reactions
- how the properties of reactants and products are different

F Where do all the atoms go?

Look at the picture below. How many atoms of hydrogen (H) are there before and after the reaction? Count the atoms of oxygen (O) before and after. What does this show?

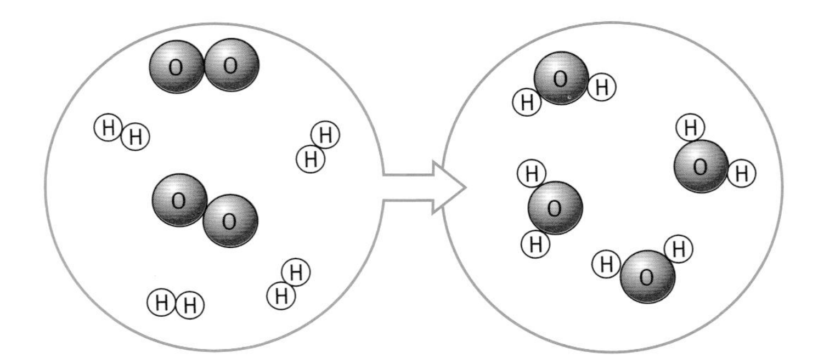

A picture showing the reaction of hydrogen and oxygen to form water.

When you have had a bonfire, some of the atoms that made up the rubbish are in the ashes left on the ground. The others are in the products released into the air.

Conservation of atoms

All the atoms present at the beginning of a chemical reaction are still there at the end. No atoms are destroyed and no new atoms are formed. The atoms are conserved. They rearrange to form new chemicals but they are still there. This is called **conservation of atoms**.

For example, when a car engine burns fuel the atoms in the petrol or diesel are not destroyed. They rearrange to form the new chemicals found in the exhaust gases.

Look again at the picture of hydrogen reacting with oxygen to form water. *Two* molecules of hydrogen react with just *one* molecule of oxygen. This produces *two* molecules of water. We can represent this change by:

Notice that there are the same numbers of each kind of atom on each side of the equation. All the atoms that are in the reactants end up in the products. The atoms are conserved.

Properties of reactants and products

The **properties** of a chemical are what make it different from other chemicals.

For example, some chemicals are solids, some are liquids, and some are gases at normal temperatures. Some are coloured, some burn easily, some smell, some react with metals, some dissolve in water, and so on. Each chemical has its own collection of properties.

The table compares the properties of the reactants and products of the reaction between sulfur and oxygen.

Chemical	Properties
sulfur (reactant)	yellow solid
oxygen (reactant)	colourless gas; no smell; supports life
sulfur dioxide (product)	colourless gas; sharp, choking smell; harmful to breathe; dissolves in water to form an acid

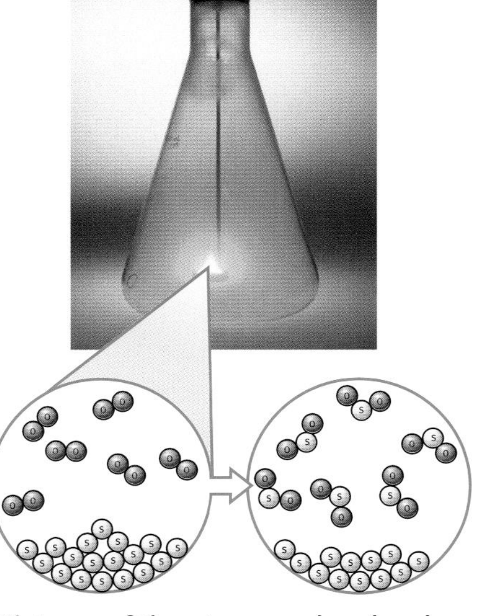

Pictures of the atoms and molecules involved in the burning of sulfur.

In any chemical reaction all the atoms you start with are still there at the end. But they are combined in a different way. So the properties of the products are different from the properties of the reactants.

This is very important for air quality. You can have a piece of coal which is a harmless black stone.

But the coal may contain a small amount of sulfur. When it burns the sulfur will change to the gas sulfur dioxide.

The sulfur dioxide escapes into the atmosphere. It will harm the quality of our air. It will dissolve to form acid rain. This is harmful to plants, animals, and buildings. The harmless piece of coal has produced a harmful gas.

Questions

1 Burning rubbish gets rid of it forever. Is this a true statement? Think about the atoms in the rubbish. Fully explain your answer.

(2) You have learned that atoms are conserved during a chemical reaction. Work out how many molecules of CO_2 and H_2O will be produced when one molecule of methane (CH_4) is burnt. Draw a picture to show the atoms and molecules in the reaction. Work out how many molecules of O_2 will be used.

3 Water (H_2O) is made by reacting molecules of hydrogen and oxygen together. Make a list of some properties of water that are different from the properties of the reactants it is made from.

Key words
conservation of atoms
properties

Find out about:

- what happens to pollutants when they are released into the atmosphere

G What happens to atmospheric pollutants?

Human activity adds pollutants directly to the atmosphere. These are called **primary pollutants**. Examples are:

- particulate carbon (C)
- carbon monoxide (CO)
- nitrogen monoxide (NO)
- sulfur dioxide (SO_2)
- hydrocarbons such as methane (CH_4) and hexane (C_6H_{14})

Some pollutants can chemically react in the air. They produce other chemicals which are called **secondary pollutants**. Nitrogen dioxide (NO_2) is an example of a secondary pollutant.

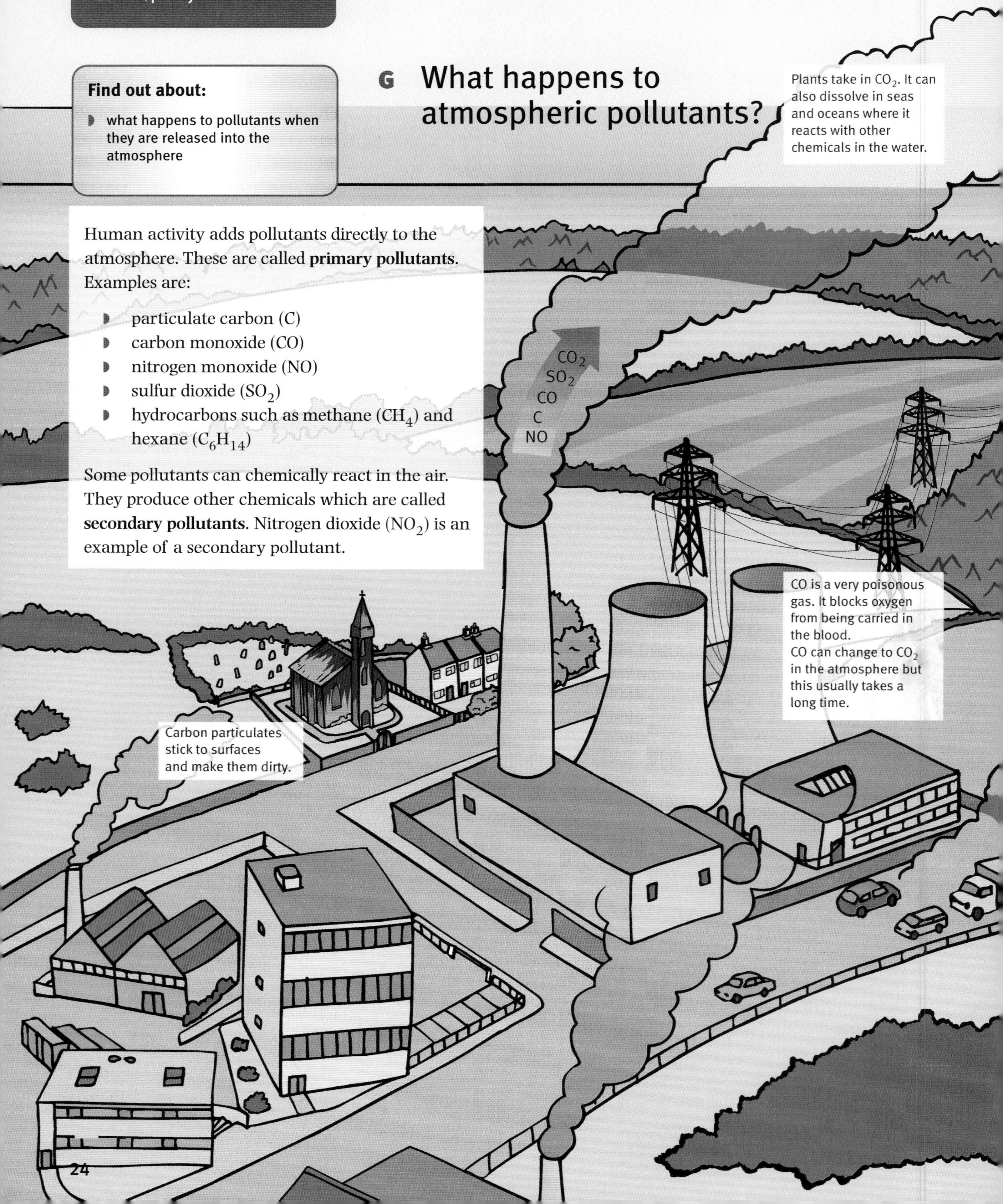

Plants take in CO_2. It can also dissolve in seas and oceans where it reacts with other chemicals in the water.

CO_2
SO_2
CO
C
NO

CO is a very poisonous gas. It blocks oxygen from being carried in the blood.
CO can change to CO_2 in the atmosphere but this usually takes a long time.

Carbon particulates stick to surfaces and make them dirty.

CO_2 SO_2 CO NO_2 O_3

SO_2 and NO_2 react with water vapour in clouds to form 'acid rain'. When it falls, the acid rain can damage plants. It can also make lakes too acidic for fish.

Ozone (O_3) is a secondary pollutant. It forms in the lower atmosphere when sunlight triggers chemical reactions between other pollutants. This happens slowly. Winds may have moved the pollutants to rural areas before the O_3 is formed. O_3 high in the atmosphere helps to shield us from damaging ultraviolet rays. But low level O_3 is a harmful pollutant. It can weaken our immune system and damage our lungs.

Key words
primary pollutants
secondary pollutants

Questions

1 Why are NO and SO_2 called primary pollutants? Why are NO_2 and acid rain called secondary pollutants?

2 Use pictures of atoms and molecules to represent the chemical changes when
a CO reacts with O_2 to form CO_2
b NO reacts with O_2 to form NO_2

3 Read all the information on these two pages. Make a note of some properties of these chemicals: CO, CO_2, SO_2, NO_2.

(4) Do you think CO_2 is an atmospheric pollutant? Give reasons for your answer.

NO comes from vehicle exhaust fumes. It reacts quickly with oxygen in the air to form nitrogen dioxide (NO_2). This happens within a few metres of the vehicle's exhaust pipe. NO_2 is harmful and is an example of a secondary pollutant.

CO_2 is used during photosynthesis. It is essential for plant growth and the start of the food chain. So, in a way, we rely on CO_2 for our food.

Human activity is increasing the amount of CO_2 in the atmosphere. This could lead to global temperatures rising too high. It may have dangerous effects like climate change. It may also cause cause sea levels to rise.

Find out about:

- how to look for links between air quality data and the symptoms of an illness
- how pollen causes hay fever
- the link between asthma and air quality

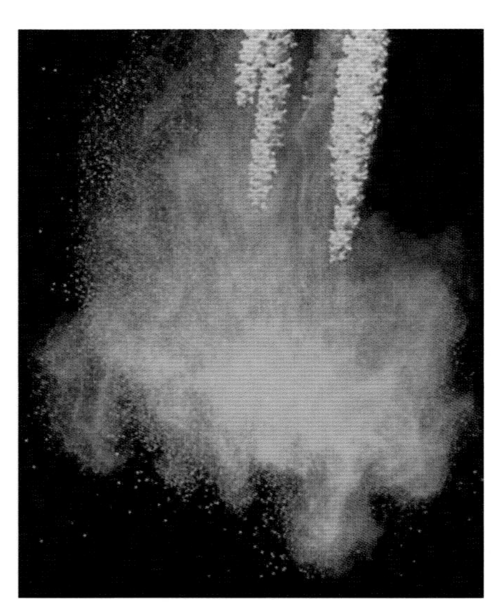
Pollen is released by plants and may travel many kilometres on the wind. Pollen grains are in the air that we breathe.

H How does air quality affect our health?

Hay fever

Do you suffer from a runny nose, sneezing, and itchy eyes in the summer? This could be hay fever.

Hay fever got its name because people noticed that it happens in the summer. This is when grass is being cut to make hay. It is also the time when pollen from plants is at its highest.

Pollen traps collect pollen grains so that they can be counted using a microscope. This gives the 'pollen count'. Newspapers, radio, and television report the pollen count during the summer.

Is there a link between hay fever and pollen?

A **correlation** is a link between two things. In this case, does hay fever increase when the pollen count increases?

Looking at thousands of people's medical records show that hay fever is highest in the summer months. This is also when most pollen is in the air.

This evidence shows that there is a link, or correlation, between pollen levels and hay fever attacks. But does this mean that pollen is the **cause** of hay fever?

Pollen grains under the microscope. Different plants release different types of pollen. (mag: × 1360 approx)

Does pollen cause hay fever?

An increase in two things could be caused by a third factor that has not been measured. Or it could be a coincidence that the two things increase at the same time.

Think about ice cream. Most ice cream is sold in the summer months but nobody would say that ice cream causes hay fever. The link may be just a coincidence. Or both increases may be caused by some other factor.

To claim that pollen causes hay fever you need some supporting evidence. You need to be able to explain how pollen causes hay fever.

Some people have hay fever at the same time each year. Their hay fever could be linked with the particular type of pollen that is released during that month. This is strong extra evidence for the link between pollen and hay fever.

Skin tests show that people who suffer from hay fever are allergic to pollen. Hay fever is an allergic reaction caused by pollen. So, there is a correlation between hay fever and pollen because pollen causes hay fever.

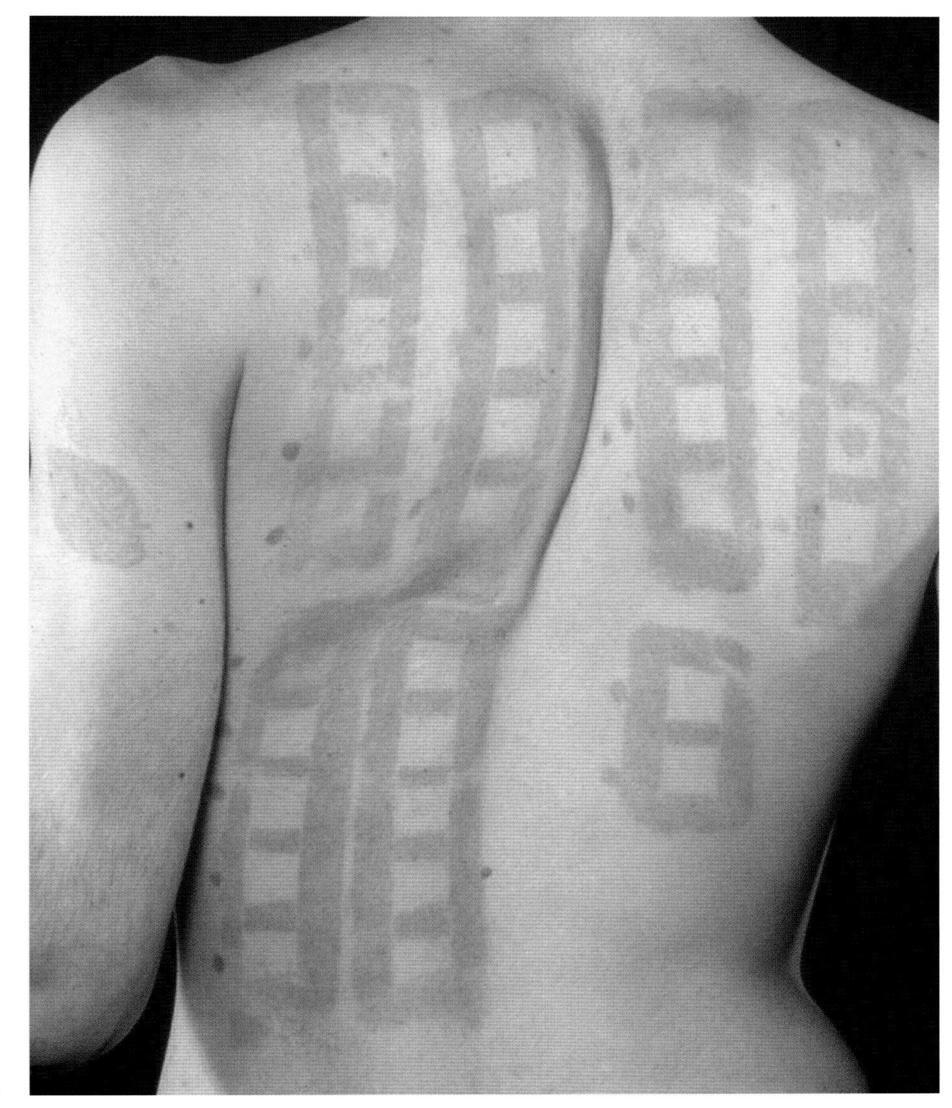

In a skin test, small disks of different chemicals are held on the skin by plasters. If you are allergic to a chemical, it will leave a round mark. This patient is also allergic to chemicals in the plaster.

Key words
correlation
cause

Questions

1 Suggest why it is useful to report the levels of pollen in the air during the summer months.

2 Write a note to a friend explaining:

- **a** what is meant by 'there is a correlation between pollen count and hay fever symptons'
- **(b)** why you need to look at medical records of a large number of people to be sure there is a correlation
- **c** why a correlation between ice cream sales and hay fever does not mean that ice cream causes hay fever

Asthma and air quality

Asthma is a common problem, especially in young adults. During an asthma attack, a person's chest feels very tight. They find it difficult to breath. It can be very frightening. A severe asthma attack can be very dangerous, especially for older people.

Elaine (14):

'I use my inhaler before I go swimming, or when it is very cold. When I first noticed my asthma, I used to feel very panicky and frightened. I felt as though I couldn't breathe. But now I have an inhaler, it isn't so bad.'

Asthma attacks are treated with inhalers. These contain medicines which help the lungs to 'open up' and breathe more freely.

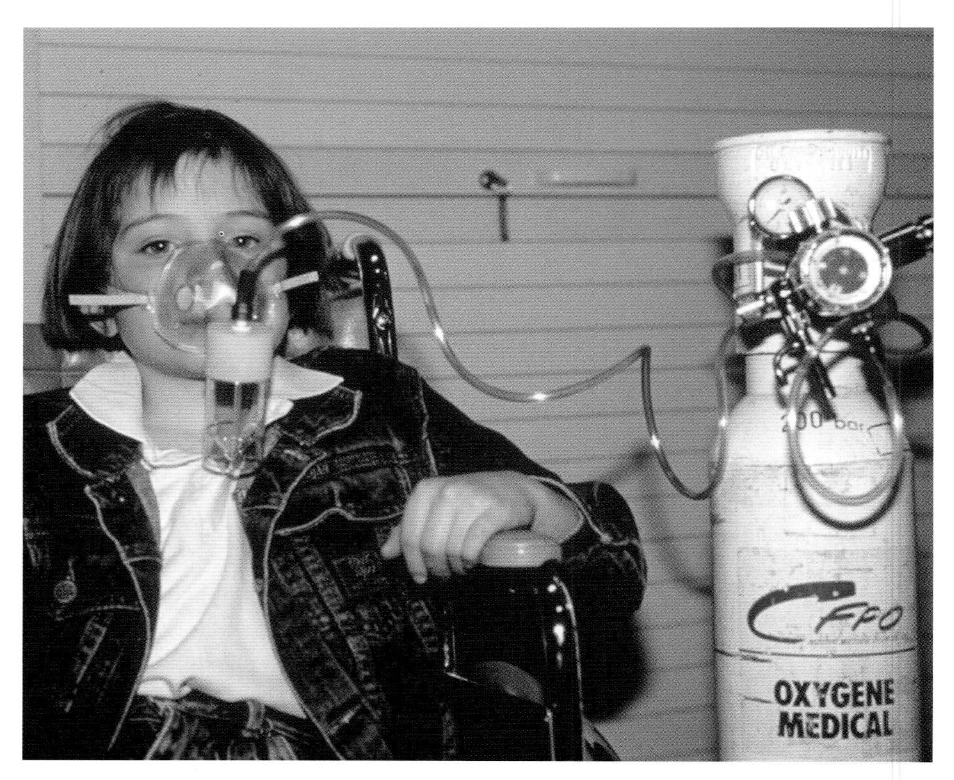

Inhalers are used to treat asthma attacks. Oxygen may be needed in severe cases.

Causes of asthma

Medical evidence shows that asthma attacks may be triggered by many different factors.

There is no clear link between concentrations of air pollutants and people starting to get asthma.

However, people who already have asthma may have sensitive lungs. Air pollutants may irritate a person's lungs. This may make their lungs sensitive to things that could trigger an attack.

Nitrogen dioxide and asthma

Because so many factors can affect asthma, studies of the effect of air quality are complicated. They must be carefully designed to eliminate the effect of other factors.

Studies have shown that NO_2 does have an effect on people who suffer from asthma. This air pollutant comes mainly from traffic exhausts.

If the concentration of NO_2 stays high for several days there is an increased number of asthma attacks. This may cause more people to die. So, although there is a correlation between NO_2 levels and asthma attacks, there is no clear evidence that it causes asthma.

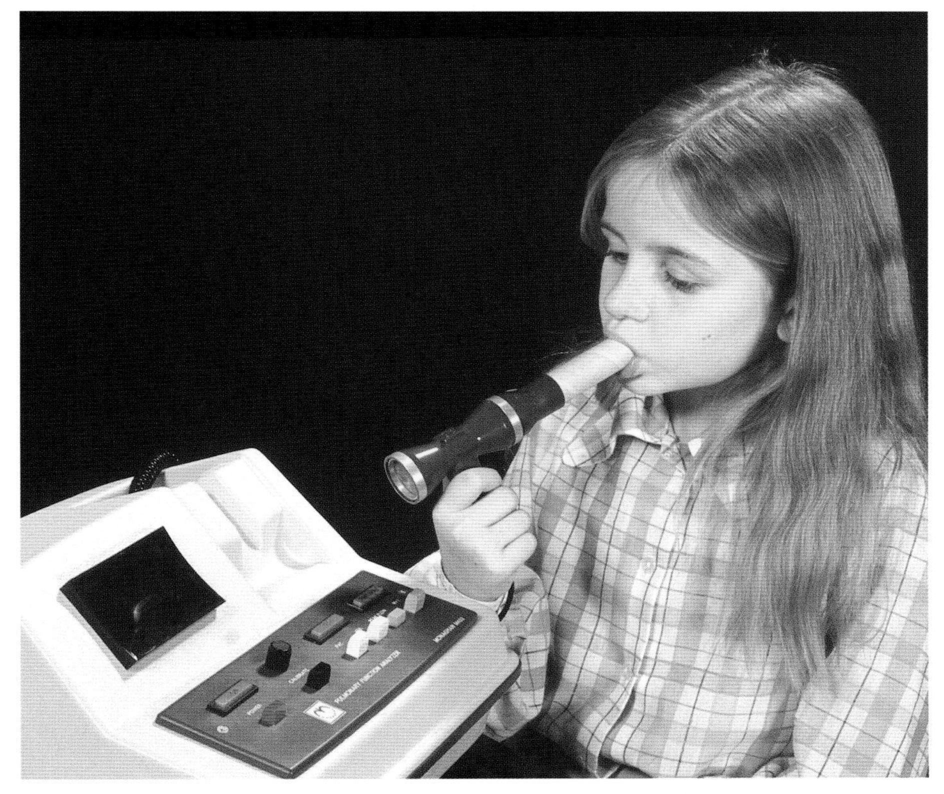

The efficiency of the lungs can be measured using a peak flow meter. This measures how quickly air can be breathed out. It can be used to measure the effects of air pollution on how well the lungs work.

Things that can trigger asthma:

- tree or grass pollen
- animal skin flakes
- dust-mite droppings
- air pollution
- nuts, shellfish
- food additives
- dusty materials
- strong perfumes
- getting emotional
- stress
- exercise (especially in cold weather)
- colds and flu

Questions

3. Explain why investigations into the effects of air quality on asthma are complicated.
4. Explain why the correlation between high levels of NO_2 and asthma attacks does not mean that NO_2 causes people to start to suffer from asthma.

Find out about:

- how new technology can reduce the harmful emissions from cars and power stations

I How can new technology improve air quality?

Scientists and technologists can find new ways of reducing the amounts of pollutants that escape into the air.

Polluted water can be purified and delivered to people. Air is all around us. You do not get it out of a tap. So, everyone should try to reduce the pollutants getting into the air.

Efficient engines and catalytic converters

Technological developments have made modern car engines more efficient than engines in old cars. They use less fuel and so produce less air pollution.

Catalytic converters have been added to car exhaust systems. The waste gases pass through a metal honeycomb structure. It has a very large surface area. This is coated with a thin layer of platinum. This metal acts as a **catalyst** (it speeds up the chemical reaction without being used up itself).

Chemical reactions happen in the exhaust gases as they pass through the catalytic converter. These reactions convert the air pollutants CO and NO to less harmful gases.

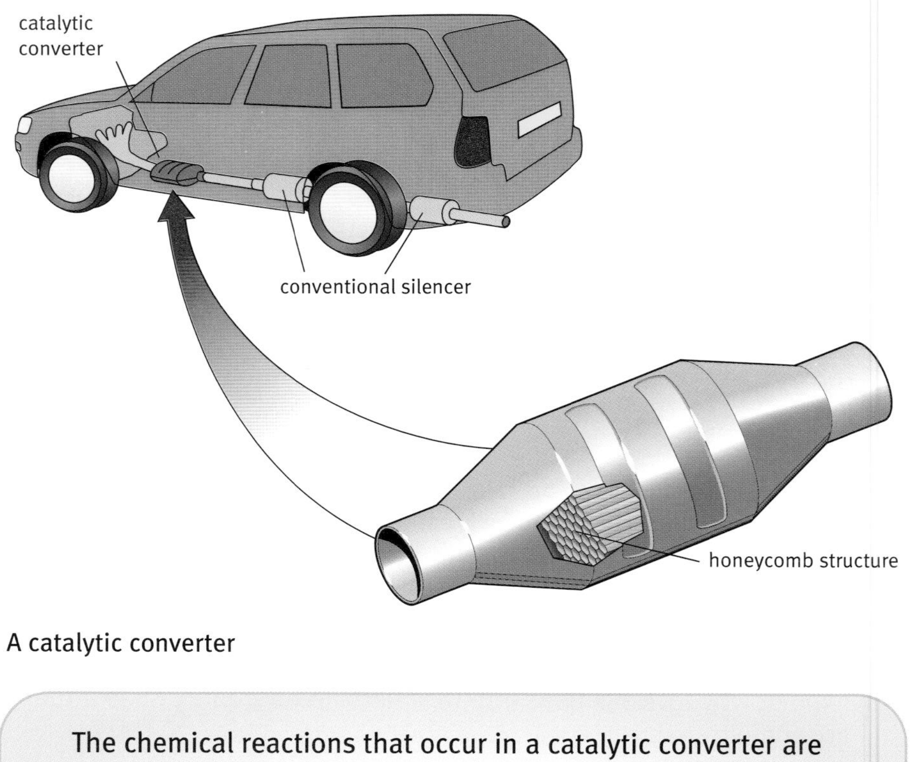

A catalytic converter

The chemical reactions that occur in a catalytic converter are

carbon monoxide + oxygen → carbon dioxide

nitrogen monoxide + carbon monoxide → nitrogen + carbon dioxide

Questions

1 Catalytic converters remove harmful CO and NO by converting them to less harmful chemicals. What are the chemicals they are converted to?

Key words

technological developments
catalyst

Reducing pollutants from power stations

When coal and natural gas burn the main product is CO_2. However, fossil fuels often have impurities that contain sulfur. When they burn, these impurities produce SO_2. This can cause acid rain if it escapes into the atmosphere. But the SO_2 can be removed before the waste gases escape from the power station chimney.

Waste gases pass through a spray of powdered lime (calcium oxide) and water. The SO_2 in the gases reacts with this mixture and oxygen from the air. Together they form a new chemical called calcium sulfate. This is a solid that has trapped the SO_2 before it can escape to the air.

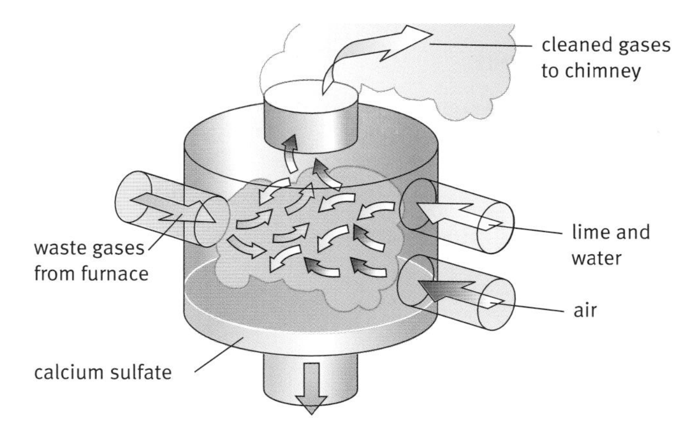

Removing sulfur dioxide to prevent it escaping from power station chimneys.

Cleaner fuels

Either existing fuels can be improved or new fuels can be developed.

Sulfur compounds can be removed from natural gas before it is used for power stations and domestic heating. Also low-sulfur petrol is now available. These improvements reduce the SO_2 in the waste gases.

When hydrogen is used as a fuel the only product is water so it would be a very clean fuel. But you would have to be careful that the electricity used to produce the hydrogen was not generated by a method that produced atmospheric pollutants.

Hydrogen-fuelled cars are being developed. Their only waste product is water. They do not add CO_2, CO, or particulates to the atmosphere.

Electric trams and trains help to remove air pollution from congested cities. However, remember that some of the electricity they use will be produced by power stations that burn fossil fuels.

Questions

2 Sulfur dioxide can be removed from the waste gases before they escape from a power station chimney. What chemical is the SO_2 converted to?

3 Write a letter to a newspaper explaining why in the long term air quality could be a bigger problem than water quality.

Find out about:

- how laws and regulations can help to improve air quality

J How can governments and individuals improve air quality?

Governments can pass laws and **regulations** that aim to improve air quality. Some of the early ones seem a little funny today. But they have all helped to tackle the problems of air pollution.

- **1845** A limit was put on the amount of smoke released by steam train engines.
- **1847** The amount of smoke that factories could give out was reduced.
- **1863** The 'Alkali Act' controlled emissions from early chemical factories that were making sodium hydroxide.
- **1956** The 'Clean Air Act' introduced smokeless zones in cities. People inside these zones had to burn 'smokeless coal'.
- **1991** Limits were set to control the emissions of carbon monoxide and particulates from vehicle exhausts.
- **1997** The National Air Quality Strategy set targets for a reduction in UK emissions.

The London smog of 1952 killed 4000 people. It led to the Clean Air Act which reduced pollution from coal fires.

Local emissions but global problems

There are international agreements that aim to reduce the emission of atmospheric pollutants.

In 1997, an international meeting in Kyoto, Japan, agreed to set targets for a reduction in carbon dioxide emissions.

Air pollution spreads around the world. Some pollutants react quickly to form other chemicals. These cause local or regional problems. Other pollutants are less reactive. They may travel long distances without changing and cause global problems.

Living more sustainably

Sustainable development means meeting people's needs without making the environment worse for future generations. This includes producing less air pollution.

Governments try to encourage people and industries to produce less air pollution. They use **financial incentives** (or taxes) such as:

- Car tax
  In general, bigger cars, and cars with bigger engines, produce more air pollutants. Owners of these cars could pay a higher vehicle excise duty (car tax).
- Fuel duty
  Higher fuel prices could discourage people from using their cars. It would also encourage people to buy more fuel-efficient vehicles. Both of these results would help to reduce emissions.
- Energy efficiency grants
  These grants might include payments for fitting wall or roof insulation. They might also include the fitting of a high efficiency boiler, such as a condensing boiler.

Ordinary people, through the choices they make, can influence the air quality.

Bad air quality would affect your everyday life.

Questions

1 Explain why reducing the distance travelled each year by people in cars would help to reduce air pollution.

2 List examples of ways the UK Government has tried to make people live more sustainably. For each one, say whether you think it will result in:
   a an increase in costs for people
   b a decrease in the convenience and quality of life
   c no change in costs or quality of life

3 Suggest three ways that you could encourage your friends and family to use less energy and so be responsible for the production of less air pollution.

Key words
regulations
sustainable development
financial incentives

C1 Air quality

Science explanations

The atmosphere supports life and controls the temperature on the surface of the Earth. Human activity affects air quality. Understanding what happens in chemical reactions will help you to explain this.

You should know:

- the gases that make up the Earth's atmosphere
- human activities add small amounts of chemicals to the atmosphere
- some of these chemicals are harmful and are called air pollutants
- power stations and vehicles that burn fossil fuels add:
  - small amounts of the gases carbon monoxide, nitrogen oxides, and sulfur dioxide to the atmosphere
  - small amounts of very small particles, such as carbon, called particulates, to the atmosphere
  - extra carbon dioxide that contributes to global warming
- primary pollutants are released directly into the atmosphere
- secondary pollutants, such as nitrogen dioxide and acid rain are produced by chemical reactions in the atmosphere
- that the fossil fuel coal is mainly carbon
- fuels such as petrol, diesel and natural gas are hydrocarbons - these are chemicals made up from carbon and hydrogen
- when a fuel burns the oxygen atoms from air combine with:
  - carbon atoms to form carbon dioxide
  - hydrogen atoms to form water
- in a chemical change/reaction atoms separate and recombine to form different chemicals
- chemical changes can be shown by pictures of the atoms and molecules involved
- during a chemical reaction, the number of atoms of each kind is the same in the products as in the reactants. The atoms are conserved.
- the conservation of atoms means that combustion reactions affect air quality
- the properties of the reactants and products of chemical changes are different
- technological developments such as catalytic converters and flue gas desulfurization can reduce amounts of pollutants released into the atmosphere

Ideas about science

Scientists need to collect large amounts of data when they investigate the causes and effects of air pollutants. They can never be sure that a measurement tells them the true value of the quantity being measured. Data are more reliable if they can be repeated.

If you make several measurements of the same quantity, the results are likely to vary. This may be because:

- you have to measure several individual examples, for example, exhaust gases from different cars of the same make
- the quantity you are measuring is varying, for example, pollen levels
- the limitations of the measuring equipment or because of the way you use the equipment

Usually the best estimate of the true value of a quantity is the mean (or average) of several repeat measurements. The spread of values in a set of repeated measurements, the lowest to the highest, gives a rough estimate of the range within which the true value probably lies.
You should:

- know that if a measurement lies well outside the range within which the others in a set of repeats lie, then it is an outlier and should not be used when calculating the mean
- be able to calculate the mean from a set of repeated measurements

When comparing information on air quality from different places you should know that:

- a difference between their means is real if their ranges do not overlap

Investigating the link between air pollution and illnesses:

- a correlation shows a link between a factor and an outcome, for example, as the pollen count goes up the number of people suffering from hay fever goes up
- a correlation does not always mean the factor causes the outcome
- scientists have evidence to explain how pollen causes hay fever
- but although poor air quality can make people's asthma worse, there is no clear evidence that it causes people to suffer from asthma

Making decisions about improving air quality:

- official regulations such as the MOT test for motor vehicles can be used to improve air quality
- using less electricity and burning less fuel will improve air quality

C2

Why study materials and their uses?

All the things we buy are made of 'stuff'. That stuff must come from somewhere. When you have finished with it, it has to go somewhere. The products people use every day are made of many different kinds of materials. Materials are chosen to do a job because of their properties. Everyone can make better choices about uses of materials if they understand more about these properties.

The science

Scientists use their knowledge of molecules to explain why different materials behave in different ways. This gives them the ability to design new materials with just the right properties to meet everyday needs.

Ideas about science

Scientists test products to check that they can do the job, are good value and safe. You can use data from these tests when you buy a product. So you need to be able to judge whether or not the results can be trusted.

Science can also help us to save money and cut down waste. Scientists make a careful analysis of the energy used, and materials needed, for each stage in the life of a product.

Material choices

Find out about:

- the testing and measurement that helps people to make good choices when buying products
- some of the explanations scientists use to design better materials
- ways to weigh up the costs and benefits of using different materials
- the choices people can make to reduce waste

Find out about:

- materials and their properties
- natural and synthetic materials
- long-chain polymers

A Choosing the right stuff

The newest fashion

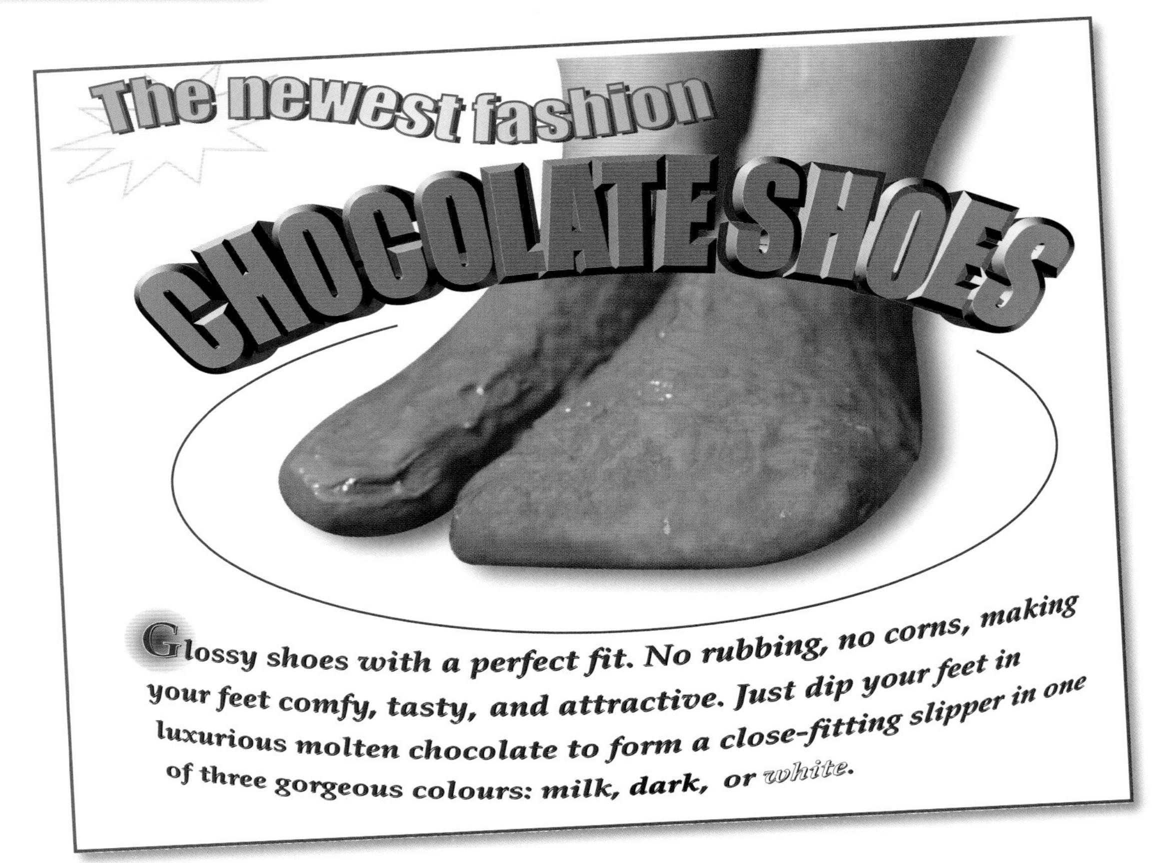

Latex is a natural polymer that can be tapped from rubber trees. After treatment, it is used in a wide variety of products, including the soles of shoes.

What the advertising agency didn't tell you

Of course chocolate shoes are a joke. Chocolate is not a good **material** for making shoes. Here are some reasons:

- chocolate would crack
- it would melt in warm weather
- dogs would follow you and lick your feet
- it would wear away
- it would leave a mess on the carpet

Maybe not chocolate

Although chocolate does not have the right **properties**, the idea of moulded shoes is not new. South American Indians used to dip their feet in liquid latex straight from the rubber tree. They would sit in the sun to let the latex harden, forming the first, snug fitting, wellies. So latex is more suitable than chocolate for making shoes. Let's see what properties it has that make it better.

Fantastic elastic

The most obvious difference between latex and chocolate is that latex is **flexible**. Any material chosen to make our shoes needs to be flexible.
It also needs to be:

- hard wearing because you will walk on it
- waterproof
- a solid at room temperature
- elastic so it keeps its shape
- flexible so you can bend your feet
- tough so that it won't crack when it bends

Latex has all these properties whereas chocolate does not.

Fit for purpose

Latex is not the only material for making shoes. Once they know the properties they need, shoe designers can a choose a number of different materials. As well as latex, they can use **natural** materials like cotton or leather. Or they can use a **synthetic** material like nylon, neoprene, or Gore-tex.

What's in a name?

Sometimes words can have more than one meaning. Take the word material for example – to some people this means cloth (or fabric) for making clothes. For a scientist, the word material means any sort of stuff you can use to make things from.

Most of the materials used to make shoes are **polymers**.

What are polymers?

All polymers have one thing in common. Their molecules are very long chains of atoms. This is true for natural polymers such as cotton, silk, and wool and for synthetic polymers such as polythene, nylon, and neoprene.

You find natural and synthetic polymers all around you.

Key words

material	natural
properties	synthetic
flexible	polymers

Questions

1 Look at the picture of young people in a car. Identify items that could be made from:
a natural polymers **b** synthetic polymers.

(2) The word 'synthetic' can mean different things at different times. Write down up to four words that come to mind when you see or hear the word synthetic.

(3) Suggest reasons why

a the natural material leather is very suitable for making smart shoes

b the synthetic material polythene is suitable for making the straps on sandals for the beach but not for making walking shoes.

Find out about:

- synthetic polymers made to meet our needs
- examples of plastics and their uses

B Polymers everywhere

Plastics can bring benefits to people by meeting their needs. These include:

- physical needs for shelter, warmth, and transport
- bodily needs for food, water, hygiene, and health care
- social and emotional needs for human contact, leisure, and entertainment
- needs of the mind to stimulate thinking and creativity

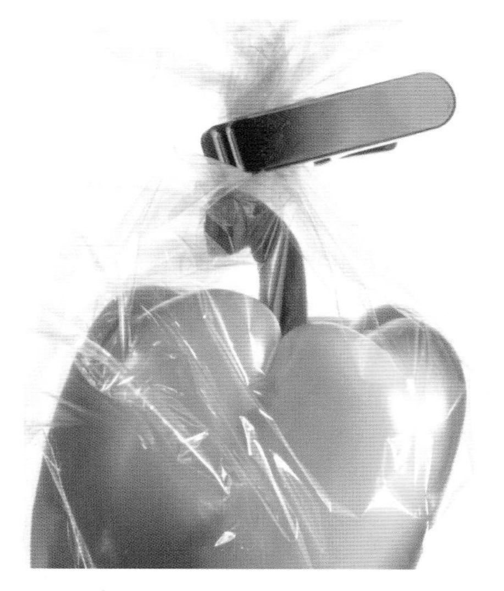

Polythene bags help people to protect, store, and carry food.

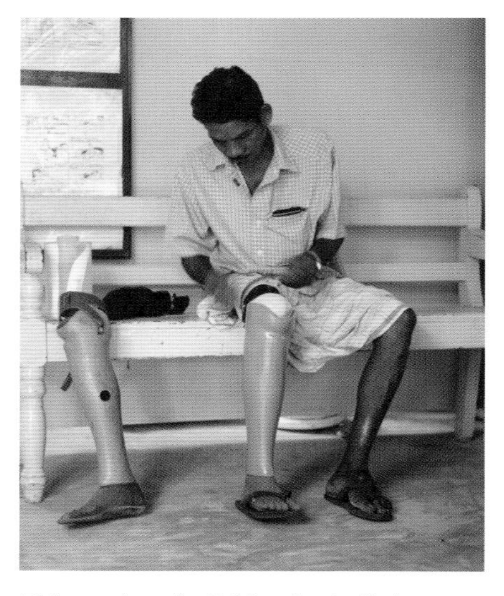

This patient in Sri Lanka is fitting a new leg made of polypropylene.

The world's first inflatable church made from PVC.

PET is a polyester used to make soft-drinks bottles and other food containers.

Polyester is used to make hulls and sails.

This acrylic painting was on show in a shop in Zanzibar.

Manchester City stadium roof is made from polycarbonate 'glass'.

Kevlar helmets have saved many soldiers lives.

Austrians on a sleigh in traditional woollen dress with their old wooden skis. Wood and wool are both made from natural fibres. they are now often replaced with synthetic polymers.

A wet suit made from neoprene offers warmth and protection.

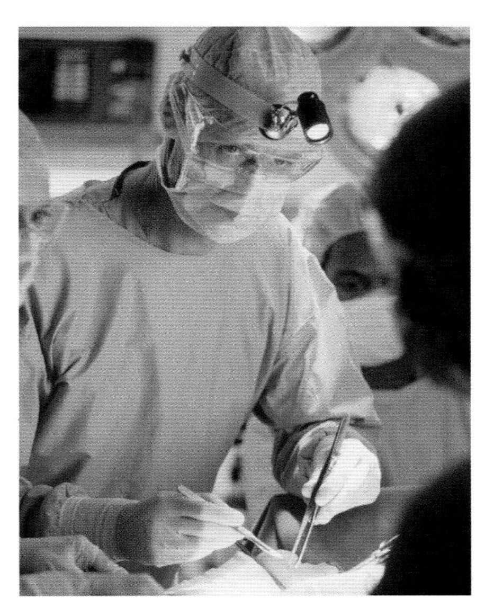

Doctors and other health workers wear gloves made of natural rubber (latex) for protection and to prevent infection.

Questions

1 Create a chart, diagram, or table to show how plastics can meet our needs. Use the examples on these pages and any other examples that you know of.

2 Give examples of objects that are now made of synthetic polymers but which were once made of metals, glass, pottery, or natural polymers such as wood.

Find out about:

- words scientists use to describe materials
- testing materials to ensure quality and safety

C Testing times

Getting the right material

Manufacturers and designers have to choose the right materials to make their products. They decide which materials to use based on their properties and cost. In many products the materials include polymers.

Modern materials with special properties

For example, the soles of shoes have to be flexible, hard wearing, and strong. Also, they must not crack when they bend – they have to be tough. A synthetic rubber is a good choice.

The case of a computer has to be very different from shoe soles. It needs to be stiff, strong, and tough. People want a case that resists scratches and keeps its appearance. So the polymer has to be hard.

Nylon ropes must be light and strong. Climbers want to be sure that their ropes have been tested.

Material words

When scientists describe the properties of materials, they use special words. Some of these, like strong, have everyday meanings that are similar to their technical meaning. However, some are a little different.

A material is **strong** if it takes a large force to break it. Some materials are strong when stretched. Examples are steel and nylon which are strong in **tension**. Concrete tends to crack when in tension but it is very strong in **compression**. This makes it useful for pillars and foundations.

Stiff is the opposite of flexible. It is difficult to stretch or bend a stiff material. High stiffness is very important in many of the materials that engineers use to make aeroplanes, bridges, and engines.

Hard and **soft** are also opposites. The softer a material, the easier it is to scratch it. A harder material will always scratch a softer one.

In many applications it is also important to know how heavy a material is for its size. Materials such as steel and concrete have a high **density**. Other materials are very light for their size and have a low density. Examples are foam rubber and expanded polystyrene.

A testing machine for plastic packaging. Measuring the force needed to crush the container gives a value for the strength of the pack.

Measuring the words

Technical words help to describe materials. There are times when more than a description is needed. Accurate measurements of properties are important when it is important to compare materials and test their quality.

For example, rope makers need to find strong, stiff fibres. Their engineers test small samples to find the ones that are strong enough. The force that breaks each fibre tells them its strength.

They can also measure how much the fibres stretch. A stretchy rope is not suitable for climbing mountains but might be just right for bungey jumping.

Quality control

John Fletcher is quality manager for Coates. This company makes sewing threads. He takes samples from every batch that leaves the factory. He tests them to ensure that they have the correct strength and stiffness. This means that his customers can be sure that the threads will always be the same.

This machine measures the force needed to break threads made at Coates. John Fletcher also measures the length of the broken thread to see how much it has stretched.

Key words

strong	compression	hard	density
tension	stiff	soft	

Questions

1 Look at the picture of rollerbladers. Identify items which are:

a flexible **b** stiff **c** strong **d** hard

2 Look at the two pictures of material testing. Which is a test of strength in tension? Which is a test of strength in compression?

③ Suggest reasons for measuring the strength of packaging materials.

Find out about:

- materials under the microscope
- molecules and atoms in materials
- models of molecules

D Zooming in

A woollen jumper is very different from a silk shirt. The shirt is more formal and less stretchy than the jumper. They are both made from natural polymers but they are very different. Their properties depend on their make-up, from the large scale to the invisibly small:

- the visible weave of a fabric
- the microscopic shape and texture of the fibres
- the invisible molecules that make up the polymer

The visible weave

The fabric of a woven shirt is tightly woven but even so it is possible to see the criss-cross pattern of threads. The fabric is hard to stretch because the strong threads are held together so tightly.

On the other hand, a knitted jumper is soft and stretchy. The loose stitches allow the threads to move around.

The weave and the stitches are visible to the naked eye. They are **macroscopic** features. However, the properties of a fabric also depend on smaller structures.

Silk

Levels of structure and detail. A millimetre is a thousandth of a metre. A micrometre is a thousandth of a millimetre. And a nanometre is a thousandth of a micrometre.

Taking a closer look

A microscope can show details of the individual fibres in a fabric. Silk, for example, has smooth, straight fibres that slide across each other.

Wool fibres have a rough surface that is covered in scales. The wool fibres tend to cling to each other in the thread and also make the threads cling together.

The invisible world of molecules

It is difficult to look much further into the structure of materials using microscopes of any kind. Scientists explain the differences between silk, wool, and other fibres by finding out about their molecules. Molecules are very small indeed, so small that it needs a giant leap of the imagination to think about them.

Scientists measure the sizes of atoms in nanometres. One **nanometre** is 1 000 000 000 times smaller than a metre. Many molecules, such as the small molecules in air, are even smaller than one nanometre but some are bigger.

The molecules in fibres are big on the nanometre scale. They are very long – 1000 nanometres or more. The shape and size of the **long-chain molecules** in a fibre make the material what it is. The length of the molecules gives polymers their special properties.

Computer model of a protein molecule. No-one knows what atoms and molecules look like. It helps to use models to understand what they do. In the real world the atoms are not coloured. In this computer image the atoms are colour-coded: carbon (green), sulfur (yellow), nitrogen (blue), hydrogen (grey), and oxygen (red).

Model molecules

Even the largest molecules and atoms are invisible. So in the nanoworld of molecules scientists build models based on the results from their experiments.

Models of molecules can be compared to the familiar map of the London tube system. The tube map does not look like an underground railway. But it has lots of useful information about the way the stations are connected. In a similar way, models of molecules do not look like real molecules. But they show what scientists have discovered about the atoms in the molecules and how they are joined together.

Key words

macroscopic
nanometre
long-chain molecules

Questions

1 a Put these in order of size. Start with the largest: fibre, fabric, atom, thread, molecule.

b Use the words in part **a** to write four sentences that describe the decreasing structures. The first sentence might be: Fabrics are made by weaving together threads.

2 a How many chemical elements are there in silk?

b Is silk a hydrocarbon?

3 A polymer molecule is about 1000 nanometres long. An atom is about 0.1 nanometres across.

a How many atoms are there along the chain?

b How many molecules would fit into a millimetre?

Find out about:

- polymer discoveries
- polymers as long-chain molecules

E The big new idea

The 1930s was the decade of the first polymers. The world was a tense place and war was on its way. Governments were looking for scientific solutions to give them an advantage. This speeded up many scientific developments. Some of these used the big new idea: polymers. However, the first synthetic polymer was discovered by accident.

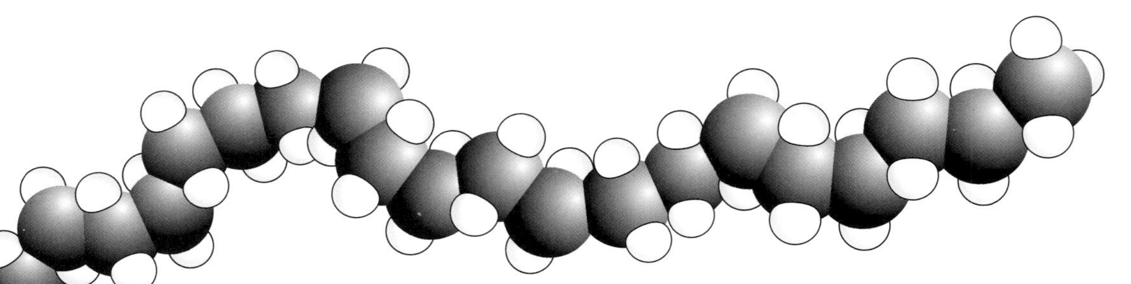

The accidental discovery of polythene

In 1933, two chemists made polythene thanks to a leaky container. Eric Fawcett and Reginald Gibson were working for ICI. Their job was to investigate the reactions of gases at very high pressures. They had put some ethene gas into the container and squashed it to 2000 times its normal pressure. However, some of the ethene escaped. When they added more ethene, they also let in some air.

Two days later, they found a white waxy solid inside the apparatus. This was a surprise. They decided that the gas must have reacted with itself to form a solid. They realized that, in some way, the small molecules of ethene had joined with each other to make bigger molecules.

They worked out that the new molecules were like repeating chains. The chains were made from repeating links of ethene molecules.

Later they understood that oxygen in the air leaking into their apparatus had acted as a catalyst. The oxygen speeded up what would otherwise have been a very, very slow reaction to join the ethene molecules together.

What are polymers?

Polymers all have one thing in common: their molecules are long chains of repeating links. Each link in the chain is a smaller molecule. It connects to the next one to form the chain. This is true for natural polymers such as cotton, silk, and wool and for synthetic polymers such as polythene, nylon, and neoprene.

Fawcett and Gibson called their material poly-ethene and we now call it polythene. The word poly means 'many'; a poly-ethene molecule is made from many ethene molecules joined together.

ethene gas under pressure

The original high-pressure container used by Fawcett and Gibson is on display at the Science Museum. The diagrams show what was happening to the small ethene molecules as they joined up in long chains to make polythene. This is polymerization.

A polymer pioneer

Wallace Carothers was an American chemist who discovered neoprene and invented nylon. Neoprene was another accidental discovery. A worker in Carothers' laboratory left a mixture of chemicals in a jar for five weeks. When Carothers had a tidy up, he discovered a rubbery solid in the bottom of the jar. Carothers realized that this new stuff could be useful. He developed it into neoprene. This synthetic rubber first came on the market in 1931 and is still used today, to make wet suits, for example. This discovery helped Carothers to work out a theory of how small molecules can **polymerize**.

The discovery of nylon

America and Japan were on bad terms in the years before World War II. Trade was difficult and the supply of silk was cut off. It became rare and expensive. Carothers started looking for a synthetic replacement. In 1934, his team came up with nylon. This is a polymer made from two chemicals. The different molecules join together as alternate links in the chain.

Sadly, Carothers died before he could see the effects of his discoveries. Nevertheless, they are both still in use today.

Key words
polymerize

Questions

1 a Write down the names of two polymers that were discovered by accident.

b Write down any other accidental discoveries that you know about.

c Many scientists have made accidental discoveries. All of these words might be used to describe these scientists:

lucky, skilful, foresight, inventive, creative.

i Choose two of these words to describe the scientists.

ii In each case, explain why you have chosen that word.

2 Draw a timeline for the years from 1930 to 1950. Draw it running down the middle of a page. Put dates every five years.

a Put on the dates of the discovery of polythene, nylon, and neoprene. Mark these on the right of the timeline.

b Mark the dates for major world events in this period. You should include World War II. Do this on the left of the timeline.

c Mark on any other dates that you think are important to the stories of these polymers.

Find out about:

- long and short polymer chains
- polymers that are crystalline
- explaining polymer materials

F Molecules big and small

The longer the stronger

The properties of a polymer depend on the length of its molecules. The molecules in candle wax are very similar to those in polythene. However, wax is weaker and more brittle than polythene. This is because the wax molecules are much shorter. They contain only a few atoms; polythene molecules contain many thousands. The longer molecules make a stronger material.

Two different bonds

The molecules are made of atoms. The bonds between the atoms are strong. So it is very hard to pull a molecule apart. The molecules do not break when the materials are pulled apart.

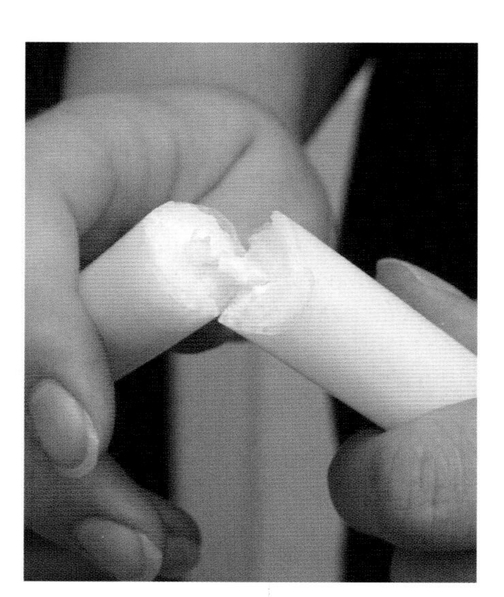

The molecules of candle wax are about 20 atoms long. Wax is weak and brittle.

The molecules of polythene are similar to those of candle wax. But they are about 5000 times longer. Polythene is much stronger and tougher than candle wax.

The forces between molecules are very weak. It is much easier to separate molecules. They can slide past each other.

Breaking wax and polythene

Stretch or bend a candle and it cracks. This is because separating the small molecules is quite easy. They slip past each other quite easily.

Breaking a lump of polythene is much more difficult. Its long molecules are all jumbled up and tangled. It is very hard to make them slide over each other. The long molecules make polythene stronger than wax.

Polythene original

Eric Fawcett and Reginald Gibson (see pages 46 – 47) made polythene under high pressure. Chemists now know that in this polythene the long polymer chains have branches. The branches stop the molecules packing together neatly.

Compare a bonfire pile with a log pile. In a bonfire the twigs and side branches are sticking out all over the place and the bonfire pile is full of holes. But the pile of straight logs is neatly stacked. It is the same with polymers but on a much smaller scale. If there are side chains sticking out of the structure it is messy and full of holes. This is the case with polythene made under pressure which has **branched chains**.

A stronger denser polythene

Chemists realized that a **crystalline polymer** might be stronger and denser. They set out to find a way of making polythene molecules in neat piles of straight lines. After the war, in the 1950s, they found a way of doing just that. It was an international effort by the German Karl Ziegler and the Italian Giulio Natta.

The scientists used special metal compounds as catalysts. These metal compounds act in a similar way to the oxygen in the high-pressure process. They speed up the rate at which the ethene molecules join together.
The growing polymer chains latch onto the solid catalyst.
The regular surface of the solid allows the molecules to build up more regularly.

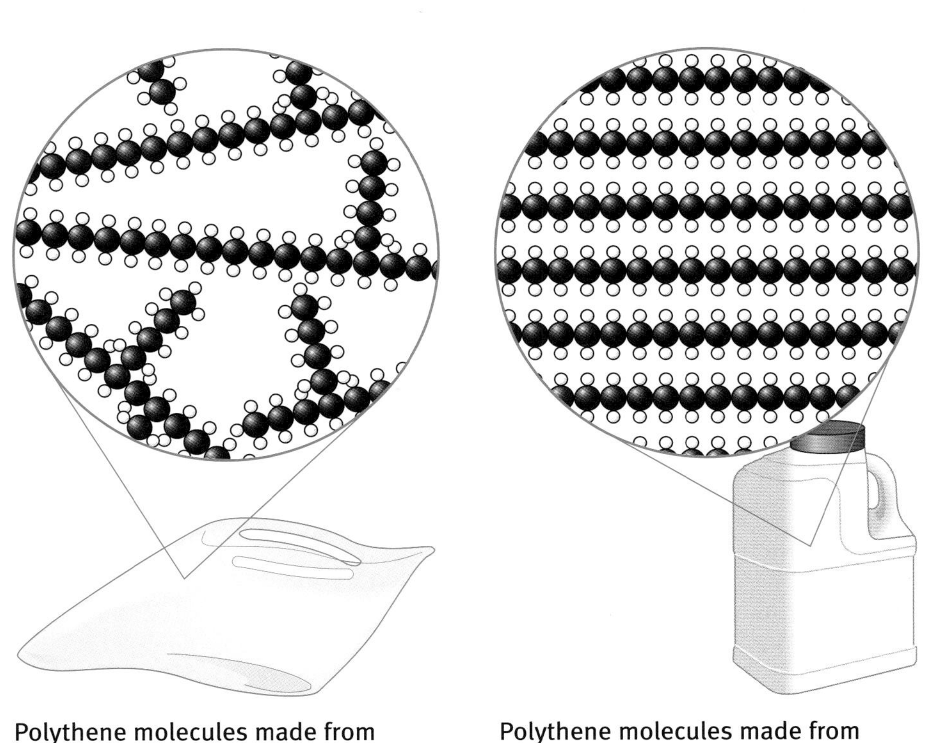

Polythene molecules made from ethene under pressure have side branches. This stops the polymer molecules lining up neatly. This type of polythene has a slightly lower density and is not crystalline.

Polythene molecules made from ethene with a special catalyst do not have side branches. The polymer molecules line up neatly. This type of polythene has a slightly higher density and is crystalline.

In this new form of polythene the molecules are more neatly packed together. It is slightly stronger and denser than the older type. It softens at a higher temperature too. Both types are still made – the old branched-molecule version is called low–density polythene (LDPE). The newer straight-molecule version is high–density polythene (HDPE).

Key words
branched chains
crystalline polymer

Questions

1 Bowls of pasta can be used as an analogy to explain the difference between wax and polythene. One bowl contains cooked spaghetti. The other bowl contains cooked macaroni (or penne).

a In the analogy, what represents a molecule?

b Which kind of pasta represents wax and which represents polythene?

c Show how this analogy can help to explain why polythene is stronger than wax.

2 Why is HDPE slightly denser than LDPE? Suggest an explanation based on the structure and arrangements of molecules.

③ LDPE starts to soften at the temperature of boiling water. HDPE keeps its strength at 100 °C. Suggest examples of products better made of HDPE rather than LDPE and give your reasons.

Find out about:

- using science to change polymer properties
- cross links to make polymers harder
- plasticizers to make polymers softer

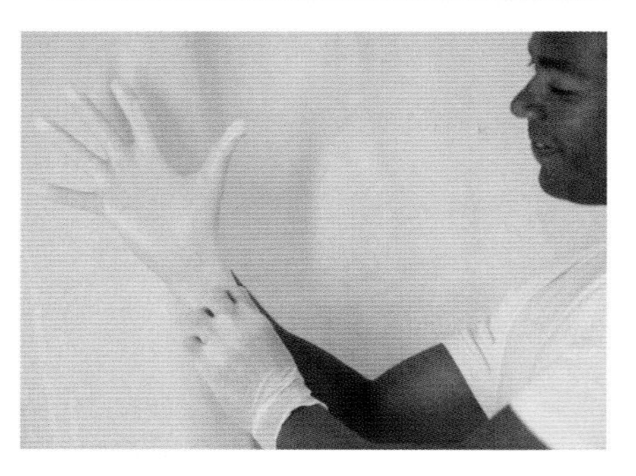

Vulcanizing natural rubber produces gloves that are strong enough not to tear.

G Designer stuff

Hardening rubber

Natural rubber is a very flexible polymer. But it wears away easily. This makes it good at rubbing away pencil marks. But not for much else.

Sometime around 1840 an American inventor called Charles Goodyear was experimenting with mixing sulfur and rubber. He was trying to improve the properties of the natural material. He accidentally dropped some of his mixture on top of a hot stove. He did not bother to clean it off, and the next morning it had hardened.

It took two more years of research to find the best conditions for the new process which Goodyear called **vulcanization**. It made rubber into a stronger material that was more resistant to heat and wear. At that time no-one knew why this happened. They just knew that it worked and that it made rubber an excellent material for car tyres.

Goodyear started a business making tyres. He began with tyres for bicycles and prams. Now the business makes tyres for cars, motorcycles, and aeroplanes.

Cross-links

Goodyear was the first person to alter the properties of a polymer. He did not know why vulcanization worked – only that it did. Now that we understand more about molecules, we know what's going on.

The sulfur makes **cross-links** between the long rubber molecules. This stops them from slipping over each other. The molecules are locked into a regular arrangement. This makes the rubber less flexible, stronger, and harder.

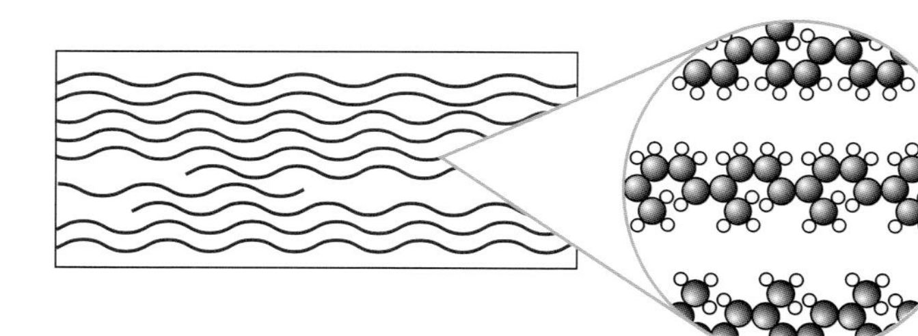

Each line represents a polymer molecule. Without cross linking, the long chains can move easily, uncoil and slide past each other.

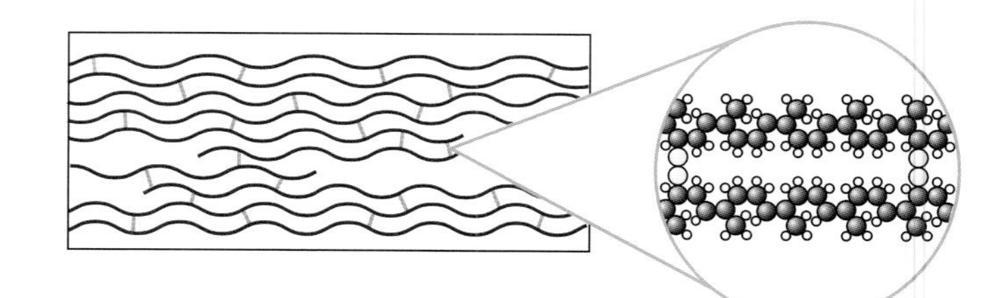

The sulfur atoms form cross-links across the polymer chains. This stops the rubber molecules uncoiling and sliding past each other

Softening up

PVC is a polymer often used for making window frames and guttering. These need to be **durable** and hard.

PVC is also a good, safe polymer for making children's toys. However, toy manufacturers often need to make it a bit softer and more flexible.

To do this, they add a **plasticizer**. This is usually an oily liquid with small molecules. The small molecules sit between the polymer chains.

The polymer chains are now further apart. This weakens the forces between them. Therefore, they can slide over each other more easily. This makes the polymer softer and more flexible.

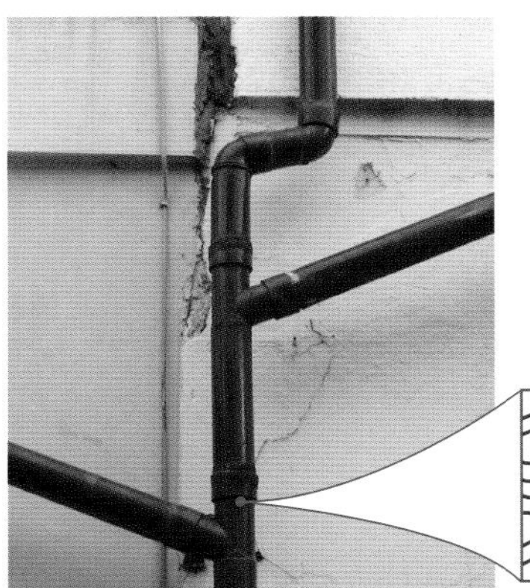

This PVC is unplasticized. It is called uPVC.

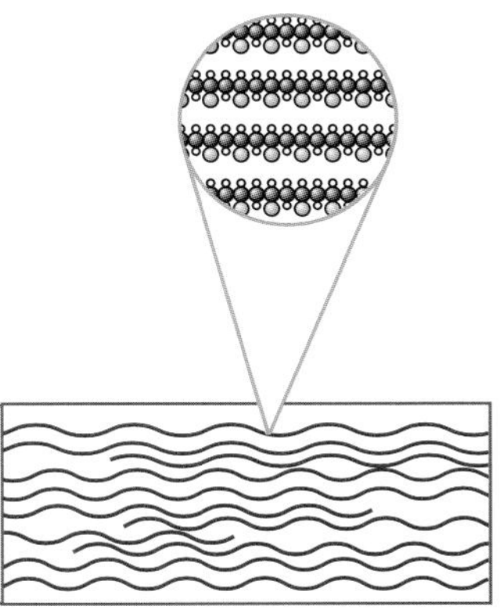

The lines represent PVC molecules. The chains of PVC lie close together. The closer they are, the stronger the forces.

This PVC has been plasticized to make it soft.

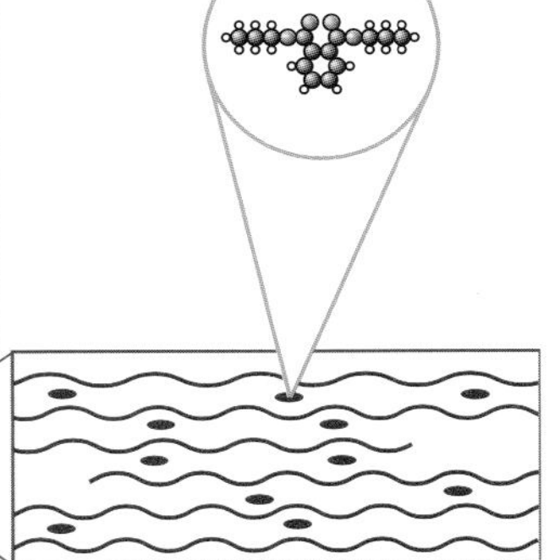

The molecules of plasticizer hold the PVC chains apart. This weakens their attraction and makes it easier for them to slide past each other.

Cling film

Cling film was first made from plasticized PVC. Unfortunately the small plasticizer molecules were able to move through the polymer and into the food – especially fatty foods such as cheese. Some people were worried that the plasticizer might be bad for their health. The evidence that plasticizers are harmful to health is controversial and strongly challenged by the plastics industry.

There is stronger evidence that the regular use of plastic foodwrap can cut down on food poisoning which is a serious and growing risk to health.

Some cling film is now made using PVC and plasticizers that are much less likely to move from the polymer to food. Alternatively cling film made from polythene is available. This is just as flexible but does not cling so well.

Key words

vulcanization
cross-links
durable
plasticizer

Questions

1 There are four sections on these pages. For each, decide what the most important point is. Write a sentence that summarizes this point.

2 Chemists can vary the extent of cross-linking between chains in rubber. How would you expect the properties of rubber to vary as the degree of cross-linking increases.

3 **a** Make a table to list the benefits of using cling film to wrap food in one column and the risks involved in a second column.

b Comment on whether or not you think that the benefits outweigh the risks.

Smart materials

Ingenious layers

Sometimes layers of different polymers with different properties are sandwiched together. An interesting example is the waterproof fabric Gore-tex, named after its inventor Bob Gore. He was working with a polymer called PTFE. This is the plastic coating for non-stick pans.

Gore discovered that if a sheet of PTFE is stretched it develops very small holes and becomes porous. A single water molecule can pass through the small holes. But a whole water droplet is too large to get through. This got Gore thinking. His idea was that vapour evaporating from someone's skin would pass through the polymer sheet, but that rain drops would not.

Gore-tex has a layer of PTFE sandwiched between two layers of cloth. The wearer stays dry and comfortable no matter how energetic they are or what the weather is like. Sweat can always pass out through the fabric but no water can get in.

Gore-tex is waterproof and windproof yet it allows the moisture from sweat to pass through.

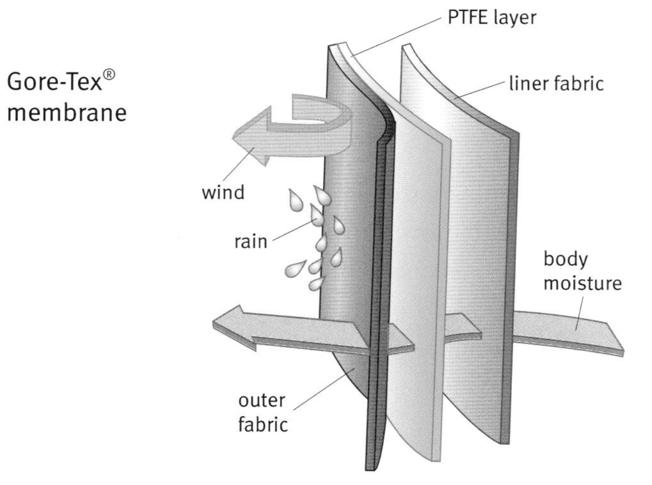

There are billions of tiny holes in the film of PTFE. The holes are 20 000 times smaller than a raindrop but 700 times larger than a water molecule.

Kevlar

Nearly all the early synthetic polymers were discovered by accident. But once chemists started to understand how polymerization works, they could predict how reactions might take place. This meant they could plan to make a polymer with certain properties.

Du Pont is a huge multinational company with a special interest in polymers. The company wanted to make a very strong but lightweight polymer with a high melting point. The chemists designed and made a polymer with very long molecules, linked together in sheets. These sheets were themselves tightly packed together in a circular pattern.

One of the scientists involved in the research was an American, Stephanie Kwolek. Her job was to make small quantities of the new polymer and turn it into a liquid. Once it was liquid the polymer could be forced through a small hole to make fibres. The problem was that the polymer would not melt nor would it dissolve in any of the usual solvents.

Stephanie Kwolek experimented with many solvents. She eventually found that the new polymer would dissolve in concentrated sulfuric acid. This is a highly dangerous chemical that can cause severe burns. But, fibres of the new polymer were manufactured in this way. This was the origin of Kevlar which is five times stronger than steel. It is used for bullet-proof vests and to reinforce tyres. A similar polymer called Nomex is used in protective clothing for racing drivers.

Stephanie Kwolek wearing protective gloves made of KEVLAR®. She discovered how to turn this polymer into fibres.

Copying nature

The inventor of Velcro, George de Mestral, was copying seed pods that he found stuck to his socks when he was out walking. The pods were covered with minute hooks that attached themselves round threads in the socks.

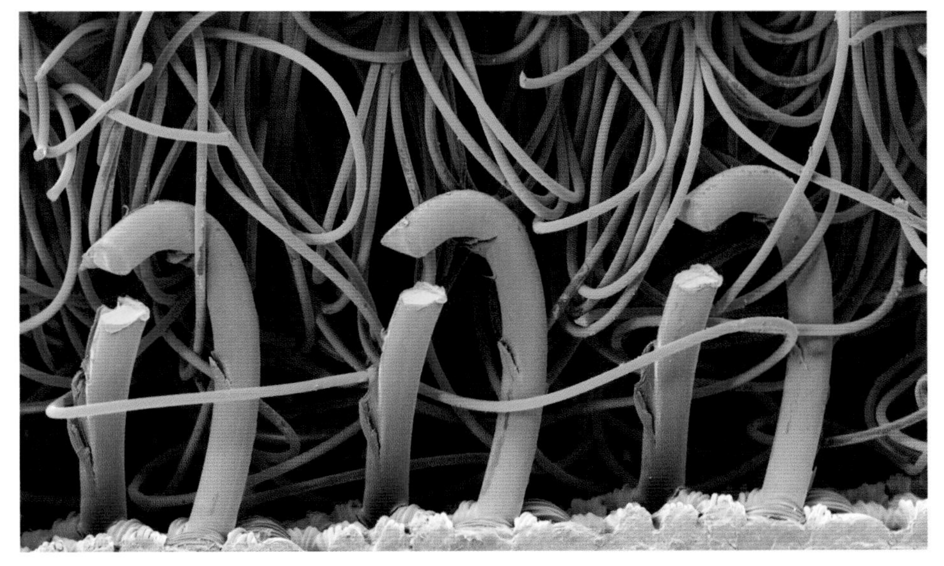

A magnified view of the nylon hooks and loops in Velcro material. This is a false colour image taken with an electron microscope. The loops are loosely woven strands. The hooks are loops woven into the fabric and then cut. When the two surfaces are brought together they form a strong bond, which can be peeled apart. Magnification $\times$ 30

Questions

4 Skim read Pages 46 – 53 and identify two examples each of polymers, or polymer products: **a** discovered by accident **b** developed by design.

5 What words describe the properties of the polymer needed to make: **a** the hooks in Velcro **b** the loops in Velcro.

Find out about:

- unsustainable development on Easter Island
- wood from sustainable forests

These amazing carved heads are all around the coast of Easter Island. They stare sadly out to sea from their treeless landscape.

Wangari Maathi won the 2004 Nobel Peace Prize for her work promoting sustainable development. Her efforts have encouraged women in poor communities to plant over 300 million trees in Kenya.

H Is it sustainable?

Modern lifestyles depend on **natural resources**. Some of these resources provide us with warmth and light. Some of them make products. Either way, using them affects our future.

We have to think whether we can replace them and whether we can sustain this lifestyle. This is a lesson that was learnt the hard way by the people of Easter Island.

The Easter Island story

Easter Island is a remote place in the middle of the Pacific Ocean. It is famous for its gigantic rock heads. These were carved by the Polynesian people.

The Polynesians arrived by boat in about 600 AD. They found a lush, palm-covered island. It was an ideal place to live. They settled there and the population grew to several thousand.

However, when Captain Cook landed there in March 1774, he found the statues toppled and just a few islanders. They were barely managing to survive on the barren island.

What went wrong?

The islanders' main resource was wood. They used this to build houses, boats, and fires. It gave them a good, comfortable life. However, they were using the trees more quickly than they could re-grow. So the stocks were being depleted. Eventually, the last tree was cut down.
They could not even build a boat to escape.

If only they had not used up all the wood – their main resource. If only they had managed their forests by planting more trees. If only they had lived in a way that was **sustainable**.

It is easy to look back on their rashness and make judgements. But there are now signs that we are doing the same with natural resources in places all over the world.

Are we sustainable?

Even now, there are forests that are dwindling. Tropical hardwood trees are being cut down to make furniture. These trees can take a hundred years to grow. This means that not many new trees reach maturity each year, not as many as are being cut down. This is not a sustainable use of timber.

It took the Easter Islanders about a thousand years to run out of wood. The hardwood forests of South America may be big but they are not infinite.

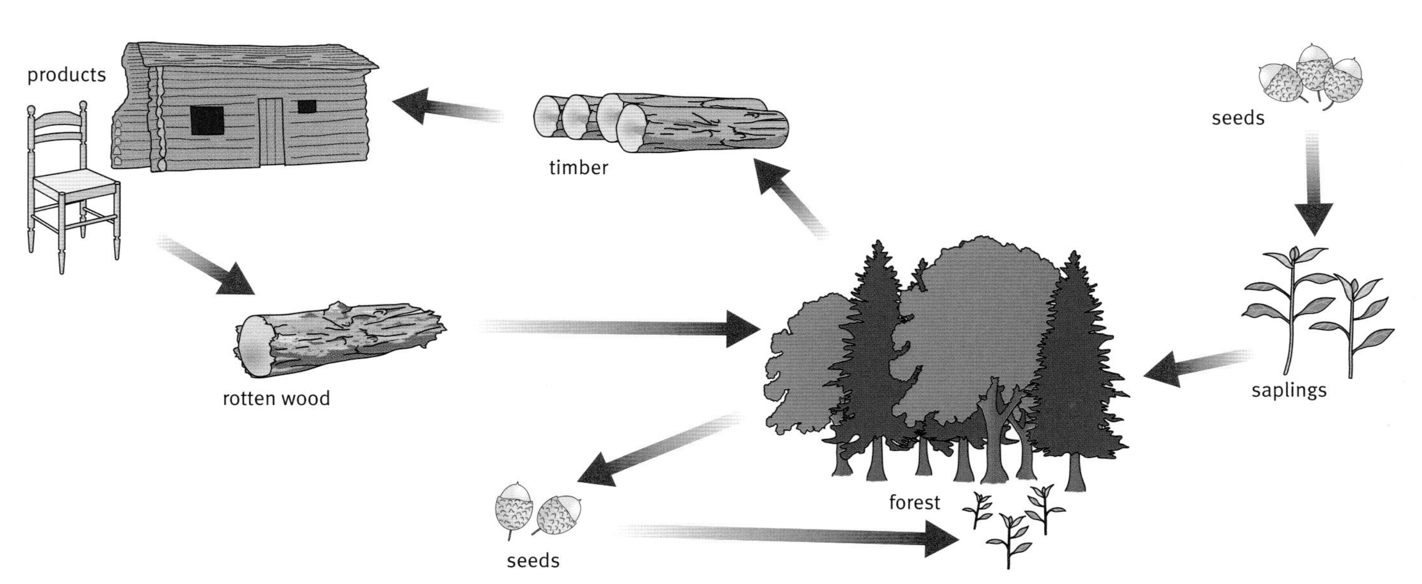

The life cycle of a forest works over hundreds of years.

Is it just trees?

Trees are not the only natural resource that is being depleted. We use materials made from metals, rocks, coal, and oil. These will not last for ever. We have to try to use them in a sustainable way. Otherwise, our isolated planet will run out.

Key words
natural resources
sustainable

Questions

1 Mahogany is a hardwood. Mahogany trees take 100 years to reach a useful size. A furniture company wants to start a sustainable forest.

a **i** They plant one new tree every year. How long do they have to wait until their first tree is ready?

ii By then, how many trees will there be in the forest?

b They decide to plant five trees every year.

i They wait the same time as in part **a i**. How many trees will there be in the forest?

ii How many trees should they cut down each year?

c Imagine they find a forest with 5000 trees. How many should they cut down each year?

② Look at these things that people do: use hardwood trees for furniture; farm vegetables; use limestone for buildings; fish for cod in the North Sea; use wool for clothes; rely on crude oil.

a In each case:

i write down whether you think it is sustainable or not;

ii explain your reasons.

b Choose one example that you think is unsustainable. Describe how to make it sustainable.

③ Farming and forestry both involve growing and harvesting plants. They can both be sustainable.

a Draw a diagram for the life cycle of a field of wheat. Include the timescale.

b It is more difficult to make forestry sustainable. Explain why.

c We use oil, coal, and stones.

i Is it easier or more difficult to make this sustainable?

ii Explain your answer.

Find out about:

- the life of products from cradle to grave
- assessing the impact of all the materials we use

1 Life cycle assessment

In our homes we are surrounded by manufactured goods including furniture, clothes, carpets, china and glass, TV sets, and CDs. The life of each of these products has three distinct phases:

1. a manufacturer makes them
2. people use them and then
3. they throw them away

Each phase uses resources.

CRADLE

- The raw materials for making the product
- The energy used to manufacture it

USE

- The energy needed to use it (for example, petrol in a car)
- The energy needed to maintain it – cleaning, mending etc
- The chemicals needed to maintain it

GRAVE

- The energy needed to dispose of it
- The space needed to dispose of it

Lives or life cycles?

Imagine a television that was bought in 1970 and thrown away in 1981. It contains glass, metals, plastics, and wood. It is now buried under 50 tonnes of rubble in a **landfill**. This is its grave.

The wood will eventually rot because it is biodegradable. But the rest of the materials are stuck there. This is not sustainable. The materials had a life but not a life cycle.

Once the life of a product is over, its materials should go back into another product. This is **recycling**.

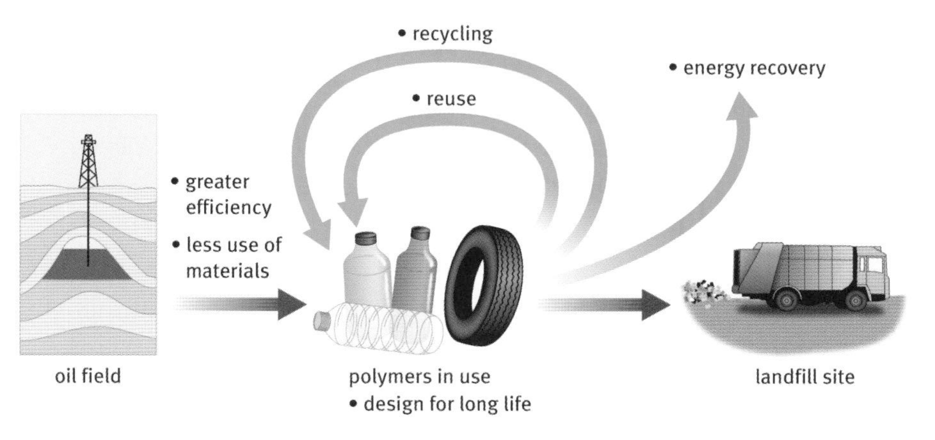

Oil and products from oil, such as polymers, have high value. They lose value as they are used up and end as waste. The aim now is to slow down the journey of materials from natural resources to landfill sites or **incinerators**.

Key words
landfill
recycling
incinerators
life cycle assessment

Life cycle assessment

Manufacturers are now assessing what happens to the materials in their products. This **Life Cycle Assessment** (LCA) is part of legislation to protect the environment.

The aim is to cut the rate at which we use up natural resources that are not renewable. One approach is to find ways to slow down the flow of materials from resources to waste.

One example of legislation to improve sustainability deals with Waste Electrical and Electronic Equipment – known as WEEE.

A WEEE problem

Manufacturers have to pay the costs of dealing with WEEE. They can recycle, burn, or bury it. But the most cost-effective solution is recycling – especially if they make their products easy to recycle. The more they recycle, the more they save. And this keeps the price of the product down. It also reduces the impact on the environment.

Also, they are likely to make products degrade easily. The casing of Sony's latest DVD player is made from a vegetable plastic. This is biodegradable. When it is buried, it will rot and its energy and chemicals will go back into the soil. Not only is it less expensive than landfill, it is part of a natural cycle for these chemicals.

Weeeman is made from electronic waste. Its size shows the amount of waste that one person is likely to produce in a lifetime, from electronic toys to mobile phones.

Questions

1. Suggest examples of attempts to slow down the flow of materials from resources to waste. Include examples of: **a** reuse; **b** recycling; **c** recovering energy
2. Choose a product that has been designed to reduce its impact on the environment. **a** Describe the product. **b** Explain how its environmental impact has been reduced.

Find out about:

- the birth, life and death of polymer products
- how to get rid of plastic waste

J Life cycle of a synthetic polymer

All manufactured goods have a cost to the environment. Manufacturers are now being asked to analyse the life cycles of their products – from cradle to grave. A polythene bottle is a typical example.

Oil companies extract millions of tonnes of oil every day.

Cradle

Polythene is a polymer made from ethene. This comes from crude oil. So the story starts underground.

Getting the oil

Oil companies extract crude oil from wells under the ground or under the sea bed. It has taken millions of years to form from the dead remains of plankton.

Crude oil is a mixture of lots of **hydrocarbons**. Most of them are used to provide energy for transport, homes, and manufacturing industry. Only about 4% of the crude oil is used for **chemical synthesis** to make polymers.

The hydrocarbons in oil have varying amounts of carbon and hydrogen in their molecules. Within this mixture there are hydrocarbons that are useful as fuels and lubricants.

Plant for processing chemicals from oil.

Making the polymer

After oil is pumped from the ground it is taken to a refinery. There, it is distilled in a tower which separates the molecules according to size. This is possible because the lighter molecules boil and turn into gas before the heavier ones.

The refinery takes the hydrocarbons with 20 or more carbons atoms and breaks them down into smaller molecules. One of these products is ethene. This is piped to a **petrochemical plant** where it is turned into polythene (see Pages 130–131).

Blow-moulding is a way of making plastic bottles. The machine extrudes a short tube of hot plastic into a mould. Then compressed air forces the plastic to take the shape of the mould. This process needs energy to heat the plastic and run the machinery. It needs water to cool the mould.

Making the product.

The raw polythene is sent to a factory where it is moulded. It is heated and forced into a bottle-shaped mould. Now the bottle can leave its cradle.

Use

The polythene bottles are transported to a filling plant, filled with a sports drink and sent to supermarkets and shops. People buy the drink, consume it, and throw the bottle away.

Graves

From here on, the story can follow different routes.

Recycling

Most plastics can be melted down and re-moulded into something else. So recycling seems to be the easy and obvious answer. Unfortunately it is not always as easy at it seems. There are a number of problems:

- *Sorting* – there are so many different types of polymer and it is difficult to separate them either in the home or at a recycling plant.
- *Cleaning* and *separating* – old containers may have food stuck to them. Some products may have more than one material. For example, trainers consist of several polymers tightly glued together.
- *Money* – all stages of recycling cost money. Collecting and sorting rubbish is expensive. Transport costs may be high. Processing recycled material may cost more than making the polymer from oil.

Recovering the chemicals

It is sometimes possible to convert the polymers in plastics back to simple molecules. This produces new raw materials for the chemical industry and is a kind of recycling. It can be possible to recover between 80 to 90 per cent of the chemicals this way, with 10 to 20 per cent being burnt to provide the energy for the process.

Recovering the energy

Some polymers can be burnt. This reduces the need to use fresh fuel from crude oil.

They have to be burnt at a very high temperature to make sure they are fully combusted. This is done in special incinerators.

Landfill

Unfortunately, most polymers still end up being tipped into holes in the ground. We call this landfill. This really is a waste.

Most of our rubbish ends up being tipped into holes called landfill.

Key words

crude oil
hydrocarbons
chemical synthesis
petrochemical plant

Questions

1. Plastics make up only 8% by weight of domestic waste but they take up 20% of the volume of rubbish collected. How do you account for this?
2. Draw and label a diagram to show a possible life cycle for a plastic soft-drinks bottle.
3. It might seem that the best solution would be to reuse articles made of plastic instead of throwing them away or recycling them. Why is this often impossible or even undesirable?
4. Burning a plastic such as polythene gives out energy. What are the other products of burning this polymer?
5. Is burning waste better than dumping it in landfill? Why is there often opposition to proposals to build waste incinerators?

Find out about:

- inventing a more sustainable product
- life cycle assessment to test claims

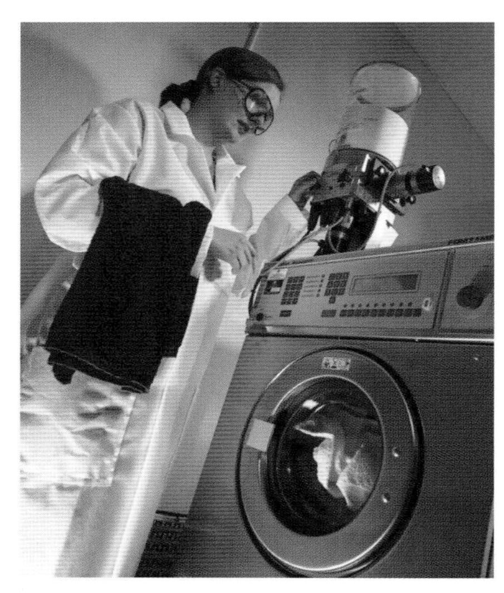

Rebecca Fay has the job of washing the towels over and over again to see if the finish will wear off.

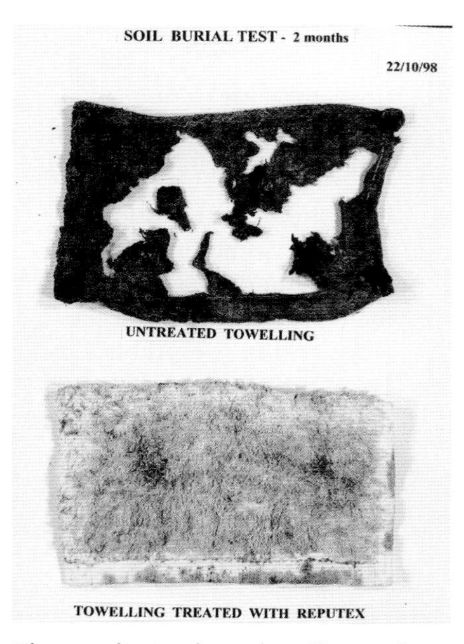

The results to show the effects of burying towels in soil. This compares the speed with which treated and untreated towels rot away. Anything which rots when buried in soil is biodegradable.

K Anti-bacterial towels – a more sustainable alternative?

You know how it is with gym towels. You have to wash them every day, which uses up water, energy, and washing powder. If you don't, they go manky and smelly in the bottom of a gym bag.

A company in Manchester called Avecia have discovered a polymer which kills bacteria. This polymer can be added to cotton. The polymer has positive electrical charges along its chain. Cotton has negative charges along its molecular chains. Positive and negative charges attract, so the two can easily be stuck together. Finished towels can be treated with the polymer, which will make them stay fresher longer.

The question is whether these new towels will save energy, water, and detergent by needing less washing. Avecia asked Richard Blackburn of Leeds University to carry out a life cycle assessment on both processed and unprocessed towels.

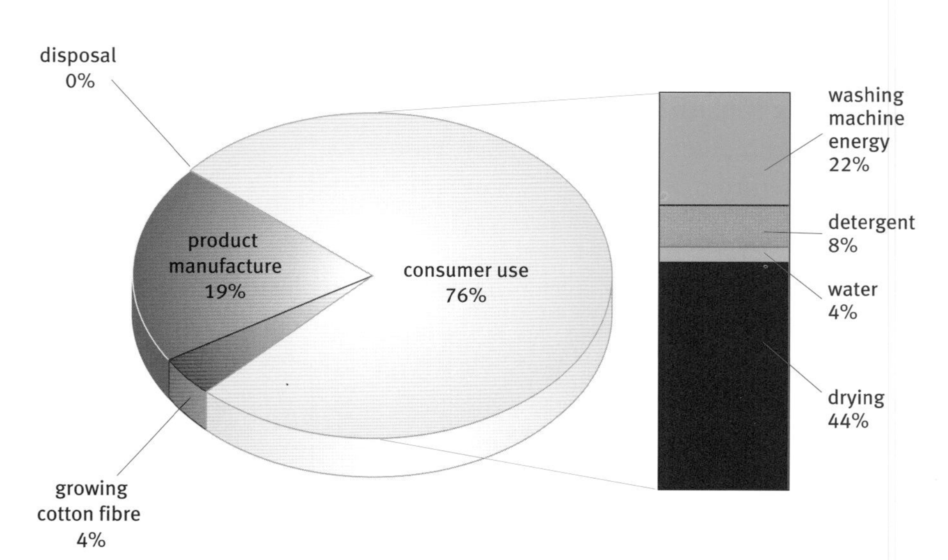

The pie chart show the energy consumption over the life of the towels. The bar chart shows the breakdown of the energy consumption during the years when the towels are in use.

First of all, Richard looked at towels in general. He worked out that most of the energy use in the lifetime of a towel happens in the home when the towel is being washed. If energy can be saved at this stage, it could be seen as beneficial to the environment.

The table shows the energy used in the production, use, and disposal of plain and treated towels. Richard assumed that the towels would be tumble-dried for half their washes. He calculated the washing costs over a year.

To find out how the towels would biodegrade after they are thrown away, they were buried and dug up.

Untreated Towels	Energy (kWh)	Water consumption (l)	Chemical/Detergent (kg)
growing cotton fibre	9.35	0	0
product manufacture	44.7	32.67	0.22
washing machine energy	47.82	0	0
detergent	18.25	0	2.16
water	6.28	1460.00	0
drying	98.55	0	0
consumer use total	170.89	1460.00	2.38
disposal	0.02	0	0
lifetime total	224.97	1492.67	2.38

Treated Towels	Energy (kWh)	Water consumption (l)	Chemical/Detergent (kg)
growing cotton fibre	9.35	0	0
product manufacture	44.87	32.67	0.25
washing machine energy	19.13	0	0
detergent	7.3	0	0.79
water	2.51	584.00	0
drying	39.42	0	0
consumer use total	68.36	584.00	0.79
disposal	0.02	0	0
lifetime total	122.60	616.67	1.04

Questions

1 Which stage in the life of a towel uses most energy?

(2) Why is energy needed:

- to provide detergent
- for the water supply when washing
- for drying?

3 Draw up a bar chart to compare the energy use of treated and untreated towels to include: washing machine energy, energy needed to supply detergent, energy needed to supply water, and drying.

4 Does the treatment make the towels more or less biodegradable?

5 Work out figures to show whether or not the anti-bacterial coating saves energy: **a** in production; **b** in its use **c** over its life time?

(6) If you bought one of the new treated towels, would it affect how often the towels get washed where you live?

(7) Would you buy one of the new treated towels **a** to use all the time, or **b** to use for sport?

C2 Material choices

Science explanations

Theory can help chemists to develop new materials with useful properties. Some materials consist of very long chain molecules. One way of developing new plastics and fibres is to changing the length and arrangement of these big molecules.

You should know that:

- one way of comparing materials is to measure their properties, including:
  - melting points
  - strength (in tension or compression)
  - stiffness
  - hardness
  - density
- when choosing a material for use it helps to have an accurate knowledge of its properties
- polymers are materials which are made up of long-chain molecules
- there are natural polymers such as cotton, paper, silk, and wool
- there are synthetic materials which are alternatives to materials from living things
- there are many examples of modern materials made of synthetic polymers that have replaced older materials such as wood, iron, and glass
- crude oil is mainly made of hydrocarbons
- most of the products from oil are fuels and only a small percentage of crude oil is used to make new materials
- refining crude oil produces some small molecules which can join together to make very long-chain polymers; the process is called polymerization
- polymerization produces a wide range of plastics, rubbers, and fibres
- the properties of polymer materials depend on how the long molecules are arranged and held together
- it is possible to modify polymers to change their properties. This includes modifications such as:
  - increasing the length of the chains
  - cross-linking the molecules
  - adding plasticizers to lubricate the movement of molecules
  - crystallinity by lining up the molecules

Ideas about science

Scientists measure the properties of materials to decide what jobs they can be used for. Scientists use data rather than opinion in justifying the choice of a material for a purpose

You should be able to:

- suggest why a measurement may not be accurate

Scientists can never be sure that a measurement tells them the true value of the quantity being measured. Data is more reliable if it can be repeated. When making several measurements of the same quantity, the results are likely to vary. This may be because:

- you have to measure several individual examples, for example, several samples of the same material
- the quantity you are measuring is varying, for example, different batches of a polymer made at different time
- the limitations of the measuring equipment or because of the way you are using the equipment
- you should be able to say why you think there is or isn't a real difference between two measurements of the same quantity

Usually the best estimate of the value of a quantity is the average (or mean) of several repeat measurements. The spread of values in a set of repeated measurements give a rough estimate of the range within which the true value probably lies. You should:

- know that if a measurement lies well outside the range within which the others in a set of repeats lie, then it is an outlier and should not be used when calculating the mean.
- be able to calculate the mean from a set of repeated measurements.

Making choices about the uses of materials:

- a life cycle assessment (LCA) tests:
  - a material's fitness for purpose
  - the effects of using the materials from its production from raw materials to its disposal
- the key features of a life cycle assessment include:
  - the main energy inputs
  - the environmental impact and sustainability of making the material from natural resources
  - the environmental impact of making the product from the material
  - the environmental impact of using the product
  - the environmental impact of disposing of the product by incineration, landfill, or recycling
- when making decisions about the uses of materials it is important to be able to:
  - know that some questions can be addressed using a scientific approach, and some cannot
  - identify the groups of people affected, and the main benefits and costs of a course of action for each group
  - explain whether the use of a material is sustainable
  - show you know regulations and laws control scientific research and applications
  - distinguish between what can be done from what should be done
  - explain why different decisions may be taken in different social and economic contexts

Why study food?

Today, most of us do not have to spend time growing or catching food. Modern farming needs only a few people to make all our food. It is important that food is safe to eat.
Food safety depends on the care taken at every stage in the food chain; from farm to home.

The science

Science can help to explain how farming affects the natural environment. For example, making and using fertilizers can have a big effect on the 'cycling' of elements such as nitrogen.

Science can also explain the chemical changes that take place in your body when you eat food. Research can tell us about the effects of diet on health, and it helps doctors treat diseases such as diabetes.

Ideas about science

Making the right choices about food and farming can help to make the food chain more sustainable. Governments try to protect consumers by regulating the food chain.
The decisions they make need to use scientific information so that judgements about risk are based on evidence.

Food matters

Find out about:

- the food chain from farm to plate
- farming methods and their effects on the environment
- natural and artificial chemicals in food, including food additives
- the possible links between obesity and diabetes

Find out about:

- the food chain from farm to plate

A The food chain

Bread, cakes, biscuits, and pasta are at the end of a long trail of events that starts on the farm. This is often called the **food chain**. That is the chain that links farms to your plate of food.

One example of the food chain starts with wheat and ends with a slice of bread.

Farmers use fertilizers or manures to keep the soil **fertile**. The soil must contain enough compounds of nitrogen, phosphorus, and potassium for healthy plant growth. It must also contain smaller quantities of other elements.

On the farm

Farmers plant seeds of wheat in the soil. The seeds grow to make new plants. At **harvest** time, combine harvesters cut the crop, thresh it, and separate the seeds in the ears of wheat from chaff and the straw. Flour is made by grinding the seeds.

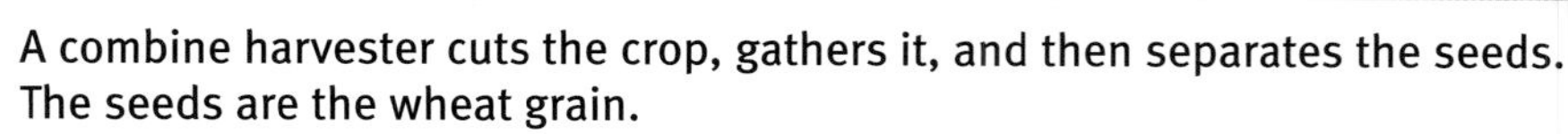

A combine harvester cuts the crop, gathers it, and then separates the seeds. The seeds are the wheat grain.

At the mill

A lot of food looks very different from the raw crop. Wheat seeds, for example, are broken up by rollers in a mill to turn them into flour.

A miller scoops wheat grains before they are milled to make flour.

On the road

Transport of food is an important part of the food chain. Much of your food now travels many miles from where it is grown before it reaches your home. More energy is needed if food has to travel a long way. Usually, the greater the distance, the less sustainable the source of food.

At the bakery

Bakers mix flour with water, fat, and yeast to make bread dough. Protein in the flour mixes with the water to make gluten. The yeast grows in the dough. As it grows, it ferments sugars from the flour. This produces carbon dioxide gas, which makes the dough rise. Gluten traps the carbon dioxide, giving bread its characteristic texture.

Bakers shape the dough to make loaves, rolls, or other products. They let the dough rise again. Next they bake the dough in a hot oven to make bread.

A baker in a supermarket mixes the ingredients to make bread dough.

Taking bread from an oven in a commercial bakery.

In the supermarket

People who buy and eat food have choices to make.

- Does it taste good?
- Is it good for you?
- Is it good for the environment?

When buying bread there are many choices: white or wholemeal? sliced or unsliced? organic or not?

Key words
food chain
harvest
fertilizers
fertile

Questions

1 Make a flow diagram to show the stages of the food chain, from the farmer's field to a piece of bread on your plate.

(2) The food industry and biologists use the term 'food chain' in different ways. Give an example to show what the term 'food chain' means in biology.

(3) Some people want us to eat more food that is grown nearer to our homes. Suggest **a** advantages and **b** disadvantages of choosing to buy food that is grown locally.

Find out about:

- methods used to keep soils fertile
- ways of protecting crops from pests
- the cycling of elements in environment

B Farming challenges

Farmers have to make sure that their crops grow well. This means that they have to keep the soil fertile. They also have to make sure that their crops are not short of water.

Farmers must also protect their crops from pests and diseases.

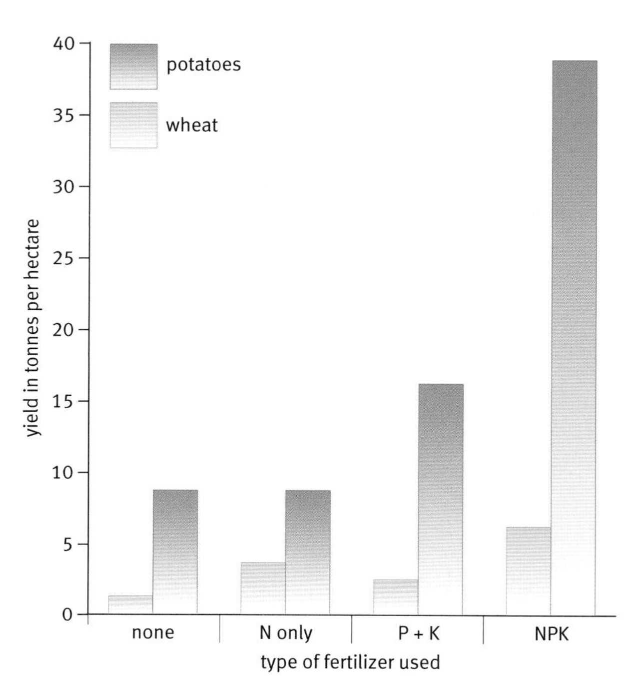

Each crop has its own nutrient needs. Fertilizer may contain one or more of the elements nitrogen (N), phosphorus (P), and potassium (K).

The nutrient challenge

Chemicals for plant growth

Plants need to make **carbohydrates**, **proteins**, and other chemicals such as oils as they grow. Carbohydrates include sugars, starch, and cellulose. These are compounds of three elements: carbon, hydrogen, and oxygen.

Plants make sugars from carbon dioxide and water. They need energy from light to do this. They take in the water from the soil and carbon dioxide from the air. The process is called photosynthesis.

Plants take in other elements from the soil to make proteins. One of these elements is nitrogen. There are nitrogen compounds dissolved in soil water. The roots of growing plants draw in soil water containing the nitrogen compounds.

Plants also need phosphorus and potassium from the soil. Phosphorus helps roots to grow better. Potassium is important for making flowers, fruits, and seeds. Plants also need traces of many other nutrients for good growth.

Cycles of nutrients

Imagine a wild apple tree growing in a hedge. It takes in nitrogen from the soil as it produces leaves and fruit in the spring and summer.

Each autumn, the apples and leaves fall to the ground and rot away. Rotting releases nitrogen compounds from the apples and leaves. Rain washes the nitrogen compounds back into the soil. In this way the nitrogen is recycled.

However, if the apple tree is growing in an orchard, people pick the apples from the tree. This means that less nitrogen can be recycled back to the soil.

Therefore, farmers use fertilizers and manures to return to the soil the nutrients removed as crops grow and are harvested.

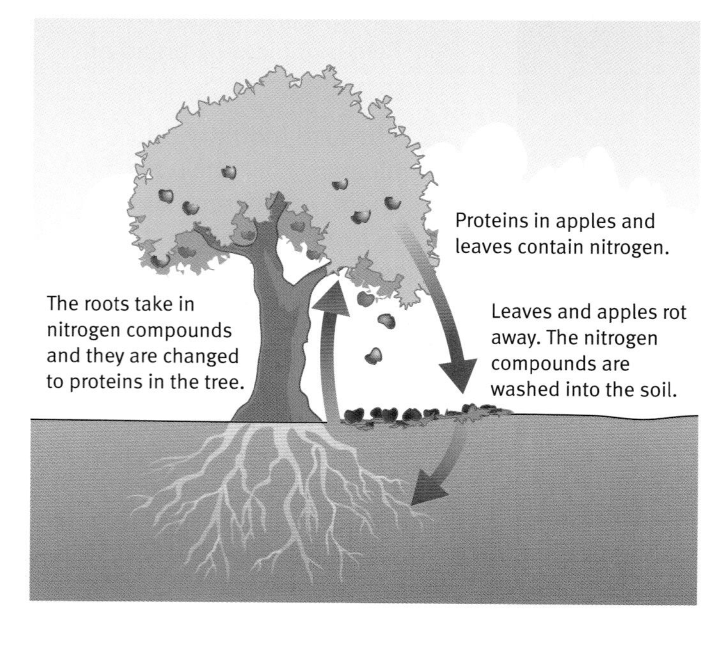

The pest challenge

Pests

Farmers try to protect their growing crops from **pests**, which include:

- insects
- weeds
- diseases caused by fungi and viruses

Insects eat the crops. Weeds compete for light, water, and nutrients. Fungal diseases make the plant sick so that it does not grow well.

Animal slurry is used to fertilize a wheat crop.

Controlling the pests

One way of controlling pests is to use chemicals to kill them. Some **pesticides** are natural chemicals. One example is pyrethrum from chrysanthemums. Other pesticides are manufactured.

Another way of controlling pests is to encourage predators that feed on the pests. All methods of pest control have advantages and disadvantages.

There are many types of pesticides, including:

- insect killers (insecticides)
- fungi killers (fungicides)
- weedkillers (herbicides)
- slug pellets (molluscicides)

Infection by a fungus can do a great deal of damage to a wheat crop. The fungus quickly spreads once one plant is infected.

Wheat crops are sprayed with a pesticide to prevent disease.

Key words

carbohydrate
protein
pest
pesticide

Questions

1 Where does the carbon come from that plants use to make carbohydrates and other chemicals as they grow?

2 Write a word equation to summarize the chemical change of photosynthesis.

3 Where do plants get the nitrogen, phosphorus, and potassium they need for healthy growth?

4 Why may soil get less and less fertile if crops are grown and harvested in the same place year after year?

5 What conclusions can you draw from the information in the bar chart showing crop yields for different fertilizers.

(6) Why do plant crops grow less well if there are lots of weeds growing in the field?

The nitrogen cycle

Nitrogen, along with carbon, hydrogen, and oxygen, is vital for life. This is because there are important nitrogen compounds in living things. The genetic code is written in DNA molecules, which contain nitrogen. The enzymes that control all living processes in cells are proteins. Proteins contain nitrogen.

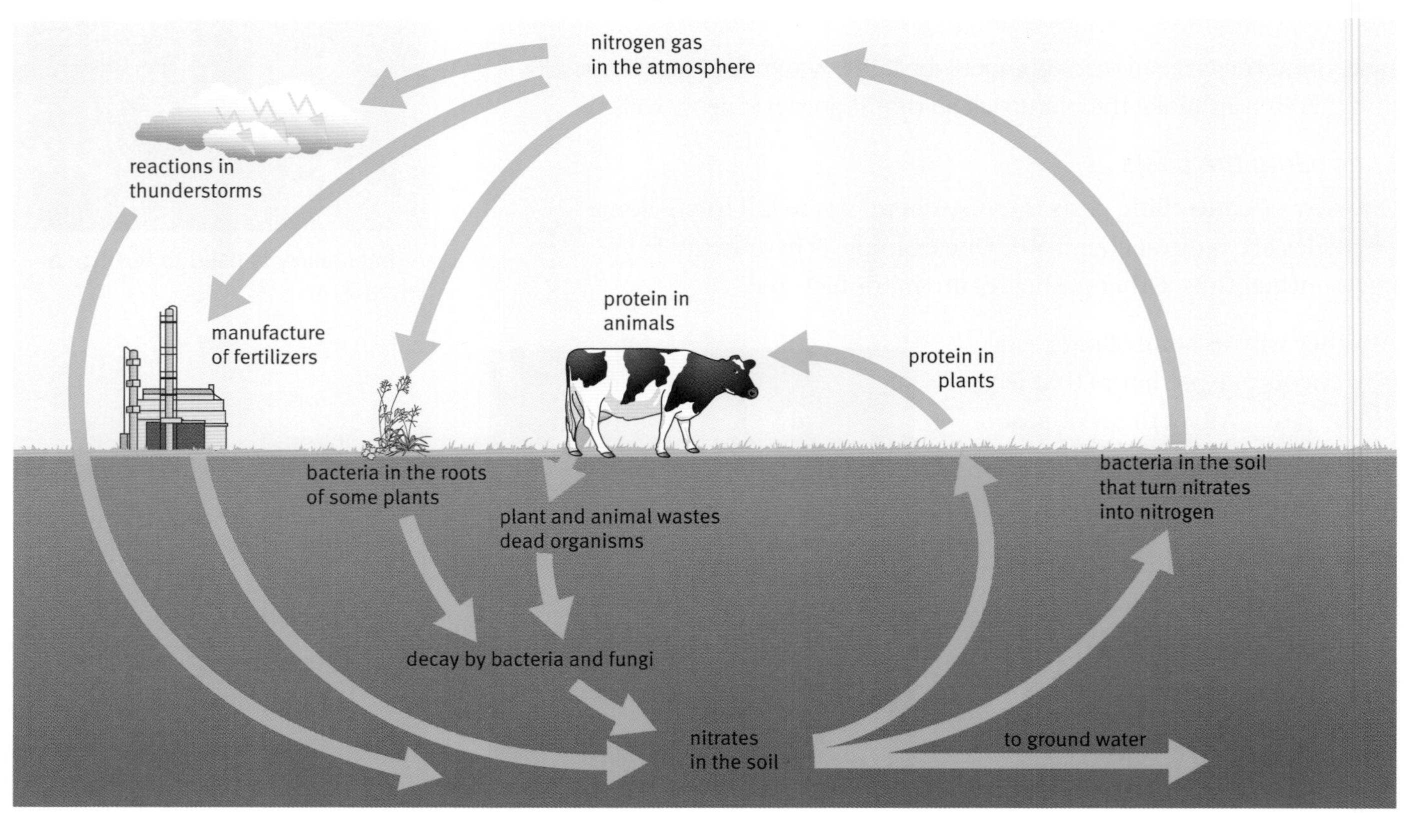

Natural and human activities contribute to the **nitrogen cycle** in the environment.

Plants in the Amazon rainforest. It appears that the soil must be rich in nutrients but this is not so. Most of the soil is clay that has no nutrients. Only the top few inches of soil are fertile. In the warm, humid conditions, fungi and bacteria rapidly recycle the chemicals from dead plants.

The nitrogen cycle in nature

Plants take in nitrates from the soil and use the nitrogen to make protein. Animals eat the plants to get the nitrogen they need from the plant proteins.

Animal urine and dung return nitrogen to the soil. Bacteria breaks down dead animals and plants to soluble nitrogen compounds. These compounds turn into nitrates again.

Adding nitrates to soil

Plants cannot use nitrogen from the air. The gas is too chemically inert. But some natural processes take nitrogen from the air and turn it into nitrates. Nitrates are inorganic salts.

Four of the natural processes which add nitrogen to the soil are:

- the decay of the remains of dead animals and plants
- the growth of bacteria in the soil which take in nitrogen gas to make nitrates
- bacteria in the roots of plants, such as peas, beans, and clover, which can also turn nitrogen into nitrates
- lightning flashes in thunderstorms which make the air hot enough for nitrogen and oxygen gases to react with each other. Then rain washes the new nitrogen compounds into the soil.

Loss of nitrates from the soil

Natural processes and human activity can remove nitrates from the soil. Some bacteria in the soil can convert nitrates back into nitrogen. Also, water trickling through the soil dissolves nitrates. The water can wash them into streams and lakes.

Farming removes nitrates from the soil. They are taken away with harvested crops and with animals used for food.

Restoring soil fertility on farms

Traditionally, farmers used manures and animal waste. These put back into the soil the nitrates removed by harvesting. But many farmers today use fertilizers manufactured by the chemical industry.

The industry uses natural gas or oil, air, and water to make inorganic nitrates on a very large scale. The first steps in the process produce ammonia. The annual production of this compound around the world is more than 130 million tonnes. Most of this goes to make fertilizers.

Making fertilizers takes energy – a great deal of energy. About half of all the energy resources used per year by agriculture in the UK is used to make fertilizers.

Questions

7 Explain why

a crop yields fall if a farmer harvests crops year by year but does not use manures or fertilizers

b a natural tropical forest has no added fertilizer, yet growth of plants does not decline

c planting peas, beans, or clover in a field one year gives an increased yield of whatever grows in the same field the next year

d there is little nitrogen in the peaty soil of watery bogs, yet insect-eating plants, such as sundews, can grow well there

The nitrogen in the air is in the form of nitrogen molecules: N N

Plants cannot use nitrogen in this form. But they can use nitrogen in nitrate compounds and ammonium compounds.

Nitrates contain this group of atoms:

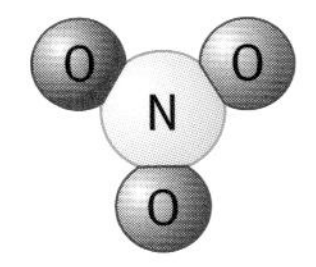

(This group of atoms has a negative electrical charge on it.)

Ammonium compounds contain this group of atoms:

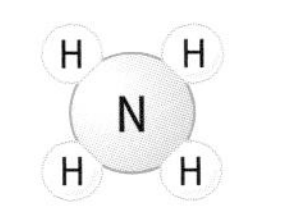

(This group of atoms has a positive electrical charge on it.)

Key words

nitrogen cycle

Find out about:

- intensive farming
- organic farming

C Farming for food

Intensive farming

On the farm

Some **intensive farms** concentrate on animals, such as cows for milk or pigs for meat. Other intensive farms mainly grow crops.

Intensive farmers aim for as large a **yield** as possible. Fields are large so it is easier to work on them with machinery.

Fertile soils

Farmers often use manufactured fertilizers to add nitrogen compounds to the soil. This can make it possible to add just the right amount of fertilizer needed at the right time.

Fighting pests and diseases

Intensive farmers use pesticides. Farmers may spray crops several times:

- to kill weeds
- to kill insects that might carry disease or damage the crop
- to stop the spread of disease

Intensive farming on a big scale.

Weeds compete with the crop for space, light, water, and nutrients. Cleavers in a ripe wheat crop.

Intensive farming and the environment

Intensive farming means that food can be produced on a smaller area of land. This could mean that there is more land for woods and other areas for wildlife. Or it could mean spare land for housing and roads.

Farmers can use intensive methods while being committed to improving the environment. These farmers control their use of fertilizers, pesticides, water, and fuels to minimize their harmful effects.

However, growing the same crop in large fields reduces the variety of wildlife. Also, pesticides kill. They do more than protect the crops. They also kill off the weeds and insects that are food for other living things.

Using too much fertilizer can also do harm. In wet weather, the nutrients can be washed into streams where they help water weeds to grow fast. This can choke the water and kill fish.

Chemicals from farmland have been washed into the Kennet and Avon canal. The canal is rich in nutrients. Algae grow fast and choke the waterway.

Large dairy farms produce a large volume of animal manure. Some of this can be spread on the land but not all. Manures leaking into streams pollute the water and kill fish.

Farming, food, and the consumer

Intensive farming can keep down the cost of food. Working on a large scale helps to bring down costs.

Using fertilizers and pesticides produces large crops with the quality that many people now expect. For example, vegetables and fruit are large. They are all about the same size and free of pests.

However, some pesticides soak right into crops to kill from the inside out. Other pesticides are sprayed on the surface. Traces of pesticides may remain in the food. Some people worry about these pesticide residues, even when the levels are well below the safety margins.

Sustainability

Over half the energy used for agriculture is used to make fertilizers. So intensive farming depends on cheap energy from fossil fuels. This type of farming may do little to recycle nutrients.

Much of the food produced travels large distances before it reaches the public. Nearly 40 per cent of the lorries on our roads carry food. About 12 per cent of the fuel burnt in the UK is for food transport and packaging.

Key words

intensive farm
yield

Questions

1 Make a table with two columns. In one column list the advantages of intensive farming. In the second column list the disadvantages.

(2) Give examples of people who

a benefit from intensive farming

b may be harmed by intensive farming

(3) Draw a simplified nitrogen cycle for an intensive farm which grows crops but has no animals.

Harvesting squash (a variety of marrow) from a field in an organic farm.

Farmer with pigs on an organic farm.

Ladybird eating aphids.

Organic farming

On the farm

On many **organic farms** the farmers keep animals and grow crops.

Fertile soils

Organic farmers use manures instead of fertilizers. So the dung from the animals is used to add nutrients to the soil.

Organic farmers also rotate their crops. This also helps to keep the soil fertile.

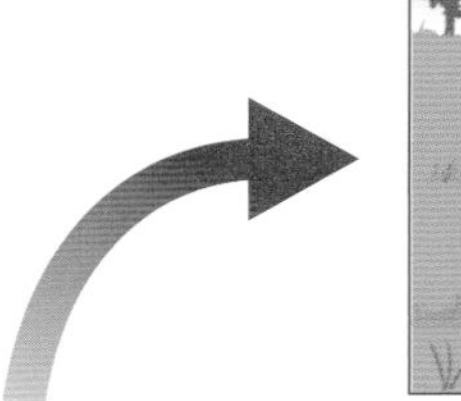

Three years of grass or clover growing in a field.

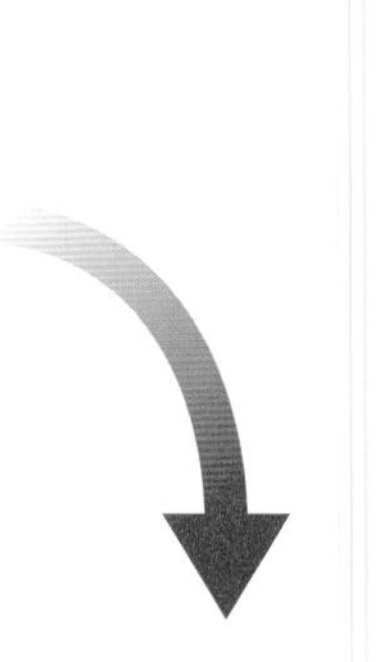

Two years of a cereal crop such as wheat.

A year of a root crops such as beet to feed animals.

Fighting pests and diseases

Organic farmers use natural predators to control pests. This is an example of biological control of pests.

Ladybirds and the larvae of hoverflies, for example, feed on greenfly and other aphids. Smaller fields mean that there are more hedges and ditches. These can be a home for insects and animals that feed on pests.

Crop rotation can also help to prevent disease by breaking the life cycle of weeds and pests. The fungi that cause disease on one crop may not survive on the next crop. So the disease can die out when a different crop is growing on the land.

Organic farmers put up with some weeds. After harvest, the weeds are ploughed into the soil to make it more fertile.

Organic farmers are allowed to use a very small number of chemical pesticides. They may need to get permission before they do so.

Organic farming and the environment

Smaller fields have more hedges round them. This helps to stop the wind blowing away soil from ploughed fields. Hedges also offer homes for wildlife. There are no pesticides killing the insects that are food for larger animals.

Manuring and ploughing can lead to nutrients being washed from the soil into streams.

Farming, food, and the consumer

Organic fruit and vegetables may be smaller and vary more in appearance. Organic food is generally more expensive, because it takes more labour to produce it.

The Soil Association is one of the organizations that sets standards for organic producers. It checks up on organic farms. A farm cannot call itself 'organic' if it does not meet the standards set nationally and internationally.

Some customers choose organic food because they think that it tastes better. Others choose it because they think that organic farmers treat their animals better.

The **Food Standards Agency** reports that there is not sufficient evidence that organic food is healthier. However, some people worry about the residues from the wider range of chemicals used by some intensive farmers.

Sustainability

Organic farmers aim to use **sustainable** resources. They recycle nutrients and produce less waste. Manures from animals fertilize the soil. Straw from cereal crops provides bedding for animals.

Organic farmers save on the cost of fertilizers and pesticides but they pay for more farm workers.

Organic farming often does not cut down on the distance that food travels. Well over half of the organic food eaten in the UK is imported. One estimate was based on a basket with 26 items of organic food. This showed that the total distance travelled by all the products was equivalent to six times round the equator.

Key words
organic farms
Food Standards Agency
sustainable

This logo can only appear on food that has been produced according to strict standards.

Questions

4 Make a table with two columns. In one column list the advantages of organic farming. In the second column list the disadvantages.

5 Give examples of people who

a benefit from organic farming

b may be harmed by organic farming

6 Explain the benefits of crop rotation on an organic farm.

(7) Give examples to show that organic farms 'rely on prevention rather than cure' when it comes to pests and diseases.

(8) Draw a simplified nitrogen cycle for an organic farm with crops and animals.

(9) Explain the meaning of the term 'sustainable development', using examples from intensive and organic farming.

Find out about:

- food additives

Strawberries can quickly turn mouldy. The large quantity of sugar in jam preserves the fruit.

D Preserving and processing food

Preserving food

Preservatives

Some foods last a long time in a kitchen cupboard. These are foods with a long shelf-life. Foods like this often contain **preservatives**.

Traditional preservatives are sugar, salt, and vinegar. These are still used to preserve some foods.

The main purpose of preservatives is to stop mould or bacteria growing in food. Examples of preservatives are:

- sulfur dioxide used to stop dried fruit going mouldy
- nitrites which help to give a longer life to bacon and ham

Antioxidants

Oxygen in the air can make foods go 'off'. Oxygen turns fats and oils rancid. Rancid food tastes horrible. Other foods change colour if they react with oxygen. **Antioxidants** stop these changes happening.

Food processors often add antioxidants to products, including:

- vegetable oils, such as cooking oil
- dairy products, such as butter
- potato products, such as crisps

Processing foods

Food manufacturers use **food additives** to create products that people want to buy and eat.

Colours

Manufacturers use colours to replace the natural colour lost during food processing or storage. They also add colour to make food products look more attractive.

Food colours brighten the coating of these sweets.

Flavourings

Many processed foods and drinks contain flavourings. These are usually added in very small amounts. They give a particular taste or smell that was lost in processing or not naturally present.

Natural flavours are a complex mixture with hundreds, even thousands, of different chemicals. It is very difficult to mimic a natural flavour exactly by mixing chemicals.

Sweeteners

Sweeteners replace sugar in products such as diet drinks and yogurt. Sweeteners, such as aspartame and saccharin, are many times sweeter than sugar. So only very small amounts are used.

Emulsifiers and stabilizers

Emulsifiers help to mix together ingredients that would normally separate, such as oil and water. **Stabilizers** help to stop these ingredients from separating again.

Emulsifiers and stabilizers also give foods an even texture. Manufacturers need them to make foods such as low-fat spreads and yogurt.

Manufacturers use emulsifiers to stop food ingredients in these foods from separating: for example, ice cream, chocolate, cakes, low-fat spread, and salad cream.

E numbers

An **E number** shows that a food additive has passed safety tests. Its use is allowed in the European Union. The numbering system is used both for additives from natural sources and for artificial additives:

- E100 series: colours
- E200 series: preservatives
- E300 series: antioxidants
- E400+ series: emulsifiers, stabilizers, and other additives

Flavourings do not have E numbers. They are controlled by different laws.

The law says that anything added to food during processing must be shown on the label. Most companies obey the law. Sometimes labels do not list everything. Very occasionally, illegal ingredients are found in food.

Questions

1 Look at the list of E numbers. Write down the reasons for adding these natural chemicals to food:

a lactic acid, E270

b pectin, E440

c cochineal, E120

d vitamin C (ascorbic acid), E300

2 Look at the list of E numbers. Write down the reasons for adding these artificial chemicals to food:

a cellulose, E461

b erythrosine, E127

c BHA (butylated hydroxyanisole), E320

d sulfur dioxide, E220

3 Some people think that adding colour makes food look more attractive. Other people think added colours are unnecessary and misleading. They worry that some colours may harm susceptible people. What do you think? Give your reasons.

Key words

preservative
antioxidant
food additive
emulsifier
stabilizer
E number

Find out about:

- the digestion of chemicals in food
- risks from harmful chemicals in food

E Healthy and harmful chemicals

Chemicals in a healthy diet

Food contains the chemicals that people need to stay alive.

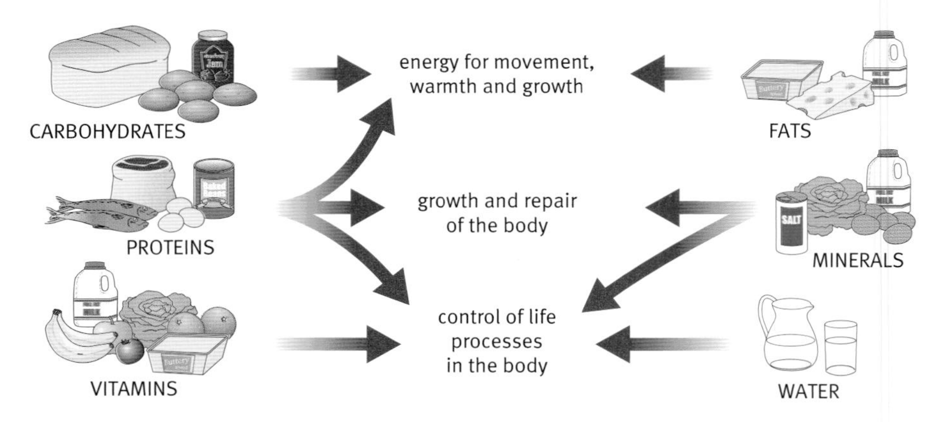

The nutrients in food and what they do.

A healthy diet must also include minerals and vitamins. Water is another vital part of the diet.

The diet should also include chemicals such as **cellulose**. Cellulose from plants makes up the fibre in the diet, which the body cannot digest.

Natural polymers

Starch and cellulose are natural polymers (see Section E in Module C2 *Material Choices*). Both are long chains of glucose molecules. The glucose molecules are linked in different ways in the two polymers. This means that their properties are not the same.

Proteins are also natural polymers (see Section D in Module C2 *Material Choices*). They are long chains of **amino acids**. There are many types of protein. Each protein has a different number of amino acids in a chain. The amino acids are also in a different order.

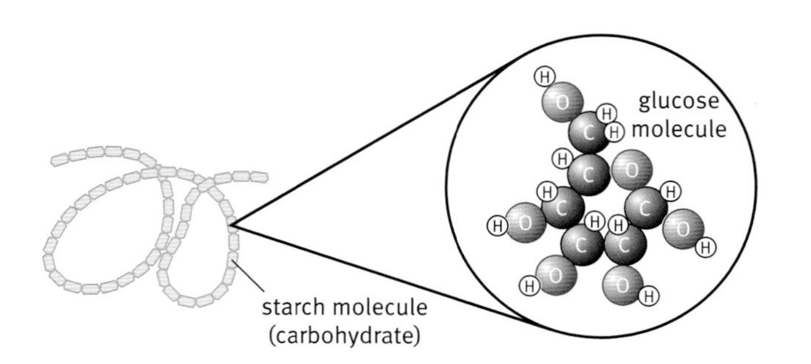

Starch is a polymer made by linking up **sugar** molecules in a long chain. The sugar is glucose. Glucose is made of carbon, hydrogen and oxygen atoms.

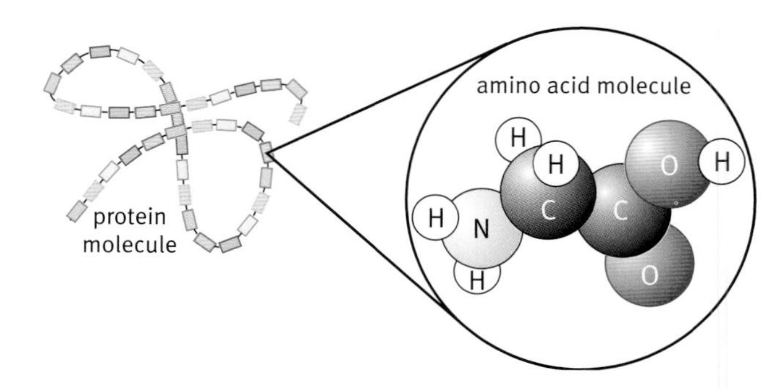

Proteins are polymers with long chains of amino acids. Amino acids are made of carbon, hydrogen, oxygen, and nitrogen atoms. There are sometimes other atoms too.

Digestion

When you swallow, food passes from your mouth to your stomach. Later it moves into your small intestine. Muscles in the gut wall squeeze the food along. They also mix the food with digestive juices. These juices contain enzymes.

The enzymes speed up the chemical reactions which break down the polymers in food into small molecules. This must happen because only small molecules can pass through the wall of the gut into your blood. This breaking down of the large molecules is called **digestion**.

The enzymes break down:

- starch into sugars and
- proteins into amino acids

Enzymes in the human body cannot break down cellulose.

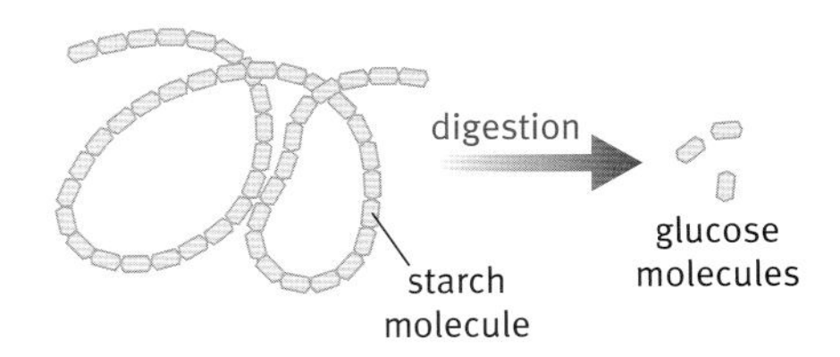

Enzymes in saliva and in the stomach break down starch into the sugar glucose.

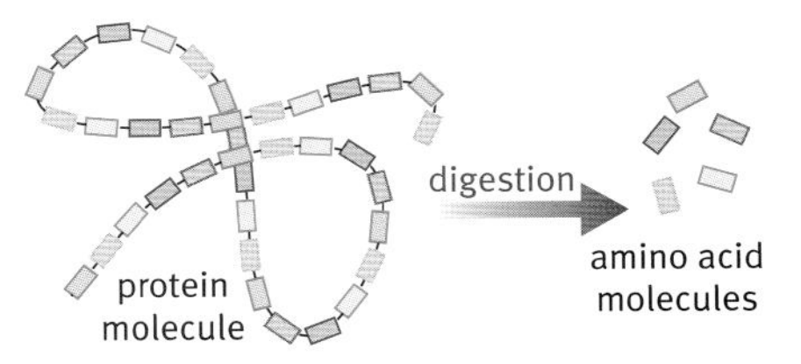

Enzymes in the small intestine break down proteins into amino acids.

Growth

Cells in the body make new cells all the time. These new cells are needed:

- for growth
- to replace worn out cells
- to replace damaged cells

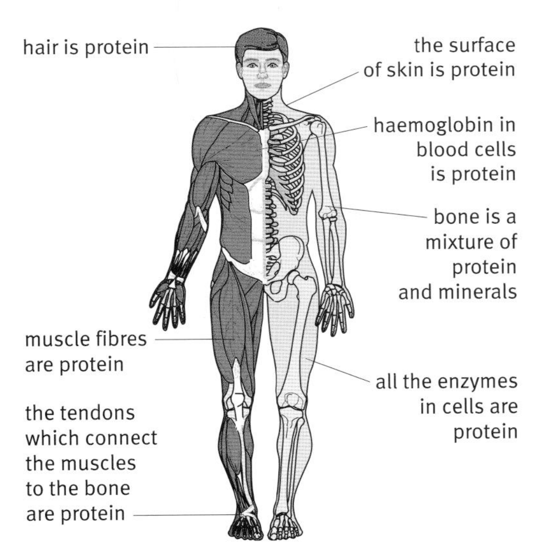

Proteins in the human body: as cells grow, they take in amino acids from the bloodstream. Cells build up the amino acids to make new proteins.

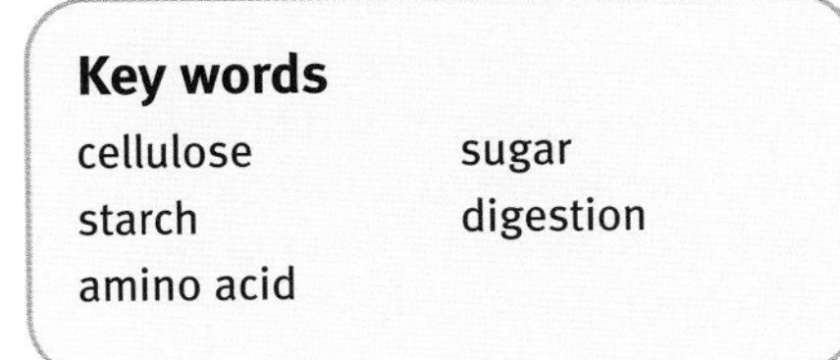

Key words

cellulose
starch
amino acid
sugar
digestion

Questions

1 a Name the three elements found in all carbohydrates.

b Name an element found in proteins that is not in carbohydrates.

2 Potatoes contain starch and cellulose. When you eat potato what happens

a to starch? **b** to cellulose?

3 The level of sugar in the blood rises quickly after eating sweets. It rises much more slowly after eating starchy foods. Why the difference?

4 Scientists estimate that there are about 100 000 different proteins in a human body. How is it possible to makes so many proteins from just 20 amino acids?

5 Copy and complete this flow diagram:

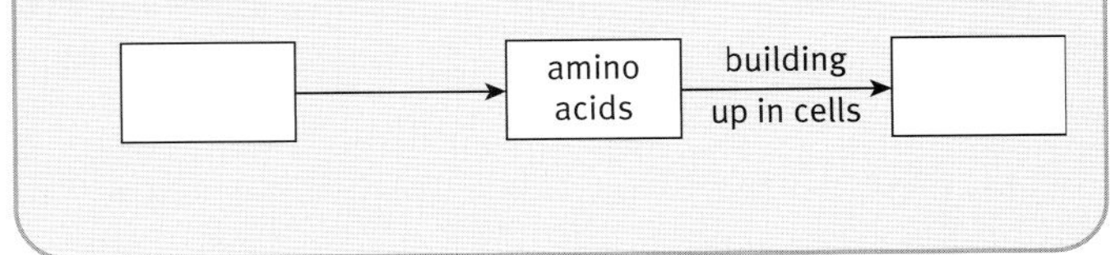

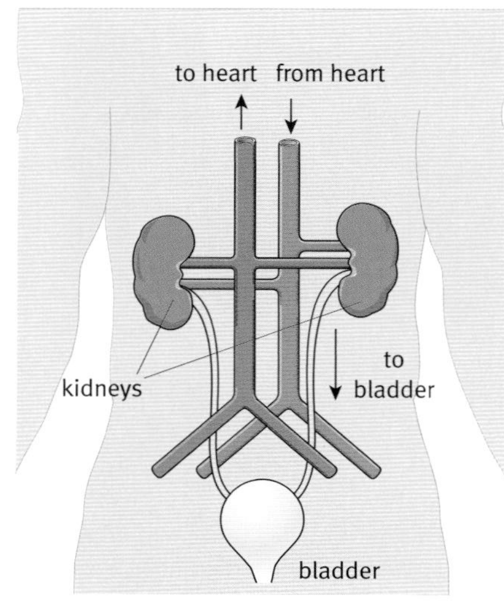

You have two kidneys which filter your blood to remove waste chemicals, such as urea. These wastes include breakdown products from bile, which are yellow.

Excretion

If you eat a lot of protein, you may have more amino acids in your blood than your body can use. The body cannot store the amino acids. It has to get rid of them if their level in the blood is too high.

The liver breaks down amino acids. The nitrogen from the amino acids can turn into poisonous (toxic) chemicals. Instead, the liver converts all the nitrogen compounds to **urea**, which is poisonous but less harmful.

Urea is a colourless chemical which is very soluble in water. The blood carries urea from the liver to the **kidneys**. The kidneys remove the urea from the blood so that it passes out with the urine.

Toxic chemicals in food and drink

Food gives pleasure and is vital for life. However, sometimes foods can be dangerous too. Everyone involved in growing, harvesting, processing, and cooking food has to take care to avoid the dangers that can make people ill. In extreme cases, food can kill.

Most of the people who die by eating mushrooms have eaten the Death Cap. **Toxins** in this variety of mushroom destroy the liver.

Moulds growing on nuts and dried fruit can produce aflatoxins. Aflatoxins can cause cancer. In the EU there are legal limits for aflatoxins in foods, to make sure that people take in as little of them as possible.

Cassava is a root crop. The roots of cassava contain poisonous compounds. The compounds release cyanide, which is very poisonous. Shredding the roots and squeezing out the juice removes most of the toxic compounds. Heating dries the flour. It also gets rid of the rest of the toxins.

Cooking starchy foods at a high temperature can produce acrylamide. The reaction involves an amino acid reacting with glucose. This was discovered in 2002. Scientists are now researching the issue. High doses of acrylamide have been found to cause cancer in some animals and so it may also harm people's health.

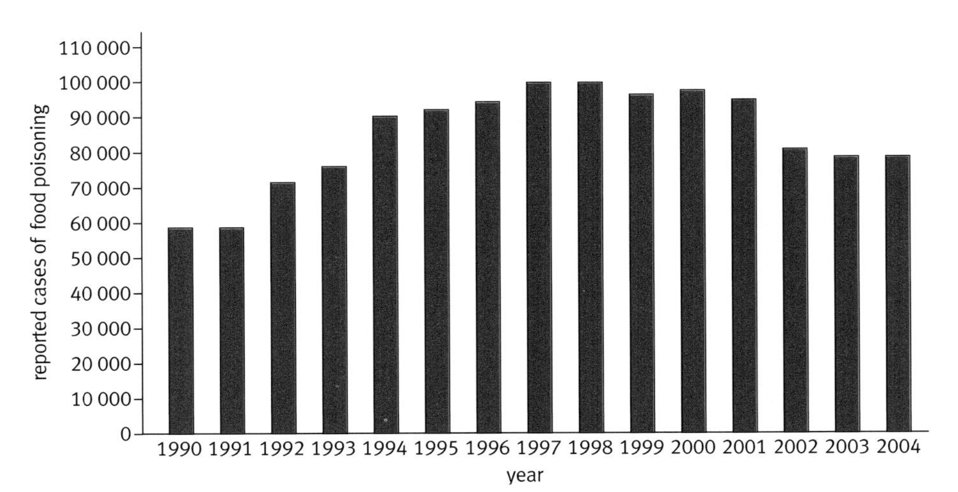

Reported cases of food poisoning in the England and Wales. Some bacteria produce toxins when they grow in food. Cooked foods can be contaminated by bacteria after cooking. Bacteria grow fast at room temperature and soon produce enough toxin to make people sick. This is one of the origins of food poisoning.

Gluten-free foods. Gluten is a protein in wheat and barley. Gluten damages the small intestine in people who suffer from intolerance to this protein. Coeliac disease is the best-known form of gluten intolerance.

Food allergies and intolerance

Allergies arise when your immune system makes the mistake of reacting to a chemical in food as if it were harmful. Most allergic reactions to food are mild, but sometimes they can be very serious.

Some people react to food because they cannot digest all the chemicals in it. Other people react because the chemicals irritate the lining of their gut. This is food intolerance. The chemical that gives rise to the most common food intolerance is lactose, from milk and other dairy products.

Some people are allergic to particular proteins found in peanuts. These proteins are not destroyed by cooking, so both fresh and cooked and roasted peanuts can cause an allergic reaction.

Questions

6 Give one example of a toxic chemical present in food because of:

- **a** the type of crop grown
- **b** the method of farming
- **c** the way the food is stored
- **d** the way the food is cooked
- **e** what happens to the food after cooking

7 Why must cooked food be kept hot or cold, but never just at room temperature?

(8) Suggest three steps that people can take to avoid being harmed by toxic chemicals in food.

(9) Suggest three questions that you would like to ask if you met a scientist doing research into the issue of acrylamide in cooked food.

Key words

urea
kidney
toxins
allergies

Find out about:

- the risk of being overweight
- the two types of diabetes
- risk factors for diabetes

F Diet and diabetes

Healthy eating

What you eat can make a big difference to your health and well-being. As well as the nutrients in a balanced diet, a healthy diet:

- contains lots of fruit and vegetables
- is based on starchy foods, such as wholegrain bread, pasta, rice, and potatoes
- is low in foods with a lot of fat, salt, and sugar, such as salty snacks, soft drinks, and confectionery.

More than half your daily energy from food should come from carbohydrates. Many processed foods contain simple carbohydrates that get into the bloodstream very quickly. They also flow through your body quickly. This means that you soon feel hungry again. It is better to eat foods with complex carbohydrate. This is digested and absorbed more slowly.

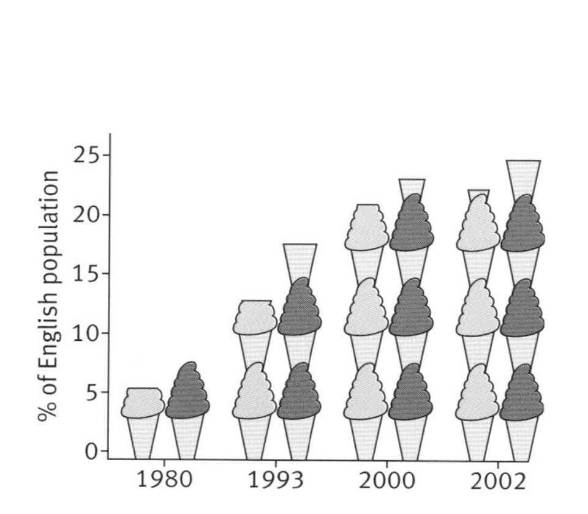

Percentage of males (blue) and females (red) in England who are obese.

Obesity and health risks

Obese people have put on so much weight that it is a danger to their health. **Obesity** is mainly caused by eating too much and not taking enough exercise. Doctors predict that, by 2020, over half of young people will be obese, if childhood obesity goes on increasing as fast as it is now.

Obesity increases the risk of heart diseases. It also increases the risk of other diseases, such as **diabetes**.

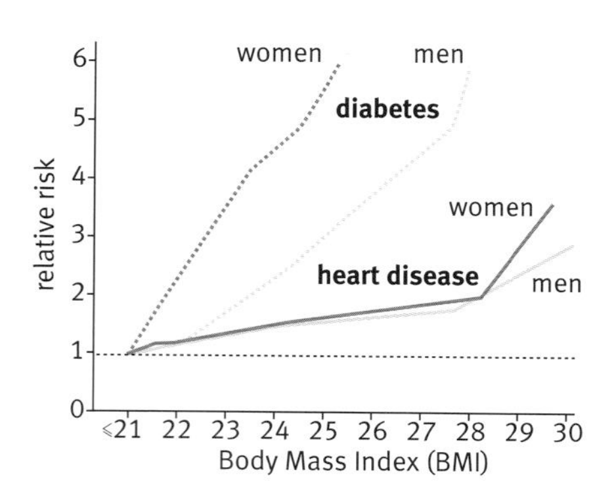

There is increased risk of diabetes and heart disease as the body mass index (BMI) rises. BMI is a number that shows a person's body mass adjusted for height.

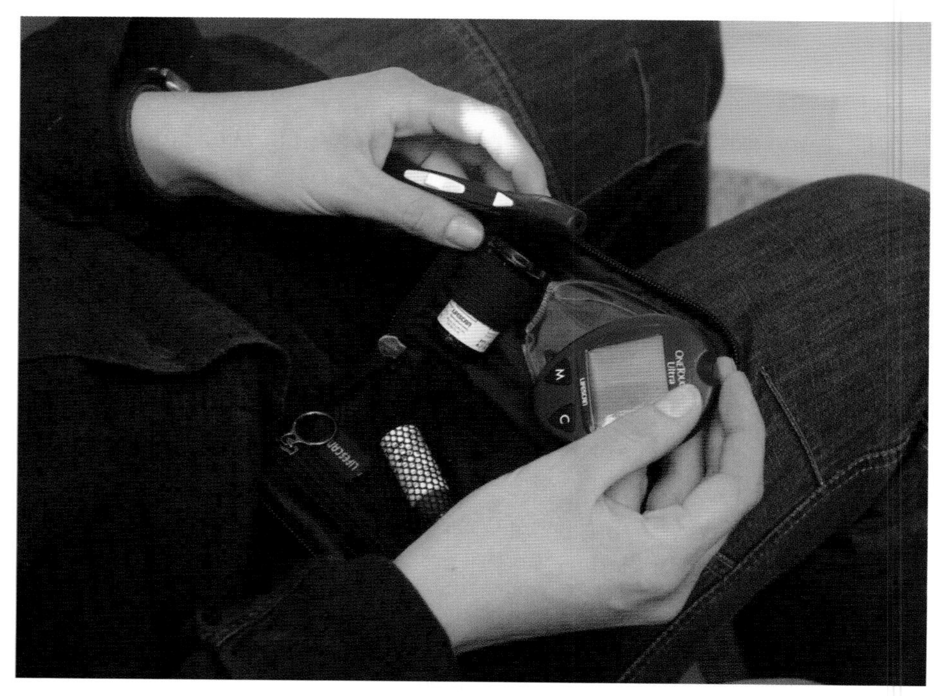

A person with diabetes checks their blood sugar levels regularly.

Diabetes

Diabetes is the third most common long-term disease in the UK, after heart disease and cancer. People with diabetes have high levels of glucose in their blood, unless they are treated. Their bodies cannot use glucose properly.

There are two types of diabetes: Type 1 and Type 2.

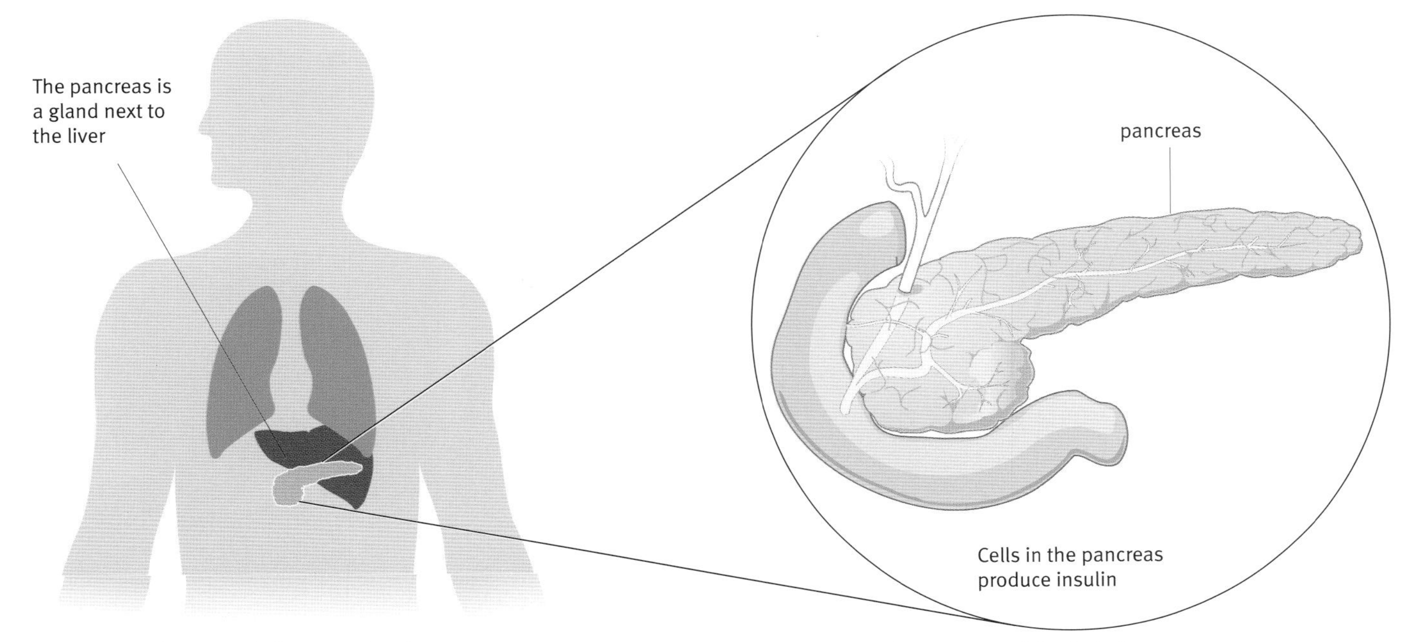

Insulin controls the level of sugar in blood. It lets sugar molecules into cells.

When the sugar levels rise the pancreas cells release insulin into the blood.

In type 1 diabetes the special cells in the pancreas are destroyed. The pancreas cannot make insulin.

In type 2 diabetes the pancreas does not make enough insulin or cells do not respond to the insulin there is.

Every cell of the body needs energy to survive. **Insulin** is a **hormone** produced in the **pancreas**. The hormone is critical for cells to take up glucose sugar and use it for energy.

Type 1 diabetes

Type 1 diabetes is more likely to start in younger people, but it can develop at any age. It develops when cells in the pancreas that produce insulin are destroyed. Insulin is a hormone that controls the levels of glucose in the blood. This type of diabetes is treated with insulin injections.

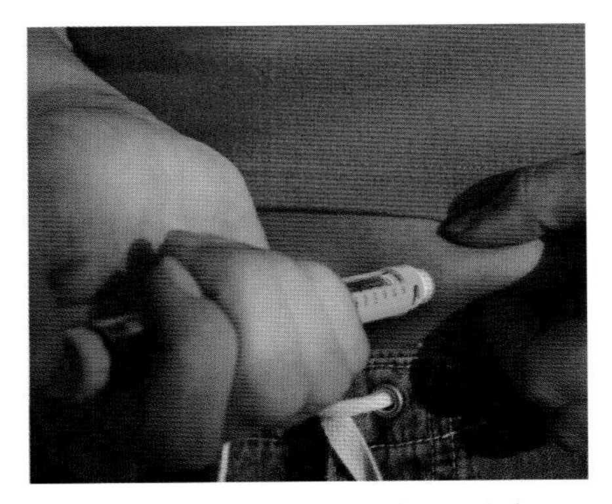

A person with type 1 diabetes injects insulin several times a day to keep blood glucose levels normal. The injection includes human insulin produced by bacteria that have been genetically modified.

Key words

obesity
diabetes
insulin
hormone
pancreas

Questions

1 What kinds of food are most likely to cause obesity if eaten in large quantities?

2 Suggest reasons why it is unhealthy to be overweight.

Type 2 diabetes

Type 2 diabetes is usually diagnosed in older people. The older you are, the greater the risk. But more young people are now developing type 2 diabetes. This type of diabetes can sometimes be treated with diet and exercise alone. But people with type 2 diabetes often need medicine, and they may need insulin too.

A dietician can advise someone with Type 2 diabetes. Choosing a healthy diet can help to control the condition.

The pancreas of a person with type 2 diabetes can still make insulin. The problem is that the cells in the body no longer respond normally to the hormone. Much more insulin than normal is needed to keep blood glucose levels at the right level.

Who gets type 2 diabetes?

Diabetes is a common health condition. There are 1.8 million people with diabetes in the UK. That is about 3 in every 100 people. There may be a million more people who have diabetes but do not know it. Over three-quarters of all the people with diabetes have type 2 diabetes.

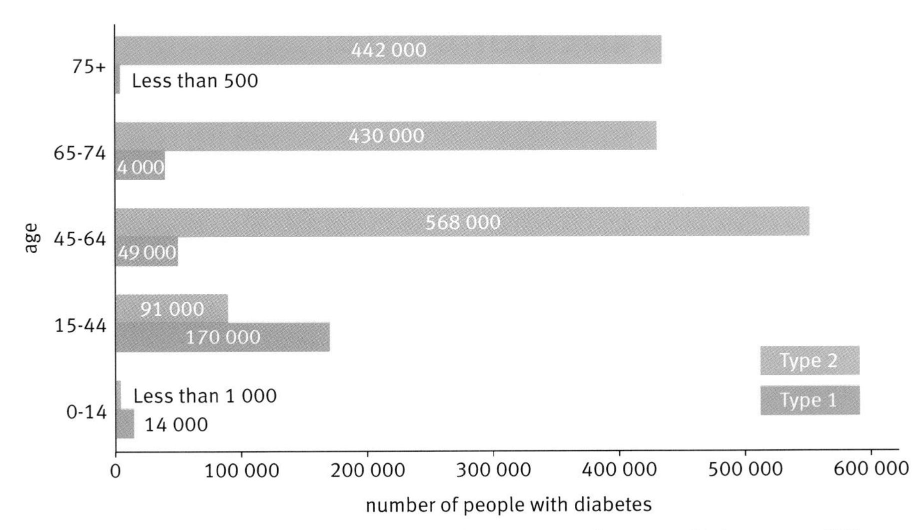

Estimates of the numbers of people with type 1 and type 2 diabetes at different ages in the population of 60 million people in the UK.

Risk factors

Being overweight is a leading **risk factor** for type 2 diabetes. Doctors classify people with a Body Mass Index (BMI) greater than 30 as obese. The risk of developing type 2 diabetes increases by up to ten times in people with a BMI of more than 30.

A lifestyle with little exercise is also a risk factor for diabetes. This is not just because people who take little exercise are often overweight. Physical activity helps the body to keep blood glucose levels in check.

Two other risk factors for type 2 diabetes are genetics and age. Type 2 diabetes tends to run in families. Also, members of some minority ethnic communities living in the UK develop type 2 diabetes at a younger age. The risk of developing diabetes is about five times higher in these communities.

Questions

3 Look at the chart showing how many people have diabetes in different age ranges. Write down the conclusions you can make from this data.

4 The number of people with type 2 diabetes is growing. Suggest a reason for this.

5 Suggest ways in which people can change their lifestyle to reduce the risk of getting type 2 diabetes.

6 What evidence is there that genetics may be a risk factor for type 2 diabetes?

7 Why is it sometimes possible to control type 2 diabetes just by careful choice of diet, but not type 1 diabetes?

Key words

risk factor

Find out about:

- monitoring and regulation of the food industry
- consumer protection in the food industry

G Food and the consumer

Governments and food safety

The European Union has passed laws covering the whole of the food chain. Officials in Brussels have the task of keeping these laws up to date.

The EU has laws regulating:

- how farmers produce food
- how food is processed
- how it is sold
- what sort of information is provided on food labels

These laws aim to encourage the food trade in the EU, while protecting the interests of consumers.

Country flags flying outside the European parliament building in Strasbourg.

National and local governments in the EU countries apply the laws. They make sure that farmers, manufacturers, and traders observe the rules.

The Food Standards Agency

The government of the UK set up the Food Standards Agency in 2000. The aims are:

- to protect the health of the public
- to defend the public interest in relation to food

The Agency aims to:

- reduce the amount of illness caused by food
- help people to eat more healthily
- promote honest and informative **food labelling**
- promote best practice in the food industry
- improve the enforcement of food law

Research and food

The Food Standards Agency wants its advice to the public to be based on the best and most up-to-date food science. It pays for scientists to do research into key issues. It also has expert committees to give advice. The Agency also carries out surveys and consults the public.

Some food issues are very controversial. The scientific evidence can be unclear, especially when a problem first comes up. So people need to be aware of the conflicting views among scientists.

Sometimes there is doubt about the level of a risk to health. Then the Agency asks one of its advisory committees for its views.

Labelling

Food labels give information about food. People can use the information to make choices about what they buy and eat.

Food labelling is controlled by law. Manufacturers cannot just print what they like on labels. This protects people from false claims and misleading descriptions.

Speaking up for the consumer

Some of the work of the Food Standards Agency is controversial. Not everyone agrees with its scientific advice. Its decisions are based on a complex mixture of factors.

There are many campaigning groups which work to make the food and farming system more sustainable.

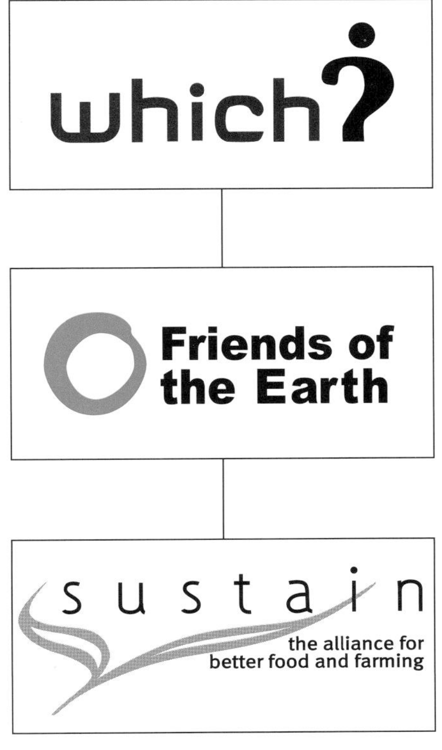

These are some of the many organizations that campaign on behalf of the public. They want a farming and food system that is better for public health, animal welfare, and the environment

Questions

1. Give examples to show why laws are needed to protect the public at each of these stages of the food chain:
   - growing and harvesting crops
   - storing food
   - processing food
   - cooking and serving food
2. You meet someone who thinks that food should be 100% safe. What arguments would you use to explain that this is impossible?

Key words

food labelling

Find out about:

- how people respond to food risks
- the application of the precautionary principle in the food industry

Today milk is pasteurized in dairies. Heating the milk to 72 °C for at least 15 seconds kills the bacteria that cause disease.

H Food hazards and risks

Changing attitudes to risk

In the 1930s, it was dangerous to drink milk in Britain. At the time, four in every ten cows were infected with tuberculosis. Every year around 50 000 people were infected and 2500 people died from tuberculosis. This was caught by drinking milk, or directly from cows with the disease.

The cows lived close to big cities and the milk was delivered directly to people's homes. The milk was untested and untreated. The untreated milk caused many deaths. But for many years the government did not think it worth the cost of processing the milk to prevent infection.

Since the 1930s, almost all milk is pasteurized before being sold. This process kills the bacteria that cause tuberculosis. Untreated cows' milk, and other dairy products such as cheese, must carry a health warning about the **risks** to health.

Now people worry more about hazards which are much less likely to cause death. Quite rightly, food labels warn people who may be allergic to ingredients such as nuts and gluten. The number of people dying from these allergies is around ten a year. Many more people die by choking on food. About 150 people a year in the UK die in this way.

Risk	Approx. number
Cancer *	66 000
Coronary heart disease (CHD)*	35 000
Food borne illness	~500
vCJD	< 20
Food allergy	~10
GMOs, pesticides, growth hormones	nil
Choking to death	151

*assumes about one-third of deaths are diet-related

Risk and approximate number of associated deaths.

The biggest risks from food come from eating an unhealthy diet and being obese. Unhealthy eating makes a big contribution to deaths from heart disease, cancer, and diabetes.

Risk from chemicals in food

Food consists of natural chemicals from plants, animals, or micro-organisms. Other chemicals may be added to food during production, processing, and preparation.

Some of the chemicals in food are hazardous. Scientists estimate the risk that people face from the known hazards. This is called risk assessment.

When carrying out a risk assessment, scientists have to decide:

- whether the chemical causes harm and how severe the harm is – based on the results of the experience of animals and people eating the food
- how likely it is that people will suffer harm – based on the amount of food eaten and how often it is eaten
- whether some people are likely to be more affected than others – depending on their age, previous illness, or genetics.

Taking a precautionary approach

Regulators, such as the Food Standards Agency, have to make judgements about levels of risk. Sometimes they have to reassess an existing risk when there is new evidence. Also, there are dozens of new food scares every year. Some of them arise from new technologies such as genetic modification. Others arise as scientists and others learn more about the effects of the food we eat.

High risk

Microorganisms: bacteria, fungi, and viruses that contaminate crops and food

Chemicals naturally present: chemicals produced by the original crop as it grows

Chemicals produced by cooking: chemicals found when food is very hot

Chemicals from pollution: contamination by hazardous waste or industrial pollution

Pesticides: chemicals added to crops to control pests, weeds, and diseases

Additives: chemicals added to preserve food or make a desirable food product

Low risk

How scientists and food-safety experts rank the level of risk of possible food hazards.

The science is often uncertain, particularly when there is a new issue. Scientists may disagree about the meaning of the data available. The **precautionary principle** says that the lack of scientific certainty should not be used as an excuse to delay action to deal with the possible risk. According to this principle, regulators and others should give priority to protecting public safety. They should not simply allow new technologies to go ahead. They must be sure that there is enough evidence that the benefits outweigh the risks.

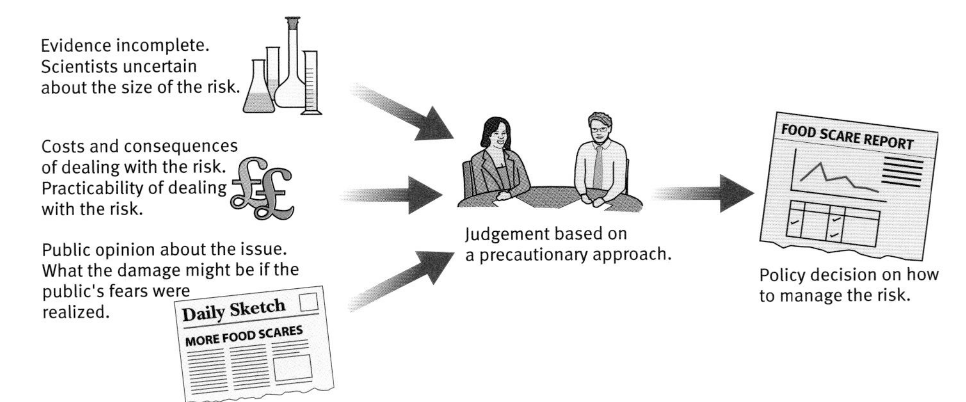

Taking a precautionary approach to making decisions about food safety. Regulators do not make judgements on their own. They have to consult both experts and the public. They have to weigh up the costs and benefits of any actions they may recommend.

Key words

risk
precautionary principle

Questions

1 There is public demand for cheap food. Suggest ways in which this demand could lead to increased food risks.

2 Some people prefer the taste of cheese made from unpasteurised milk. Cheese made this way carries a risk of contamination with harmful bacteria.

a What can be done to limit the risk from eating unpasteurised cheese?

b Who should weigh up the balance of benefit and risk?

3 Many people are putting their health at risk with too many calories, too much saturated fat, too much sugar and too much salt. What, if anything, should the government do to deal with this situation?

C3 Food matters

Science explanations

Chemicals are all around us. Their interactions govern our lives.

You should know:

- all living things are made of chemicals
- there is continual cycling of elements in the environment
- the nitrogen cycle is an example of a natural cycle
- where crops are harvested elements, such as nitrogen, potassium, and phosphorus, are lost from the soil
- land becomes less fertile unless these elements are replaced
- organic and intensive farmers use different methods to keep soil fertile for growing crops
- organic and intensive farmers use different methods to protect crops against pests and diseases
- farmers have to follow UK national standards if they want to claim that their products are organic
- farming has an impact on the natural environment
- some methods of farming are more sustainable than others
- some natural chemicals in plants that we eat may be toxic if they are not cooked properly, or they may cause allergies in some people
- moulds that contaminate crops during storage (such as aflatoxin in nuts and cereals) may add toxic chemicals to food
- chemicals used in farming (such as pesticides and herbicides) may be in the products we eat and be harmful
- harmful chemicals may be produced during food processing and cooking
- natural and synthetic chemicals may be added to food during processing
  - food colours can be used to make processed food look more attractive
  - flavourings enhance the taste of food
  - artificial sweeteners help to reduce the amount of sugar in processed foods and drinks
  - emulsifiers and stabilizers help to mix ingredients together that would normally separate, such as oil and water
  - preservatives help to keep food safe for longer by stopping the growth of harmful microbes
  - antioxidants are added to foods containing fats or oils to stop them reacting with oxygen in the air
- many chemicals in living things are natural polymers (including carbohydrates and proteins)
- cellulose, starch and sugars are carbohydrates that are made up of carbon, hydrogen and oxygen
- amino acids and proteins consist mainly of carbon, hydrogen, oxygen and nitrogen
- digestion breaks down natural polymers to smaller, soluble compounds (for example digestion breaks down starch to glucose, and proteins to amino acids)
- these small molecules can be absorbed and transported in the blood
- cells grow by building up amino acids from the blood into new proteins
- excess amino acids are broken down in the liver to form urea, which is excreted by the kidneys in urine
- high levels of sugar, common in some processed foods, are quickly absorbed into the blood stream, causing a rapid rise in the blood sugar level
- there are two types of diabetes (type 1 and type 2)
- late-onset diabetes (type 2) is more likely to be triggered by a poor diet
- obesity is one of the risk factors for type 2 diabetes
- In type 1 diabetes the pancreas stops producing enough of the hormone, insulin
- In type 2 diabetes the body no longer responds to its own insulin or does not make enough insulin
- type 1 diabetes is controlled by insulin injections and type 2 diabetes can be controlled by diet and exercise

Ideas about science

Science-based technology provides people with many things that they value, and which enhance the quality of life. Some applications of science can have unwanted affects on our quality of life or the environment.

For different farming methods you should be able to:

- identify the groups affected, and the main costs and benefits of a decision for each group
- explain how science helps to find ways of using natural resources in a more sustainable way
- show you know that regulations and laws control scientific research and applications
- distinguish from what can be done from what should be done
- explain why different decisions may be made in different social and economic contexts

New technologies and processes based on scientific advances sometimes introduce new risks. Some people are worried about the health effects arising from the use of some food additives. You should be able to:

- explain why nothing is completely safe
- suggest ways of reducing some risks

Scientific advisory committees carry out risk assessments to determine the safe levels of chemicals in food. The Food Standards Agency is an independent food safety watchdog set up by an Act of Parliament to protect the public's health and consumer interests in relation to food.

Additives with an E number have passed a safety test and been approved for use in the UK and the rest of the EU. Food labelling can help consumers decide which products to buy. You should be able to:

- interpret information on the size of risks
- show you know that regulations and laws control scientific research and applications
- explain that if it is not possible to be sure about the results of doing something, and if serious harm could result, then it makes sense to avoid it (the 'precautionary principle')

People's perception of the size of a risk is often very different from the actual measured risk. People tend to over-estimate the risk of unfamiliar things (like chemicals with strange names added to food compared with overeating and obesity), and things whose effect is invisible (like pesticides residues). You should be able to:

- discuss a particular risk, taking account both of the chance of it happening and the consequences if it did
- suggest why people will accept (or reject) the risk of a certain activity, for example, eating a diet rich in sugar and fat because they enjoy this food

Why study chemical patterns?

The periodic table is so important in chemistry because it helps to make sense of the mass of information about all the elements and their compounds. The table offers a framework that can give meaning to all the facts about properties and reactions.

The science

What fascinates chemists is that it is possible to use ideas about atomic structure to explain the periodic table and the properties of different elements.

Light and electrons in atoms can affect each other. This is the science behind spectroscopy. At first, spectroscopy led to the discovery of new elements. Today, a wide range of techniques means that spectroscopy provides the essential tools for studying chemicals and chemical reactions.

Chemistry in action

The ideas in this module are of great practical importance. Many of the most sensitive methods of chemical analysis depend on spectroscopy.

Without ionic theory there would be no aluminium metal. Ionic theory is also vital in explaining how our nerves and brain work.

Chemical patterns

Find out about:

- the chemistry of some very reactive elements
- the patterns in the periodic table
- how scientists can learn about the insides of atoms
- the use of atomic theory to explain the properties of chemicals
- the ways in which atoms become charged and turn into ions

Find out about:
- relative masses of atoms
- periodic patterns
- groups and periods

A The periodic table

A century of discovery

There were only about 30 known elements when Horatio Nelson led the British fleet to victory at Trafalgar in 1805. Nearly a hundred years later, when Queen Victoria died, scientists had discovered all but three of the stable elements found on Earth.

The discovery of so many elements encouraged chemists to look for patterns in the properties of the elements. One idea, which seemed strange at first, was to look for a connection between the chemistry of elements and the masses of their atoms.

Germanium – one of Mendeléev's missing elements. He used his version of the periodic table to predict that the missing element would be a grey metal that would form a white oxide with a high melting point. He also predicted that its chloride would boil below 100 °C and have a density of about 1.9 g/cm^3.

Relative atomic masses

In the 1800s, scientists could not measure the actual masses of atoms – they could only compare them. They chose to compare the masses of atoms with the mass of the lightest atom, hydrogen. On the **relative atomic mass** scale the relative mass of a hydrogen atom is 1, that of a carbon atom is 12, that of an oxygen atom is 16, and so on.

Johann Döbereiner, a German scientist, noticed that there were several examples of groups of three elements with similar properties (for example calcium, strontium, and barium). For each group, the relative mass of the atoms of the middle element was the mean of the relative masses of the other two elements.

This, and other early attempts to find connections between chemical properties and atomic mass, were not taken seriously at the time.

Elements in order

Dmitri Mendeléev, a Russian scientist, showed that it is possible to come up with patterns with real meaning when elements are lined up in order of the masses of their atoms (their relative atomic masses). Mendeléev's inspiration was to realize that not all of the elements had yet been discovered. He left gaps for missing elements when this was necessary to produce a sensible pattern.

When Mendeléev put the elements in order of relative atomic mass, he spotted that at intervals along the line there were elements with similar properties. Using elements known today, for example, you can see that the third, eleventh, and nineteenth elements (lithium, sodium, and potassium) are very similar.

Gold, platinum, titanium, and other transition metals are used to make jewellery. They are shiny metals which do not react with the air.

Periodicity

A repeating pattern of any kind is a **periodic** pattern. An example is the repeating pattern on a roll of fabric or wallpaper. The table of the elements gets its name from the repeating, or periodic, patterns you see when chemists line up the elements in order.

The periodic table now

In the periodic table, the elements are arranged in rows, one above the other. Each row is a **period**. The most obvious repeating pattern is from metals on the left to non-metals on the right. Every period starts with a very reactive metal in group 1 and ends with an unreactive gas in group 8.

Elements with similar properties fall into a column. Each column is a **group** of similar elements.

Key words

relative atomic mass
periodic
period
group

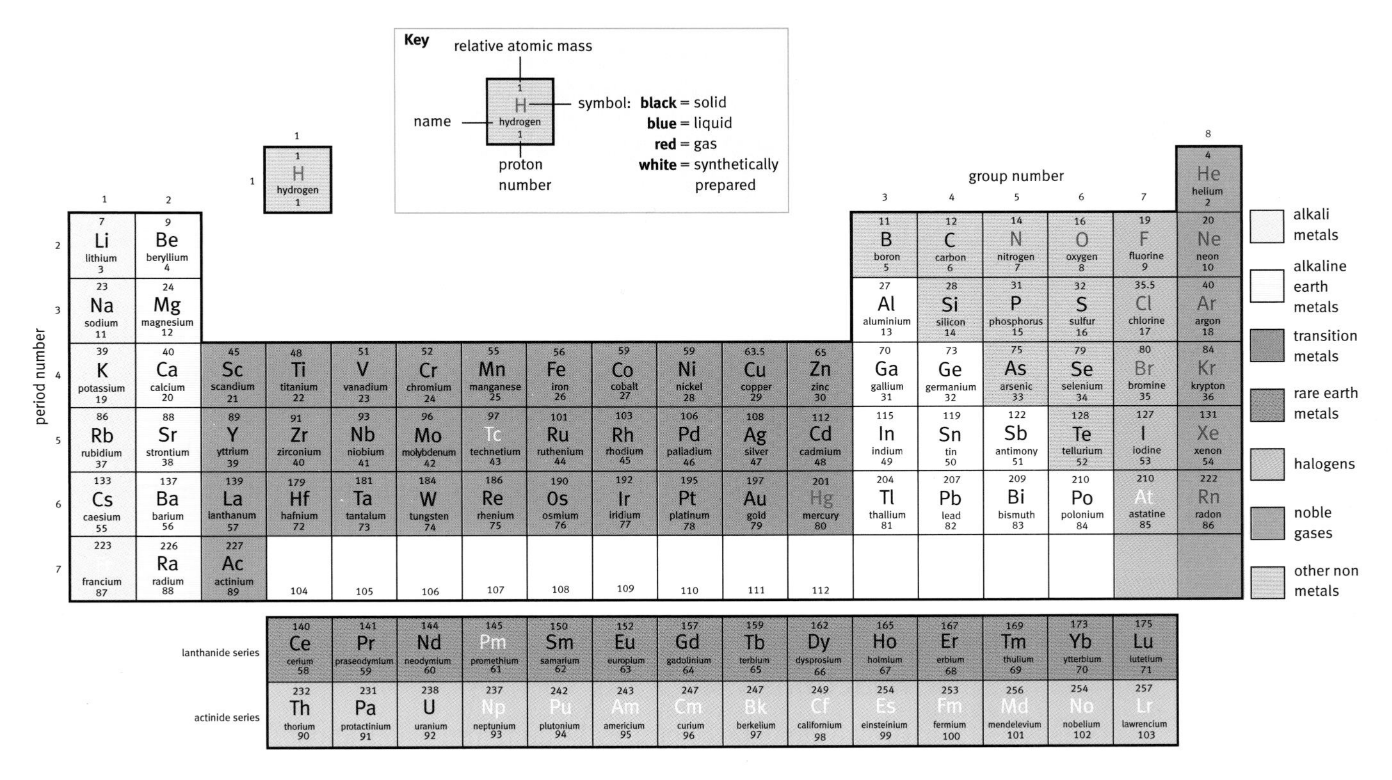

The periodic table. Over three-quarters of the elements are metals. They lie to the left of the table. Most of them are in groups 1 and 2 and the block of transition metals.

Questions

1 In the periodic table identify and name:
- **a** a liquid halogen
- **b** an alkali metal that does not occur naturally
- **c** a gaseous element with properties similar to sulfur
- **d** a solid element similar to chlorine
- **e** a liquid metal with properties similar to zinc

2 How many times heavier than a hydrogen atom are the atoms of:
- **a** carbon?
- **b** magnesium?
- **c** bromine?

3 How many times heavier is:
- **a** a magnesium atom than a carbon atom?
- **b** a sulfur atom than a helium atom?
- **c** an iron atom than a nitrogen atom?

4 Name two elements in the modern periodic table that break Mendeléev's rule that the elements should be arranged in order of relative atomic mass.

(5) Explain how Mendeléev could predict the properties of the unknown element germanium from what was known about other elements such as silicon and tin.

Find out about:
- group 1 metals
- reactions with water and chlorine
- similarities and differences between group 1 elements

B The alkali metals

The metals in group 1 of the periodic table are very reactive. They are so reactive that they have to be kept under oil to stop them reacting with oxygen or moisture in the air.

There are six elements in the group. Two of them, rubidium and caesium, are so reactive and rare that you are unlikely to see anything of them except on video. A third, francium, is highly radioactive. Its atoms are so unstable that it does not occur naturally. As a result, the study of group 1 usually concentrates on lithium (Li), sodium (Na), and potassium (K).

Chemists call these elements the alkali metals because they react with water to form alkaline solutions. Note that it is the compounds of these metals which are alkalis and not the metals themselves.

Cutting a lump of sodium to show a fresh surface of the metal

Strange metals

Most metals are hard and strong. The alkali metals are odd in this respect because it is possible to cut them with a knife. Cutting them helps to show up one of their most obvious metallic properties: they are very shiny but they tarnish quickly in the air. The shiny surface becomes dull with the formation of a layer of oxide. Group 1 elements, like other metals, are also good conductors of electricity.

Most metals are dense and have high melting points. Again the alkali metals are odd: they float on water and melt on very gentle heating.

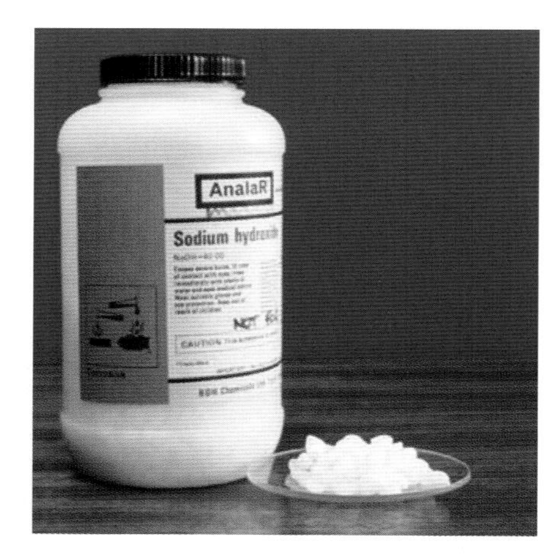

Pellets of the alkali sodium hydroxide. The traditional name is caustic soda. Anything caustic attacks skin. Alkalis such as NaOH are more damaging to skin and eyes than many acids.

Reactions with water

Drop a small piece of grey lithium into water and it floats, fizzes gently, and disappears as it turns into lithium hydroxide (LiOH).
It dissolves, making the solution alkaline. It is possible to collect the gas and use a burning splint to show that it is hydrogen.

$$\text{lithium} + \text{water} \rightarrow \text{lithium hydroxide} + \text{hydrogen}$$

The reaction with sodium is more exciting. The reaction gives out enough energy to melt the sodium, which skates around on the surface of the water. Sometimes sparks from the molten sodium ignite the hydrogen formed, which then burns with a yellowish flame. Like lithium, the sodium turns into its hydroxide (NaOH) and dissolves to give an alkaline solution.

The reaction with potassium is very violent. The hydrogen given off catches fire at once, and molten metal may be thrown from the surface of the water. The result is an alkaline solution of potassium hydroxide (KOH).

Reactions with chlorine

Hot sodium burns with a bright yellow flame. It produces clouds of white sodium chloride crystals (NaCl). This is everyday 'salt', used for seasoning food.

The other alkali metals react in a similar way with chlorine. Lithium produces lithium chloride (LiCl). Potassium produces potassium chloride (KCl). Like everyday salt, these compounds are also colourless, crystalline solids which dissolve in water.

Chemists use the term **salt** to cover all the compounds of metals with non-metals. So the chlorides of lithium, sodium, and potassium are all salts.

Sodium burning in chlorine gas

Trends

The alkali metals are all very similar, but they are not identical. There are clear **trends** in their properties down the group from lithium to sodium to potassium. These trends cover both **physical properties**, such as density and melting point, and **chemical properties**, such as the reactivity of the metals with water and chlorine.

Compounds of the alkali metals

The compounds of the alkali metals are very different from the elements. The elements are dangerously reactive. But chlorides of sodium and potassium, for example, have a vital role to play in the blood and in the way in which our nerves work.

Many compounds of alkali metals are soluble in water. Soluble sodium compounds, in particular, make up a number of common everyday chemicals. All homes contain sodium chloride ('salt'). Other important domestic products include sodium hydroxide (in oven cleaners), sodium hypochlorite (in bleach), and sodium hydrogencarbonate (as the bicarbonate of soda in antacids).

Key words
salt
trends
physical properties
chemical properties

Questions

1 Arrange the names of the alkali metals Li, Na, and K in order of reactivity with water, placing the most reactive of the metals first.

2 Predict these properties of rubidium:
- **a** How easily can it be cut with a knife?
- **b** What happens to a fresh-cut surface of the metal in the air?
- **c** What happens if you drop a small piece of rubidium onto water?

3 For the hydroxide of rubidium predict:
- **a** its colour
- **b** its formula
- **c** whether or not it is soluble in water

4 For the chloride of caesium predict:
- **a** its colour
- **b** its formula
- **c** whether or not it is soluble in water

5 Give an example to show that the trend is for the alkali metals to become more chemically reactive down the group from lithium to potassium.

(6) Suggest and explain the precautions necessary when demonstrating the reaction of potassium with water.

Find out about:
- chemical symbols
- formulae
- balanced equations

c Chemical equations

Equations are important because they do for chemists what recipes do for cooks. They allow chemists to work out how much of the starting materials to mix together and how much of the products they will then get.

Chemical models

In a **chemical change**, there is no change in mass because the number of each type of atom stays the same. The atoms regroup but no new ones appear and none are destroyed during a chemical reaction.

Hydrogen burns in oxygen to form **molecules** of water. The models in the figure below show what happens. In each water molecule there is only one oxygen atom. So one oxygen molecule reacts with two molecules of hydrogen to make two water molecules. There are equal numbers of hydrogen atoms and oxygen atoms on each side of the arrow. The term **chemical equation** can be used.

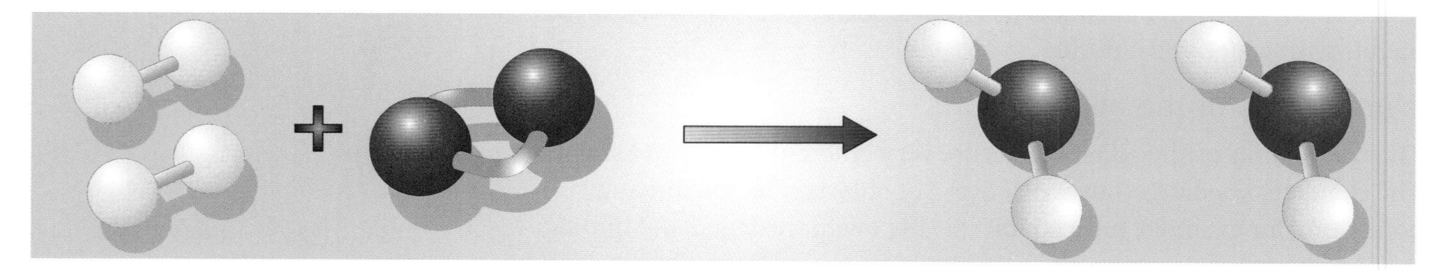

A model equation for the reaction of hydrogen with oxygen

Chemical symbols

Using drawings or photographs of models to describe every reaction would be very tiresome. Instead chemists write symbol equations to show the numbers and arrangements of the atoms in the reactants and the products.

When written in symbols the equation in the figure above becomes

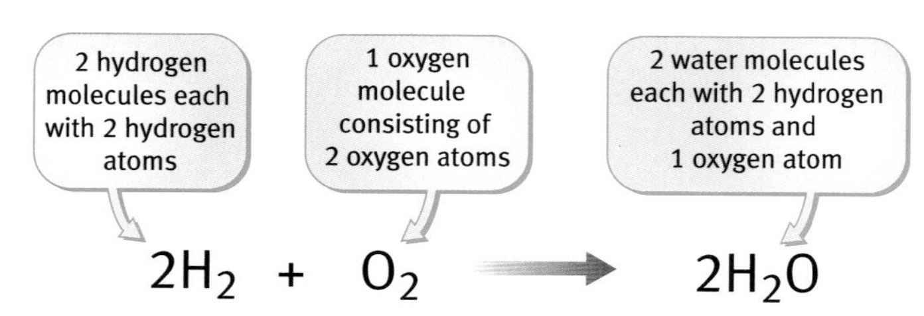

$$2H_2 + O_2 \rightarrow 2H_2O$$

The equation is 'balanced' because it has the same number of atoms of each type on the left (of the arrow) and on the right. It balances in the literal sense too. The reactants have the same mass as the products, so the chemicals on the right balance the chemicals on the left, if placed on an old-fashioned pair of scales.

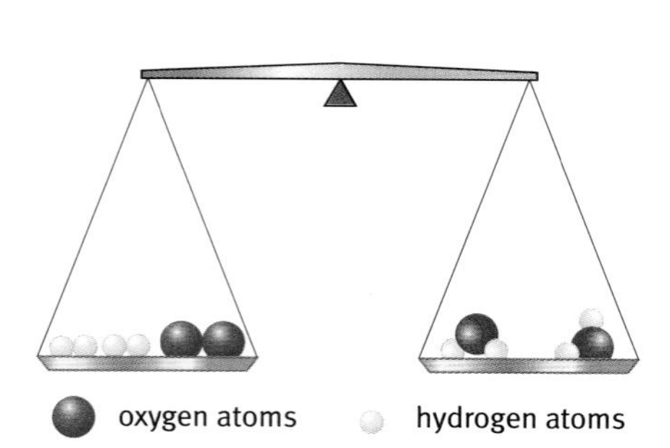

The reactants and the products of a reaction 'balance'

H

Formulae

You cannot write an equation unless you first know

- all the starting chemicals (the reactants)
- everything that is formed during the change (the products)

When writing an equation, you have to write down the correct chemical **formulae** for the reactants and products. Chemists have worked these out by experiment, and you can look them up in data tables.

If the element or compound is molecular, you write the formula for the molecule in the equation. This applies to most non-metals (O_2, H_2, Cl_2) and most compounds of non-metals with non-metals (H_2O, HCl, NH_3).

Not all elements and compounds consist of molecules. For all metals, and for the few non-metals that are not molecular (C, Si), you just write the symbol for a single atom.

The compounds of metals with non-metals are also not molecular. For these compounds you write the simplest formula for the compound, such as LiOH, NaCl, potassium carbonate (K_2CO_3).

Writing balanced equations

Follow the four steps shown in the margin to write **balanced equations**.

Example

Write down a balanced equation to show the reaction of natural gas (methane, CH_4) with oxygen.

Step 1 Describe the reaction in words:

methane + oxygen → carbon dioxide + water

Step 2 Write down the formulae for the reactants and products:

$$CH_4 + O_2 \rightarrow CO_2 + H_2O$$

Step 3 Balance the equation:

You must not change any of the formulae. You balance the equation by writing numbers in front of the formulae. These numbers then refer to the whole formula.

$$CH_4 + 2O_2 \rightarrow CO_2 + 2H_2O$$

Step 4 Add state symbols:

State symbols usually show the states of the elements and compounds at room temperature and pressure. The chemicals in an equation may be solid (s), liquid (l), gaseous (g), or dissolved in water (aq, for aqueous).

$$CH_4(g) + 2O_2(g) \rightarrow CO_2(g) + 2H_2O(l)$$

Key words

chemical change
molecules
chemical equation
formulae
balanced equation

RULES FOR WRITING BALANCED EQUATIONS

STEP 1 Write down a word equation.

STEP 2 Underneath, write down the correct formula for each reactant and product.

STEP 3 Balance the equation, if necessary, by putting numbers in front of the formulae.

STEP 4 Add state symbols.

NEVER change the formula of a compound or element to balance the equation.

Questions

1 Write balanced symbol equations for these reactions of the alkali metals:
- **a** sodium with water
- **b** potassium with water
- **c** sodium with chlorine
- **d** lithium with bromine
- **e** potassium with iodine

Find out about:
- group 7 elements
- halogen molecules
- similarities and differences between group 7 elements

Crystals of the mineral fluorite. This mineral is calcium fluoride.

D The halogens

Salt formers

Fluorine, chlorine, bromine, and iodine are all very reactive non-metals. They are interesting because of their vigorous chemistry. As elements they are hazardous because they are so reactive. For the same reason, they are not found free in nature. They occur as compounds with metals.

It is not normally possible to study fluorine because it is so dangerously reactive.

The name 'halo-gen' means 'salt-former'. These elements form salts when they combine with metals. Examples include everyday 'salt' itself, which occurs as the minerals halite (NaCl) and fluorite (CaF_2) found as the mineral Blue John (used in jewellery) in Derbyshire caves.

Non-metal patterns

Non-metals typically have low melting and boiling points. Chlorine is a greenish gas at room temperature. Bromine is a dark-red liquid which easily turns to an orange vapour. Iodine is a dark-grey solid which turns to a purple vapour on gentle warming.

The **halogens**, like most non-metals, are molecular. They each consist of molecules with the atoms joined in pairs: Cl_2, Br_2, and I_2. The forces between the molecules are weak, and so it is easy to separate them and turn the halogens into gases.

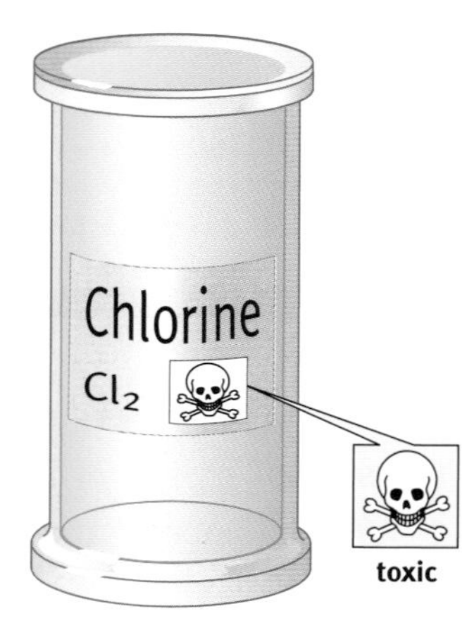

- dense, pale-green gas
- smelly and poisonous
- occurs as chlorides, especially sodium chloride in the sea

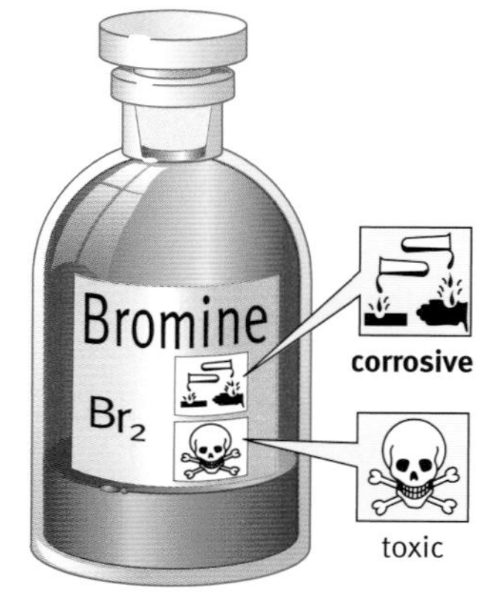

- deep red liquid with red–brown vapour
- smelly and poisonous
- occurs as bromides, especially magnesium bromide in the sea

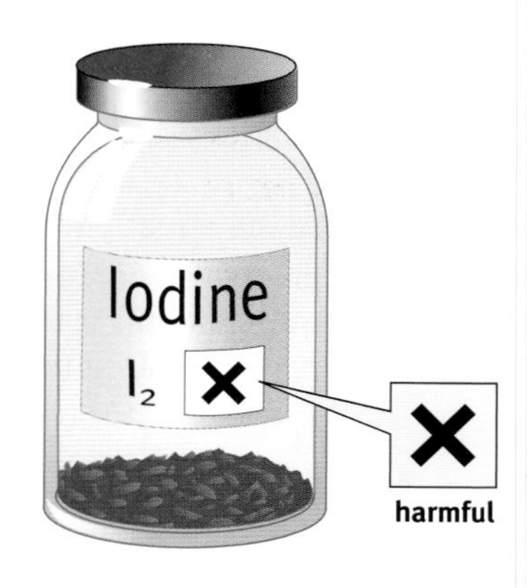

- grey solid with purple vapour
- smelly and poisonous
- occurs as iodides and iodates in some rocks and in seaweed

Chlorine, bromine, and iodine

Halogen patterns

All the halogens can harm living things. They can all kill bacteria. Domestic **bleach** is a solution of chlorine in sodium hydroxide sold to disinfect worktops and toilets. In the days before modern antiseptics, people used a solution of iodine to prevent infection of wounds.

The bleaching effect of the halogens illustrates the general trend in reactivity down the group. The usual laboratory test for chlorine shows that the gas quickly bleaches moist indicator paper. Bromine vapour also bleaches vegetable dyes such as litmus, but more slowly. Iodine has a slight bleaching effect too, but it also stains paper brown, which masks the change.

The reaction with iron also illustrates the clear trend in the reactivity of the elements down the group. Hot iron glows brightly in chlorine gas. The product is iron chloride (FeCl3), which appears as a rust-brown solid. Iron also glows when heated in bromine vapour, but less brightly. There is even less sign of reaction when iron is heated in iodine vapour.

Hot iron in a jar of chlorine

Practical importance

While the halogens themselves are too hazardous for everyday use, their compounds are of great practical importance.

The chemical industry turns everyday salt (NaCl) into chlorine (and sodium) and then uses the chlorine to make plastics such as polyvinylchloride (PVC). Another large-scale use of chlorine is water treatment to stop the spread of diseases. So chlorine compounds offer many benefits, but there are hazards too. Some chlorine compounds are so stable that they persist in the natural environment, where they can be a threat to life or bring about long-term damage. A dramatic example is the impact of chlorofluorocarbons (CFCs) and other halogen compounds on the concentration of ozone in the upper atmosphere.

Most of the bromine we use comes from the sea. Liquid bromine itself is extremely corrosive. However, the chemical industry makes important bromine compounds. These include medical drugs, and pesticides to protect food crops.

Traces of iodine compounds are essential to a healthy human diet. In regions where there is little or no natural iodine, it is usual to add potassium iodide to everyday table salt or to drinking water. This prevents disease of the thyroid, a gland in the neck. Iodine and its compounds are starting materials for the manufacture of medicines, photographic chemicals, and dyes.

Key words

halogens
toxic
corrosive
harmful
bleach

Questions

1. Which halogen is dangerously corrosive?
2. Write balanced equations for the reactions of:
   - **a** iron with chlorine to form $FeCl_3$
   - **b** potassium with chlorine
   - **c** lithium with iodine
3. Fluorine (F_2) is the first element in group 7. Predict the effect of passing a stream of fluorine over iron. Write an equation for the reaction.

Find out about:
- flame colours
- line spectra
- discovery of the elements

E The discovery of helium

A new burner for chemistry

Robert Bunsen moved to the University of Heidelberg in Germany in 1852. Before taking up the job as professor of chemistry, he insisted on having new laboratories. He also demanded gas piping to bring fuel from the gas works – this had just opened to light the city streets.

Existing burners produced smoky and yellow flames. Bunsen wanted something better. In 1855 he invented the type of burner that is still used today in laboratories all over the world.

The great advantage of Bunsen's burner was that it could be adjusted to give an almost invisible flame. Bunsen used his burner to blow glass. He noticed that whenever he held a glass tube in a colourless flame, the flame turned yellow.

Robert Bunsen (1811–1899), who discovered the flame colours of elements with the help of his new burner

Flame colours

Soon Bunsen was experimenting with different chemicals, which he held in the flame at the end of a platinum wire. He found that different chemicals produced characteristic **flame colours**.

Bunsen thought that this might lead to a new method of chemical analysis, but he soon realized that it seemed only to work for pure compounds. It was hard to make any sense of flames from mixtures. So he mentioned his problem to Gustav Kirchhoff, who was the professor of physics.

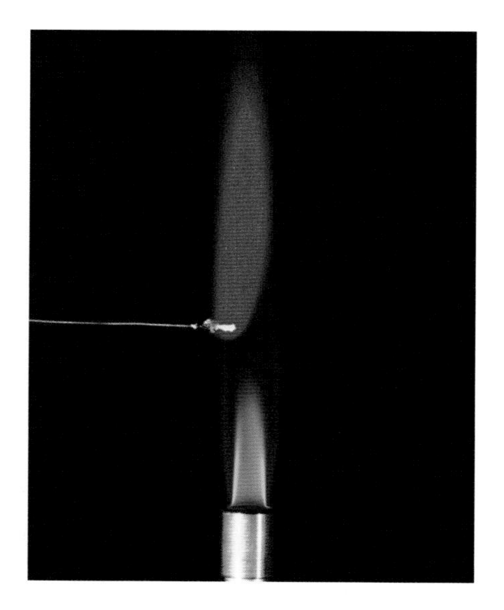

The bright red flame produced by lithium compounds. The compounds of other elements also produce colours in a flame:
- sodium – bright yellow
- potassium – lilac
- calcium – orange–red
- barium – green

Flame spectra

'My advice as a physicist', said Kirchhoff, 'is to look not at the colour of the flames, but at their spectra.'

Kirchhoff built a spectroscope, by putting a glass prism into a wooden box and inserting two telescopes at an angle. Light from a flame entered through one telescope. It was split into a spectrum by the prism and then viewed with the second telescope.

Bunsen and Kirchhoff soon found that each element has its own characteristic spectrum when its light passes through a prism. Each spectrum consists of a set of lines. With their spectroscope they were able to record the **line spectra** of many elements.

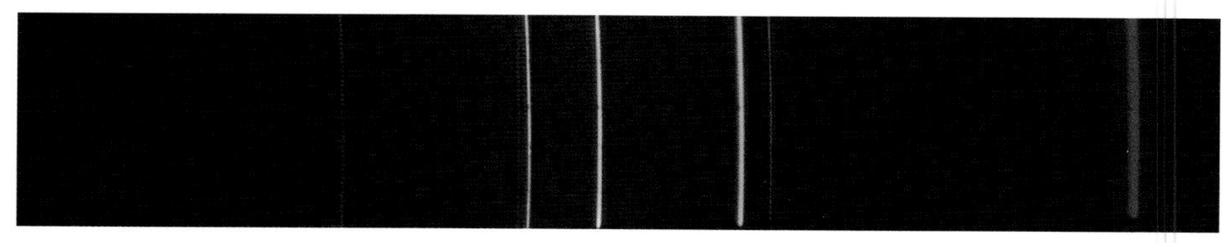

The spectrum of cadmium flame is made up of series of lines. Note the difference with the continuous spectrum of white light.

Using **spectroscopy**, Bunsen discovered two new elements in the waters of Durkheim Spa. He based their names on the colours of their spectra. He called them caesium and rubidium from the Latin for 'sky blue' and 'dark red'.

A Sun element

In 1868 there was a total eclipse of the Sun. Normally, the blinding light from the centre of the Sun makes it impossible to see the much fainter light from the hot gases around the edges of the star. During an eclipse, the Moon hides the whole bright disc of the Sun but not the much fainter light from the hot gases around the edges. This makes it possible to study this light from these gases.

Pierre Janssen, a French astronomer, took very careful observations of the Sun's spectrum during the eclipse. In the spectrum of the light, he saw a yellow line where no yellow line was expected to be.

Excited by these observations, both Janssen and an English astronomer, Joseph Lockyer, developed new methods to study the light from the Sun's gases. They worked independently, but both came to the same conclusion. There must be an unknown element in the Sun, producing the unexpected yellow line in the spectrum. Janssen and Lockyer published their findings at almost the same time – a coincidence which led to them becoming good friends.

The new element was called 'helium' from the Greek word *helios*, meaning Sun. Both astronomers were still alive in 1895 when William Ramsay, a British chemist, used spectroscopy to discover helium on Earth. The gas came from boiling up a uranium ore with acid.

A solar eclipse in 1868 helped scientists to discover helium. During an eclipse it is possible to study the spectra of the light from the hot gases around the edges of the Sun.

WARNING!

Never look directly at the Sun, even during an eclipse. You can damage your eyes or even be blinded!

Questions

1. Why was it important for Bunsen to have a burner with a colourless flame?
2. Why is it not possible to analyse chemical mixtures simply by looking at their flame colours?
3. Use your knowledge of group 1 chemistry to suggest an explanation for the fact that rubidium and caesium were not discovered until the technique of spectroscopy was developed.
4. Suggest reasons why helium was discovered on the Sun before it was discovered on Earth.
5. Why is it rare for two or more scientists who make the same discovery at the same time to end up as friends?

Key words

flame colour
line spectra
spectroscopy

Find out about:

- atomic theory
- the nuclear model of the atom
- protons, neutrons, and electrons

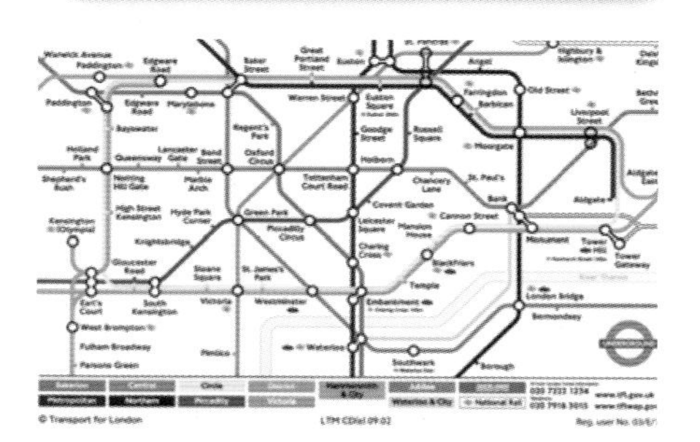

Part of the map of the London Underground

F Atomic structure

Atomic models

A picture, or model, of an atom can be used to understand how atoms join together to form compounds and how they regroup during chemical reactions.

Scientists use different models to solve different problems. There is not one atomic theory that is 'true'. Each model can represent only a part of what we know about atoms.

It is like using maps to travel through London. The usual underground rail map is a very useful guide for getting from one tube station to another. It is 'true' in that it shows how the lines and stations connect, but it cannot solve all of a traveller's problems. The map does not show how the tube stations relate to roads and buildings on the surface. For that you need a street map.

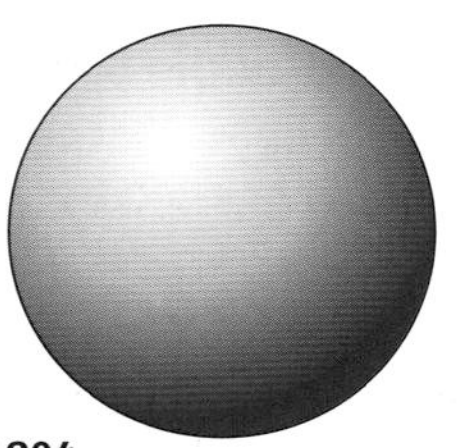

1804
Dalton's solid atom

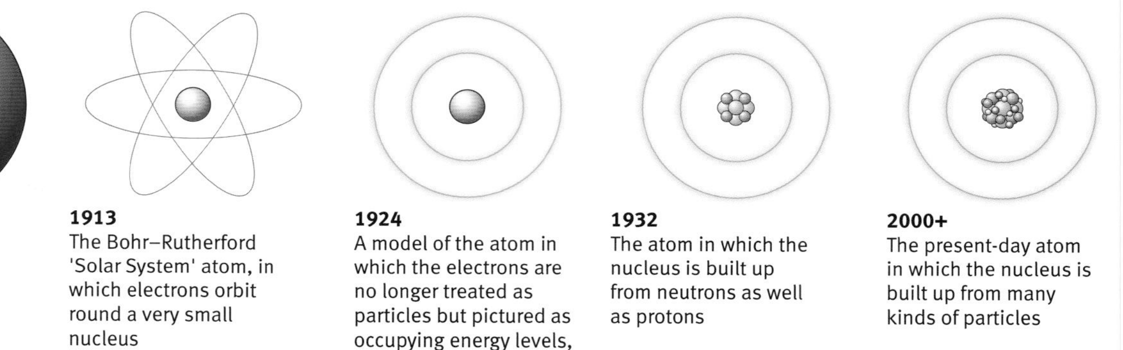

1913
The Bohr–Rutherford 'Solar System' atom, in which electrons orbit round a very small nucleus

1924
A model of the atom in which the electrons are no longer treated as particles but pictured as occupying energy levels, which give rise to regions of negative charge around the nucleus (charge clouds)

1932
The atom in which the nucleus is built up from neutrons as well as protons

2000+
The present-day atom in which the nucleus is built up from many kinds of particles

Atomic models from 1800 to the present. The diameter of an atom is about ten million times smaller than a millimetre. These diagrams are distorted. On this scale the nuclei would be invisibly small.

Dalton's atomic theory

The story of our modern thinking about atomic structure began with John Dalton at the beginning of the nineteenth century. In Dalton's theory everything is made of atoms that cannot be broken down. The very word 'atom' means 'indivisible'.

The main ideas in Dalton's theory still apply to everyday chemistry. So far as chemistry is concerned, each element does have its own kind of atom, and the atoms of different elements differ in mass. The idea that equations must balance is based on Dalton's view that atoms are not created or destroyed during chemical changes.

Even so, Dalton's theory is limited. It cannot explain the pattern of elements in the periodic table. Nor can it explain how atoms join together in elements and compounds.

Key words

nucleus
protons
neutrons
electrons
proton number

Inside the atom

In time it became clear that atoms are not solid, indivisible spheres. From the middle of the nineteenth century, scientists began to find ways of exploring the insides of atoms. Still today, scientists are spending vast sums of money to build particle accelerators (atom-smashers) which work at higher and higher energies. They hope to discover more about the fine structure of atoms.

It is possible to explain much more about the chemistry of elements and compounds with the help of a model of atomic structure that includes sub-atomic particles.

The Large Hadron Collider at CERN in Switzerland is a particle accelerator that will probe more deeply into matter than ever before. Due to switch on in 2007, it will ultimately collide beams of protons with very high energies.

A model for chemistry

In your study of chemistry you will be using an atomic model which dates back to 1932, when James Chadwick discovered the neutron. In this model the mass of the atom is concentrated in a tiny, central **nucleus**. The nucleus consists of **protons** and **neutrons**. The protons have a positive electric charge. Neutrons are uncharged.

Around the nucleus are the **electrons**. The electrons are negatively charged. The mass of an electron is so small that it can often be ignored. In an atom, the number of electrons equals the number of protons in the nucleus (the **proton number**). This means that the total negative charge equals the total positive charge, and overall an atom is uncharged.

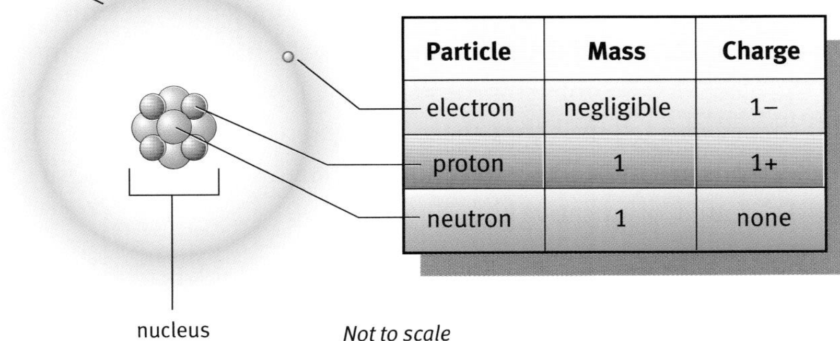

Particle	Mass	Charge
electron	negligible	1–
proton	1	1+
neutron	1	none

All atoms consist of these three basic particles. The nucleus of an atom is very, very small. The diameter of an atom is about ten million million times greater than the diameter of its nucleus.

Questions

1 With the help of the periodic table on page 95, work out:

a the element with one more proton in its nucleus than a chlorine atom

b the element with one proton fewer in its nucleus than a neon atom

c the number of protons in a sodium atom

d the number of electrons in a bromine atom

e the size of the positive charge on the nucleus of a fluorine atom

f the total negative charge on the electrons in a potassium atom

Find out about:
- evidence for energy levels
- electrons in shells
- electron configurations

G Electrons in atoms

Electrons in orbits

In 1913, the Danish scientist Niels Bohr came up with an explanation for the line spectra from atoms. He made a close study of the spectrum from hydrogen.

In the Bohr model for atoms, the electrons orbit the nucleus as the planets orbit the Sun. Bohr's idea was that heating atoms gives them energy. This forces the electrons to move to higher-energy orbits further from the nucleus. These electrons then drop back from outer orbits to inner orbits. They give out light energy as they do so. Each energy jump corresponds to a particular colour in the spectrum. The bigger the jump, the nearer the line to the blue end of the spectrum. Only certain energy jumps are possible, so the spectrum consists of a series of lines.

The line spectrum of hydrogen. Atomic theory can explain why this spectrum is a series of lines.

Bohr was able to use his theory to calculate sizes of the energy jumps. He could then deduce the energy levels of electrons in the various orbits.

Electrons in shells

The comparison with the Solar System and the use of the term orbit can be misleading. The theory has moved on since Bohr's time. Scientists still picture the electrons at a particular **energy level**. However, in the modern theory the electrons do not orbit the nucleus like planets round the Sun. All that theory can tell us is that there are regions around the nucleus where electrons are most likely to be found. Chemists describe these regions as 'clouds' of negative charge.

Think of each electron cloud as a **shell** around the nucleus. Each shell is one of the regions in space where there can be electrons. The shells only exist if there are electrons in them. Electrons in the same shell have the same energy.

Key words
energy level
shell
electron configuration

Electron configurations

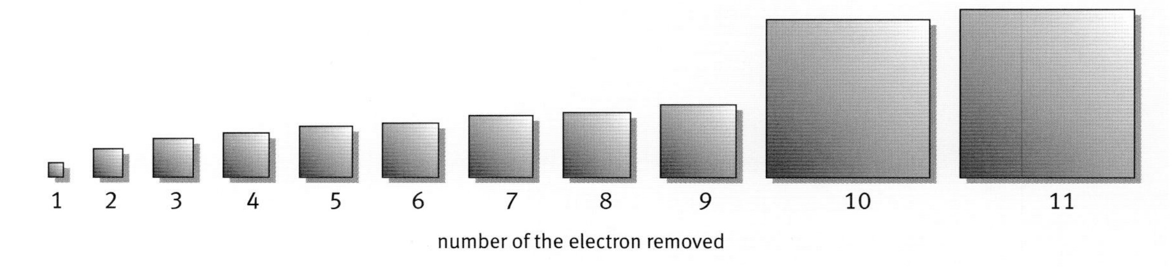

The areas of the squares are in proportion to the amount of energy needed to remove the electrons one by one from a sodium atom.

Each electron shell can contain only a limited number of electrons. The innermost shell with the lowest energy fills first. When full, the electrons go into the next shell. Evidence for this theory comes not only from spectra but also from measurements of the energy needed to remove electrons from atoms.

There are eleven electrons in a sodium atom (proton number 11). Scientists have measured the quantities of energy needed to remove these electrons one by one from a sodium atom. The relative values are represented by the areas of the squares in the picture above. You can see that it is quite easy to remove the first electron. The next eight are more difficult to remove. Finally it becomes really hard to remove the last two electrons, which are held very powerfully because they are in the shell closest to the nucleus.

This supports the idea that the electrons in a sodium atom are arranged in three shells as shown in the figure on the right. The diagram shows common representations of the **electron configuration** of the element.

The first shell that is closest to the nucleus can hold up to two electrons. The second shell can hold eight. Once the second shell holds eight electrons, the third shell starts to fill.

If there are more electrons, they occupy further shells. After the first twenty elements the arrangements become increasingly complex as the shells hold more electrons and the energy differences between shells get smaller.

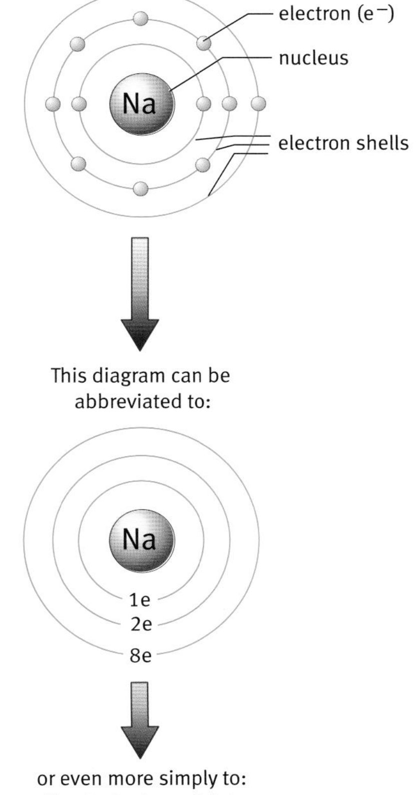

Two-dimensional representations of the electrons in shells in a sodium atom

Questions

1 Draw diagrams to show the electrons in shells for these atoms:
a beryllium **b** oxygen **c** magnesium
Refer to the periodic table on page 95 for the proton numbers, and therefore the number of electrons, in each atom.

2 How does the diagram at the top of the page support the representations of a sodium atom in the diagram underneath?

Find out about:
- atomic structure and periods
- electron configurations and groups
- explaining similarities and differences between the elements

H Electronic structures and the periodic table

The periodic table then and now

Scientists discovered electrons in 1897, which was nearly thirty years after Mendeléev published his first periodic table. Mendeléev knew nothing about atomic structure and he used the relative masses of atoms to put the elements in order.

A modern periodic table shows the elements in order of proton number, which is also the number of electrons in an atom. One of the convincing pieces of evidence for the 'shell model' of atomic structure is that it can help to explain the patterns in the periodic table.

Periods

The diagram below shows the connection between the horizontal rows of the periodic table and the structure of atoms. From one atom to the next, the proton number increases by one and the number of electrons increases by one. So the electron shells fill up progressively from one atom to the next.

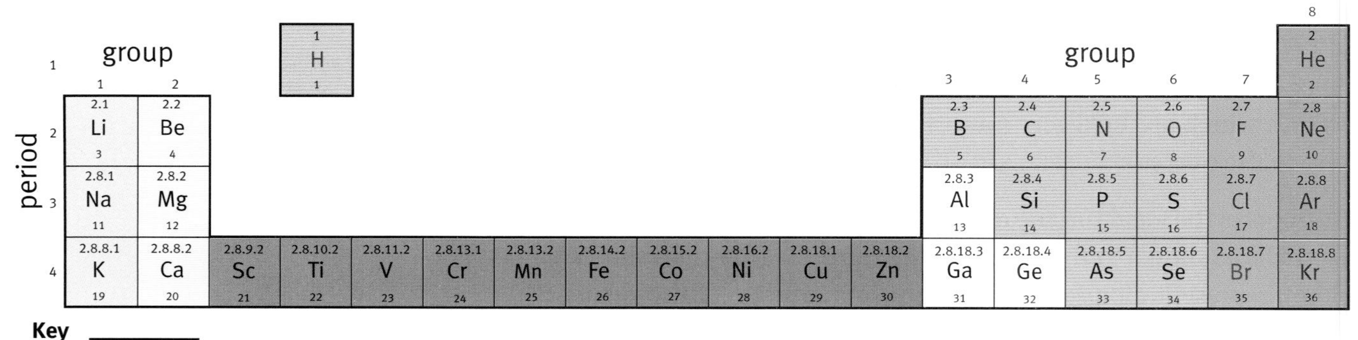

Electron configurations for the first 20 elements in the periodic table

The first period from hydrogen to helium corresponds to filling the first shell. The second shell fills across the second period from lithium (2.1) to neon (2.8). Eight electrons go into the third shell from sodium (2.8.1) to argon (2.8.8) and then the third shell starts to fill from potassium to calcium.

In fact the third shell can hold up to 18 electrons. This shell is completed from scandium to zinc, before the fourth shell continues to fill from gallium to krypton. This accounts for the appearance of the block of transition metals in the middle of the table. Why this happens cannot be explained by the simple theory described here. You will find out the explanation if you go on to a more advanced chemistry course.

Groups

When atoms react, it is the electrons in their outer shells which get involved as chemical bonds break and new chemicals form. It turns out that elements have similar properties if they have the same number and arrangement of electrons in the outer shells of their atoms.

Three of the alkali metals appear in the diagram on the right. You can see that they each have one electron in the outer shell of their atoms. This is the case for the other alkali metals too. This helps to account for the similarities in the chemistries of these elements.

The alkali metals are not all the same because their atoms differ in the number of inner full shells. A sodium atom has two inner filled shells, so it is larger than a lithium atom, and its outer electron is further away from the nucleus. As a result, the two metals have similar, but not identical, physical and chemical properties.

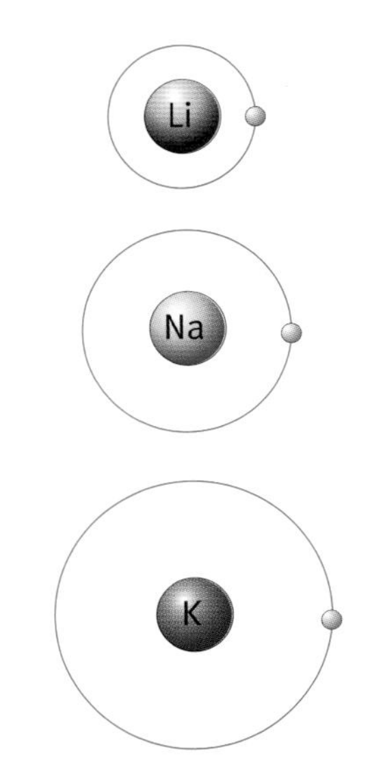

The trend in the size of the atoms of group 1 elements reflects the increasing number of full, inner electron shells down the group. Only the outer shells are shown here.

Metals and non-metals

Elements with only one or two electrons in the outer shell are metals. Elements with more electrons in the outer shell are generally non-metals, though there are exceptions to this, such as aluminium, tin, and lead. The halogens are non-metal elements with seven electrons in the outer shell.

At the end of each period there is a noble gas. This is a group of very unreactive elements. The first member of the group is helium.

The term 'noble' has been used by alchemists and chemists for hundreds of years to describe elements that are inert to most common reagents. The chemical nobility stand apart from the hurly-burly of everyday reactions.

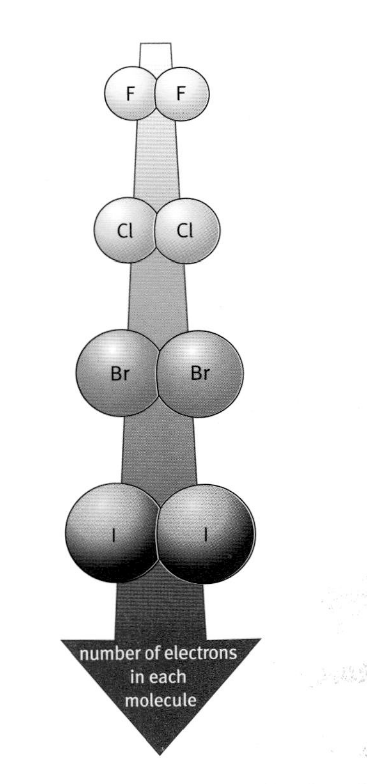

The trend in the size of the molecules of group 7 elements reflects the increasing number of full, inner electron shells in the atoms down the group.

Questions

1 Explain the meaning of this statement: the electron configuration of chlorine is (2.8.7).

2 a What are the electron configurations of the elements beryllium, magnesium, and calcium?

b In which group of the periodic table do these three elements appear?

c Are the elements metals or non-metals?

Find out about:

- salts
- properties of salts
- electricity and salts

1 Salts

Why are salts so different from their elements?

Compounds of metals with non-metals are salts. Chemists can explain the differences between a salt and its elements by studying what happens to the atoms and molecules as they react. A good example is the reaction between two very reactive elements to make the everyday table salt you can safely sprinkle on foodstuffs.

A chemical reaction in pictures: sodium and chlorine react to make sodium chloride.

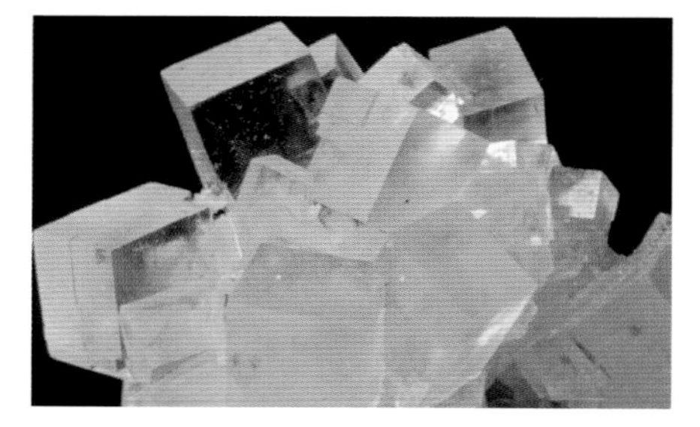

Sodium chloride crystals. Sodium chloride is soluble in water. The chemical industry uses an electric current to convert sodium chloride solution into chlorine, hydrogen, and sodium hydroxide.

Salts

Salts such as sodium chloride are crystalline. The crystals of sodium chloride are shaped like cubes. So are the crystals of calcium fluoride shown on page 100 at top left.

Salts have much higher melting and boiling points than compounds such as chlorine and bromine, which are made up of small molecules.

Chemical	Formula	Melting point (°C)	Boiling point (°C)
sodium	Na	98	890
chlorine	Cl_2	−101	−34
sodium chloride	NaCl	808	1465
potassium	K	63	766
bromine	Br_2	−7	58
potassium bromide	KBr	730	1435

Crystals of the mineral galena, which is an ore of lead. Galena consists of insoluble lead sulfide.

Sodium chloride is an example of a salt that is soluble in water. There are many other examples of soluble salts, including most of the compounds of alkali metals with halogens.

Some salts are insoluble in water. Lithium fluoride is an example of a salt which is only very slightly soluble in water. Many minerals consist of insoluble salts. Fluorite (CaF_2) is one example. Others are galena (PbS) and the brassy looking pyrites (FeS_2), sometimes called fool's gold.

Crystals of the mineral pyrites. Pyrites consist of insoluble iron sulfide.

Molten salts and electricity

The apparatus on the right is used to investigate whether or not chemicals conduct electricity. The crucible contains some white powdered solid. This is lead bromide.

At first the bulb does not light, showing that the solid does not conduct electricity. This is true of all compounds of metals with non-metals; they do not conduct when solid.

Heating the crucible melts the lead bromide. As soon as the compound is **molten**, there is a reading on the meter. This shows that a current is flowing round the circuit. As a liquid, the compound is a conductor. That is not all. The electric current causes the compound to decompose chemically. The most obvious change is the bubbling around the positive **electrode**. Puffs of orange gas appear as the bubbles burst. The orange gas is bromine.

After a while, it is possible to show that lead has formed at the negative electrode. This is done by switching off the current and pouring the liquid from the crucible into a mortar. The unchanged lead bromide quickly solidifies. Gently crushing the solid reveals a shiny lump of metallic lead. So the electric current splits the compound into its elements: lead and bromine.

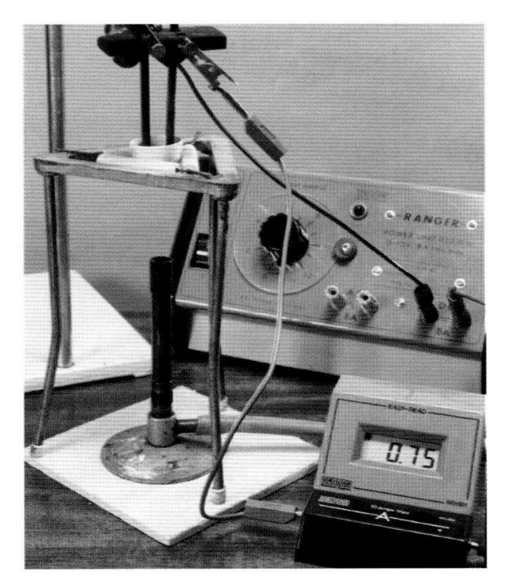

The crucible contains lead bromide. The carbon rods dipping into the crucible are the electrodes. A current begins to flow in the circuit when the lead bromide is hot enough to melt.

Salts in solution and electricity

Soluble salts also conduct electricity. This can be studied using the apparatus shown on the right. There are changes at the electrodes when an electric current flows.

The presence of water has an effect on the chemicals produced when a salt solution conducts electricity. The products are not always the same as the elements in the compound.

This apparatus is used to study the changes at the electrodes when a solution of a salt conducts electricity. In the example shown, the flow of an electric current is producing gases at the electrodes.

Questions

1 Draw up a table to compare the properties of sodium, chlorine, and sodium chloride.

2 Refer to the table of data on page 110. Which of the chemicals is a liquid:
  a at room temperature?
  b at the boiling point of water?
  c at 1000 °C?

3 Draw a two-dimensional line diagram and circuit diagram to represent the apparatus used to show that lead bromide conducts electricity when hot enough to melt.

Key words

molten
electrode

Find out about:
- ions
- ionic compounds
- explaining properties of salts

Michael Faraday lectured at the Royal Institution. He started the Christmas lectures, which continue today in the same lecture theatre.

Key words
electrolysis
ions
positive ions
negative ions

J Ionic theory

Electrolysis

An electric current can split a salt into its elements if it is molten or dissolved in water. This is the process called **electrolysis**. The term 'electro-lysis' is based on two Greek words that mean 'electricity-splitting'.

The discovery of electrolysis was very important in the history of chemistry because it made it possible to split up compounds which previously no-one could decompose. An English chemist, Humphry Davy, was Professor of Chemistry at the Royal Institution in London from 1802 to 1812. During 1807 and 1808 he used electrolysis to isolate for the first time the elements potassium, sodium, barium, strontium, calcium, and magnesium.

Faraday's theory

Michael Faraday also worked at the Royal Institution. He began as an assistant to Humphry Davy but established himself as a leading scientist in his own right. In 1833 he began to study the effects of electricity on chemicals.

Faraday decided that compounds that can be decomposed by electrolysis must contain electrically charged particles. Since opposite electrical charges attract each other, he could imagine the negative electrode attracting positively charged particles and the positive electrode attracting negatively charged particles.

The charged particles move towards the electrodes. When they reach the electrodes, they turn back into atoms. This accounts for the chemical changes that decompose a compound during electrolysis.

Faraday consulted a Greek scholar, and together they named the moving, charged particles **ions**, from a Greek word meaning 'wanderer'.

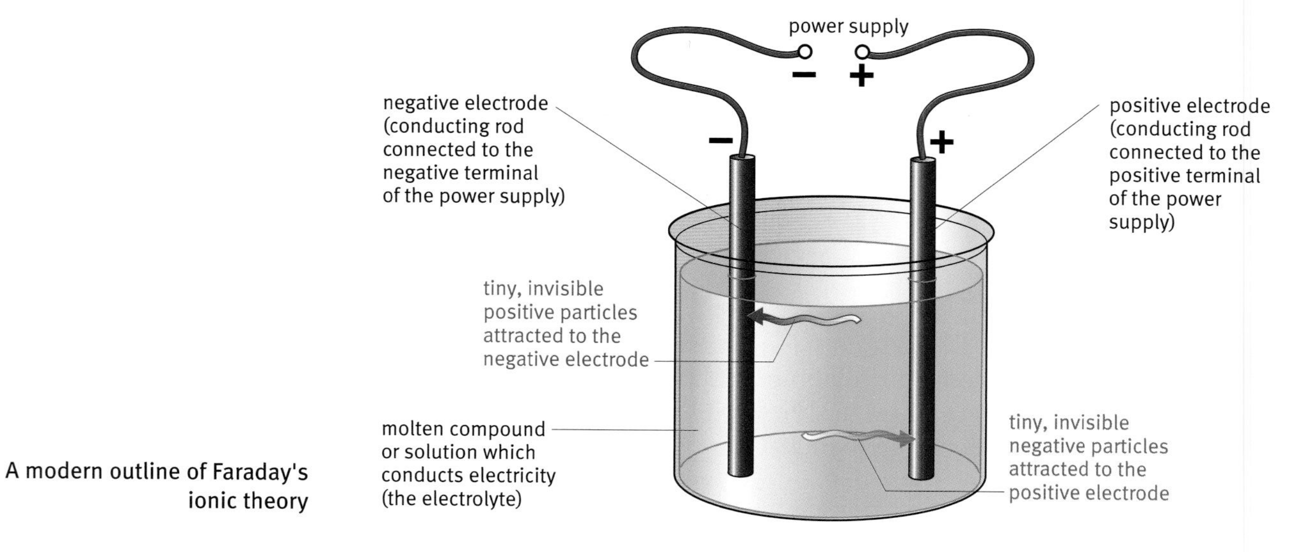

A modern outline of Faraday's ionic theory

Explaining electrolysis

Chemists continue to use ionic theory to explain electrolysis. According to the theory, salts such as sodium chloride consist of ions.

Sodium chloride is made up of sodium ions and chloride ions. Sodium ions, Na^+, are positively charged. The chloride ions, Cl^-, carry a negative charge. These oppositely charged ions attract each other.

A crystal of sodium chloride consists of millions and millions of Na^+ and Cl^- ions closely packed together. In the solid, these ions cannot move towards the electrodes, and so the compound cannot conduct electricity. The ions can move when sodium chloride is hot enough to melt or when it is dissolved in water.

During electrolysis the negative electrode attracts the **positive ions**. The positive electrode attracts the **negative ions**. When the ions reach the electrodes, they lose their charges and turn back into atoms.

Metals form positive ions, and non-metals generally form negative ions.

Elements and compounds

Ionic theory can help to explain why compounds are so different from their elements. Sodium atoms are dangerously reactive, and so are chlorine molecules. Sodium chloride is safe because its ions are much less reactive.

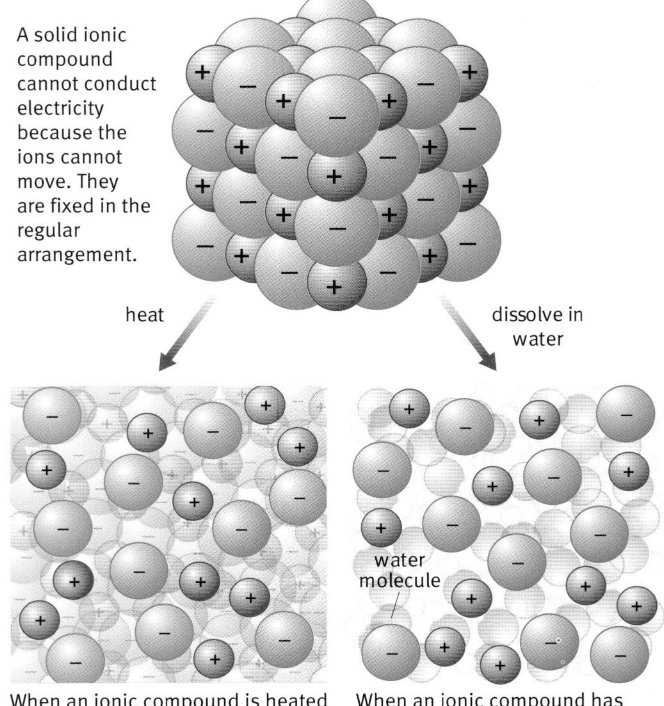

When an ionic compound is heated strongly, the ions move so much that they can no longer stay in the regular arrangement. The solid melts. Because the ions can now move around independently, the molten compound conducts electricity.

When an ionic compound has dissolved, it can conduct electricity because its ions can move independently among the water molecules.

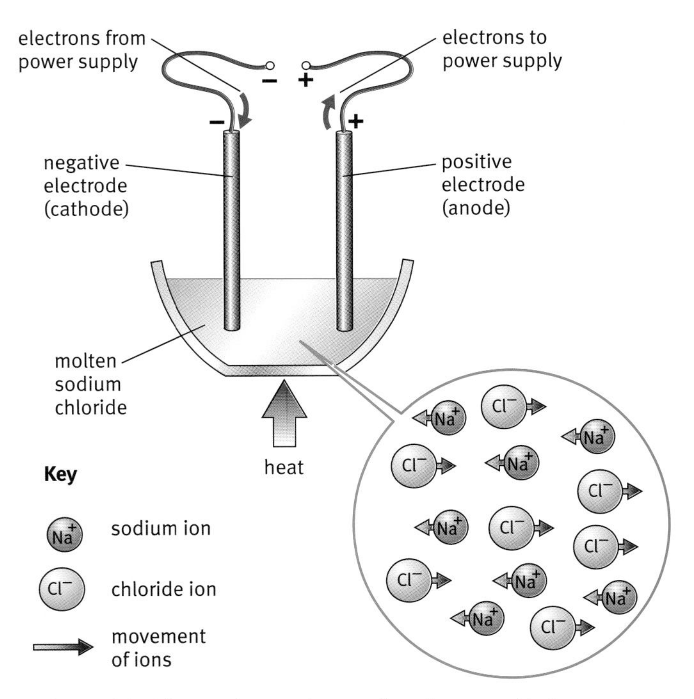

Sodium chloride conducts when molten because its ions can move towards the electrodes.

Questions

1 Why do solid compounds made of ions not conduct electricity?

2 Chemists sometimes call the negative electrode the cathode. Cations are the ions that move towards the cathode. What is the charge on a cation? Which type of element forms cations? Give an example of a cation.

3 Chemists sometimes call the positive electrode the anode. Anions are the ions that move towards the anode. What is the charge on an anion? Which type of element forms anions? Give an example of an anion.

Find out about:
- atoms and ions
- electron configurations of ions
- the formulae of ionic compounds

K Ionic theory and atomic structure

Atoms into ions

Faraday could not explain how atoms turn into ions because he was working long before anyone knew anything about the details of atomic structure. Today, chemists can use the shell model for electrons in atoms to show how atoms become electrically charged.

The metals on the left-hand side of the periodic table form ions by losing the few electrons in the outer shell. This leaves more protons than electrons, and so the ions are positively charged.

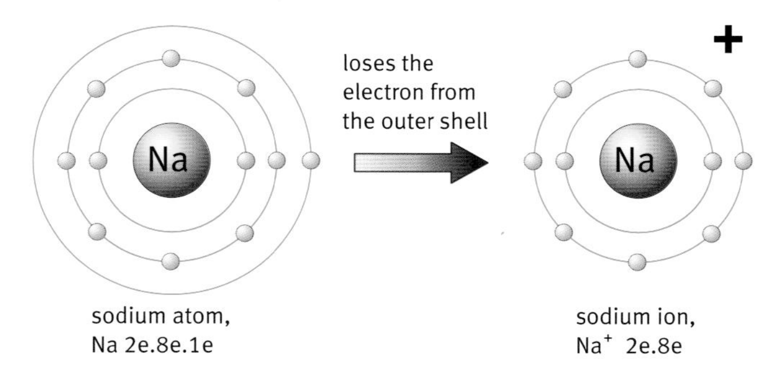

A sodium atom turns into a positive ion when it loses a negatively charged electron.

All the metals in group 1 have one electron in the outer shell. The diagram on page 107 shows that removing the first electron from a sodium atom needs relatively little energy. The same is true for the other group 1 metals, so they all form ions with a 1+ charge: Li^+, Na^+, and K^+, for example.

Chlorine gas consists of Cl_2 molecules. But it is easier to see what happens when chlorine gas forms by looking at one atom at a time, as shown in the diagram on the left. As each chlorine atom turns into an ion, it gains one electron and becomes negatively charged, Cl^-.

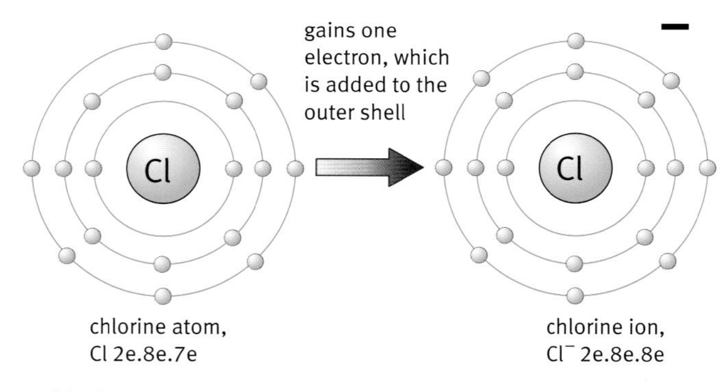

A chlorine atom turns into a negative ion by gaining an extra negatively charged electron.

Electron configurations of ions

Notice that when sodium and chlorine atoms turn into ions, they end up with the same electron configuration as the nearest noble gas in the periodic table. This is generally true for simple ions of the first 20 or so elements in the periodic table.

An explanation of why this is so requires a detailed analysis of all the energy changes when metals react with non-metals. This is something you will study if you go on to a more advanced chemistry course.

Ions into atoms

Electrolysis turns ions back into atoms. Metal ions are positively charged, so they are attracted to the negative electrode. It is a flow of electrons from the battery into this electrode that makes it negative. Positive metal ions gain electrons from the negative electrode and turn back into atoms.

Non-metal ions are negatively charged, so they are attracted to the positive electrode. This electrode is positive because electrons flow out of it to the battery. Negative ions give up electrons to the positive electrode and turn back into atoms.

Formulae of ionic compounds

The formula of sodium chloride is NaCl because there is one sodium ion (Na^+) for every chloride ion (Cl^-). There are no molecules in everyday table salt, only ions.

Not all ions have single positive or negative charges like sodium and chlorine. The formula of lead bromide is $PbBr_2$. In this compound there are two bromide ions for every lead ion. All compounds are overall electrically neutral, so the charge on a lead ion must be twice that on a bromide ion. A bromide ion, like a chloride ion, has a single negative charge (Br^-), so a lead ion must have a double positive charge (Pb^{2+}).

Compound	Ions present		Formula
	Positive	Negative	
magnesium oxide	Mg^{2+}	O^{2-}	MgO
calcium chloride	Ca^{2+}	Cl^- Cl^-	$CaCl_2$
aluminium oxide	Al^{3+} Al^{3+}	O^{2-} O^{2-} O^{2-}	Al_2O_3

Examples of formulae of ionic compounds

Ions in the periodic table

The charges of simple ions show a periodic pattern. You can see this from the diagram below, in which the ionic symbols appear in the periodic table. Many of the transition metals in the middle block of the table can form more than one type of ion. Iron, for example, can form Fe^{2+} and Fe^{3+} ions, while copper can exist as Cu^+ and Cu^{2+} ions. Why this should be so is something you will study if you go on to a more advanced chemistry course.

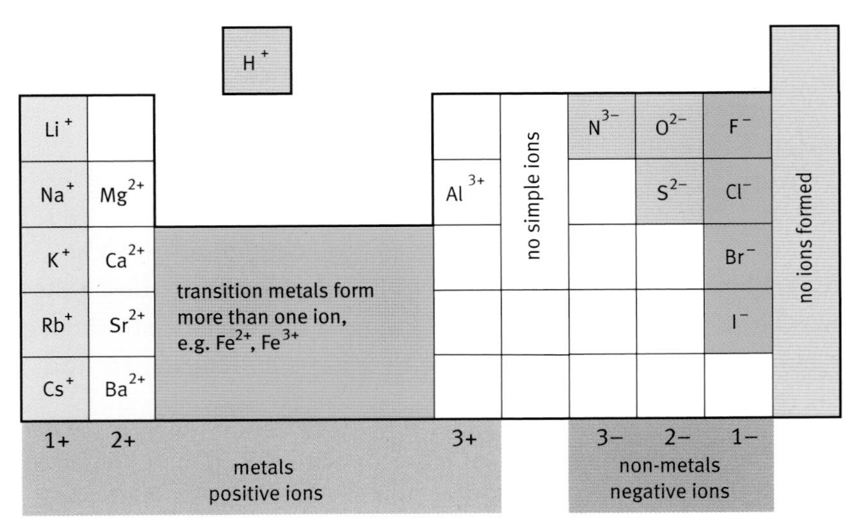

Simple ions in the periodic table

Questions

1 Draw diagrams to show the number and arrangement of electrons in a lithium atom and in a lithium ion. What is the charge on a lithium ion?

2 Draw diagrams to show the number and arrangement of electrons in a fluorine atom and in a fluoride ion. What is the charge on a fluoride ion?

3 Write down the electron configurations of:

a a fluoride ion (nucleus with 9 protons and 10 neutrons)

b a neon atom (nucleus with 10 protons and 10 neutrons)

c a sodium ion (nucleus with 11 protons and 12 neutrons). In what ways are a fluoride ion, a neon atom, and a sodium ion the same. How do they differ?

4 With the help of the table of ions, work out the formulae of these ionic compounds:

a potassium iodide

b calcium bromide

c aluminium chloride

d magnesium nitride

e aluminium sulfide

Find out about:
- atoms, molecules, and ions
- chemical species

L Chemical species

In this module you have met the idea that the same element can take different chemical forms with distinct properties. Chemists describe these different forms as **chemical species**.

Species of chlorine

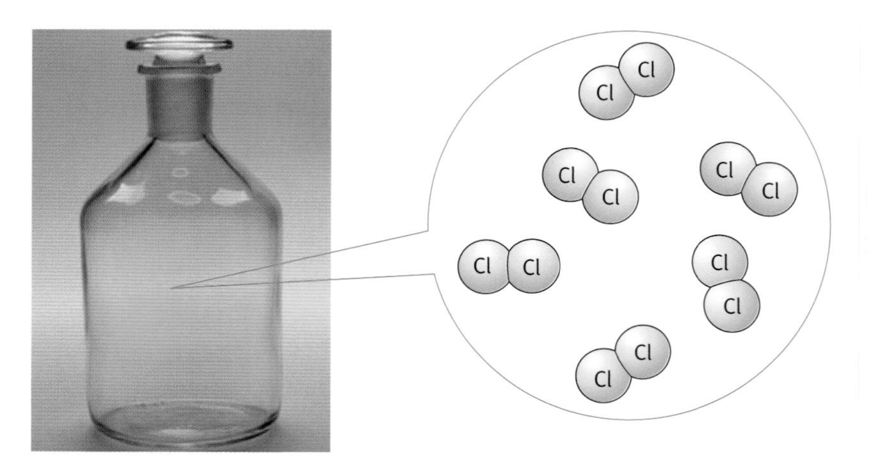

Chlorine gas consists of chlorine molecules. Chlorine molecules are chemically very reactive.

Chlorine has three simple species: atom, molecule, and ion. Each of these species of chlorine has distinct properties. Chlorine atoms (Cl) do not normally exist in a free state. They rapidly pair up to form chlorine molecules (Cl_2). However, ultraviolet radiation can split chlorine and chlorine compounds into atoms. This is what happens to CFCs such as CCl_3F when they get into the upper atmosphere. In the full glare of the Sun's radiation the molecules break up into atoms. Then the very reactive free chlorine atoms rapidly destroy ozone. Lowering the concentration of ozone creates the so-called 'hole' in the ozone layer.

Chlorine gas at room temperature consists of chlorine molecules (Cl_2). These are very reactive, as illustrated by the chemistry of the alkali metals and halogens described in section D. The chlorine molecules are reactive enough to do damage to human tissues, so the gas is given the label 'toxic'.

Chloride ions are quite different. They occur in compounds such as sodium chloride and magnesium chloride. Chloride ions in these salts are essential to life and occur in all living tissues. Chloride ions are chemically active in many ways, but they are not as reactive and harmful as the atoms or molecules of the element.

There are more complex species of chlorine with the element joined to other atoms. This includes molecules that contain chlorine and other elements, such as tetrachloromethane (CCl_4).

Key words
chemical species

Species of sodium

There are only two species of sodium: atom and ion. The atoms in sodium metal are chemically very active. In the presence of other chemicals, the sodium atoms react to produce compounds containing sodium ions.

Sodium combines with chlorine to produce the ionic compound sodium chloride. This is made up of two chemical species: Na^+ and Cl^-. These two ions are quite unreactive. Sodium chloride is soluble in water, but its solution is neutral. Water does not react with the ions.

When sodium reacts with water, it produces another ionic compound: sodium hydroxide, Na^+ and OH^-. The sodium hydroxide dissolves in the water to give a solution that is very alkaline. It is the hydroxide ions that make a solution of sodium hydroxide alkaline, not the sodium ions. You will find out more about the ionic theory of acids and alkalis in Module C6 *Chemical synthesis*.

Chemical species of the natural environment

You will learn more about the importance of being precise about chemical species in Module C5 *Chemicals of the natural environment* when you explore the chemical changes that affect elements such as nitrogen in the environment. Nitrogen gas is very different from oxides of nitrogen, which in turn are quite different in their properties from a salt containing nitrate ions.

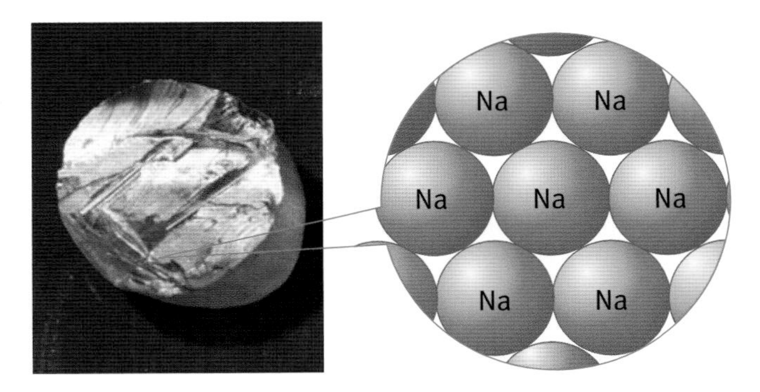

Sodium metal consists of sodium atoms. Sodium atoms are chemically very reactive.

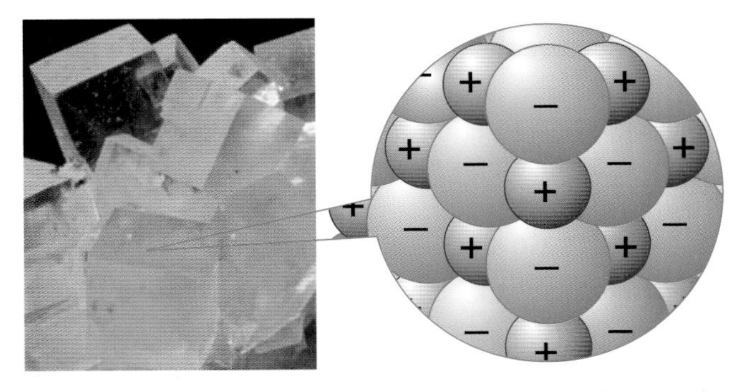

Sodium chloride consists of sodium ions and chloride ions. These ions are not very reactive.

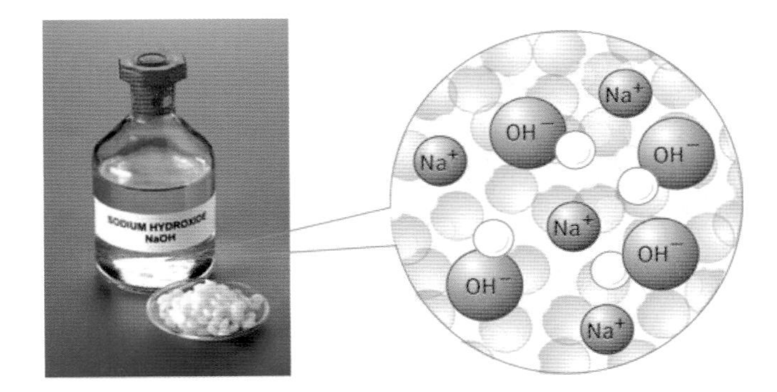

Sodium hydroxide is a strong alkali. It consists of sodium ions and hydroxide ions. In solution the ions move around separately mixed with water molecules.

Questions

1 Use the idea of chemical species to explain why the properties of sodium chloride are very different from the properties of its elements sodium and chlorine.

2 Give the name and formulae of all chemical species in:
a potassium
b bromine
c potassium bromide

③ a Identify four distinct chemical species in unpolluted air by giving their names and formulae.
b Identify three more chemical species present in the polluted air of a busy city street.

C4 Chemical patterns

Summary

You have now met some of the key patterns and theories that chemists use to make sense of the world and to explain how roughly 100 elements can give rise to such a huge variety of chemical compounds.

Atomic structure and the periodic table

- Atoms have a tiny central nucleus surrounded by negative electrons.
- The chemistry of an element is largely determined by the number and arrangement of the electrons in its atoms.
- The number of electrons is equal to the proton number of the atom.

Electrons in atoms

- Electrons in an atom have definite energies.
- The electron shell with the lowest energy fills first until it contains as many electrons as possible, then the next shell starts to fill.

Periodic table

- Arranging the elements in order of their proton numbers gives rise to the periodic table.
- For the first two rows in the table each period corresponds to the filling of an electron shell.
- After calcium the relationship between atomic structure and the periodic table becomes more complex and gives rise to the block of transition elements.

Groups

- Each column in the periodic table consists of a group of related elements.
- The elements in a group have similar chemistries because they have the same number of electrons in the outer shell.
- There are trends in the properties of the elements down a group because of the increasing number of inner full shells.

Atoms into ions

- When metals react with non-metals, the metal atoms lose electrons while the non-metal atoms gain electrons.
- This produces ionic compounds such as sodium chloride, $Na^{+}Cl^{-}$.
- The properties of an ionic compound are the properties of its ions, which behave in a different way from the atoms or molecules in the elements.

Questions

1 Copy this table and extend it to include the first 20 elements.

Element	Proton number	Number of electrons in each shell			
		First shell	Second shell	Third shell	Fourth shell
hydrogen	1	1			
helium	2	2			
lithium	3	2	1		

2 Summarize in outline how the model of atomic structure with electrons in shells can account for:

a the line spectra of elements

b the arrangement of the elements in the periodic table

c the charges on simple ions.

3 The table below shows part of the periodic table with only a few symbols included.

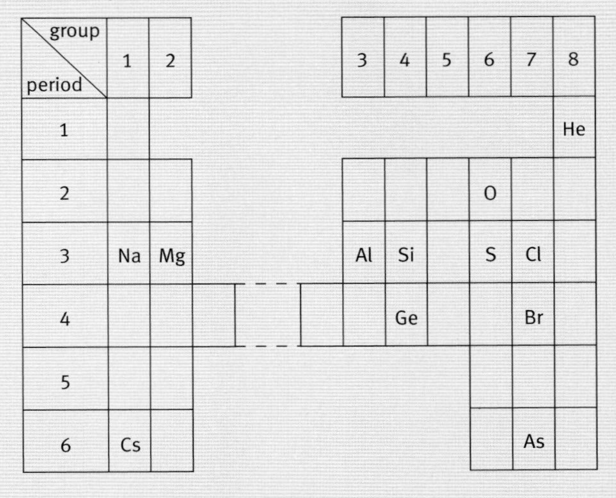

a Using only the elements in the table, write down the symbols for the following:

i a metal stored in oil

ii two non-metals that are gases at room temperature

iii an element used to kill bacteria

iv a metal that floats on water

v an element with similar properties to silicon (Si)

vi an element that has molecules made up of two atoms

vii the element with the largest number of protons in the nucleus of its atoms

viii the most reactive metal

ix an element with two electrons in the outer shell of its atoms

x an element X that forms an ionic chloride with the formula XCl

b Predict the formula of the compound of sulfur (S) with hydrogen.

c Is astatine (As) a solid, liquid, or gas at room temperature? Explain how you decide on your answer.

4 Potassium chloride (KCl) melts at 772 °C and boils at 1407 °C. Crystals of potassium chloride are colourless and shaped like cubes. The compound is soluble in water. A solution of potassium chloride conducts electricity.

a Is potassium chloride a solid, liquid, or gas at room temperature?

b Why do crystals of KCl all have the same shape?

c Why does a solution of potassium chloride conduct electricity?

d State two ways that potassium chloride differs from the element potassium.

e State two ways that potassium chloride differs from the element chlorine.

f Why are the properties of potassium chloride so different from its elements?

Why study chemicals of the natural environment?

Conditions on Earth are special. The temperature is just right for most water to be liquid. The atmosphere has enough oxygen for living things to breathe but not so much that everything catches fire. Rocks are the source of many of the chemicals that meet our daily needs.

The science

Theories of structure and bonding explain how atoms are arranged and held together in all the chemicals and materials that make up our Earth. There are three types of strong bonding that give rise to the useful materials that are metals, polymers, and ceramics. There are weaker forces of attraction that allow molecules to stick together enough to make liquids and solids.

Chemistry in action

Modern life depends on a wide range of advanced materials that have been developed as scientists understand more about structure and bonding. The science of structure and bonding is now so advanced that scientists can do engineering on an atomic scale, called nanotechnology.

Scientists are applying their understanding of chemistry in the environment to work out how society can deal with the growing impacts of human development.

Chemicals of the natural environment

Find out about:

- the chemicals in the main spheres of the Earth
- theories of structure and bonding
- the impact of human activity on the chemistry of the environment
- methods used to extract metals from their ores
- how to calculate chemical quantities

Find out about:

- naturally occurring elements and compounds
- abundances of elements in different spheres
- cycling of elements between the spheres

A Chemicals in four spheres

People who study the Earth often think of it as being made up of spheres (see the diagram below). Starting from the middle, first comes the core, then the mantle, then the crust.

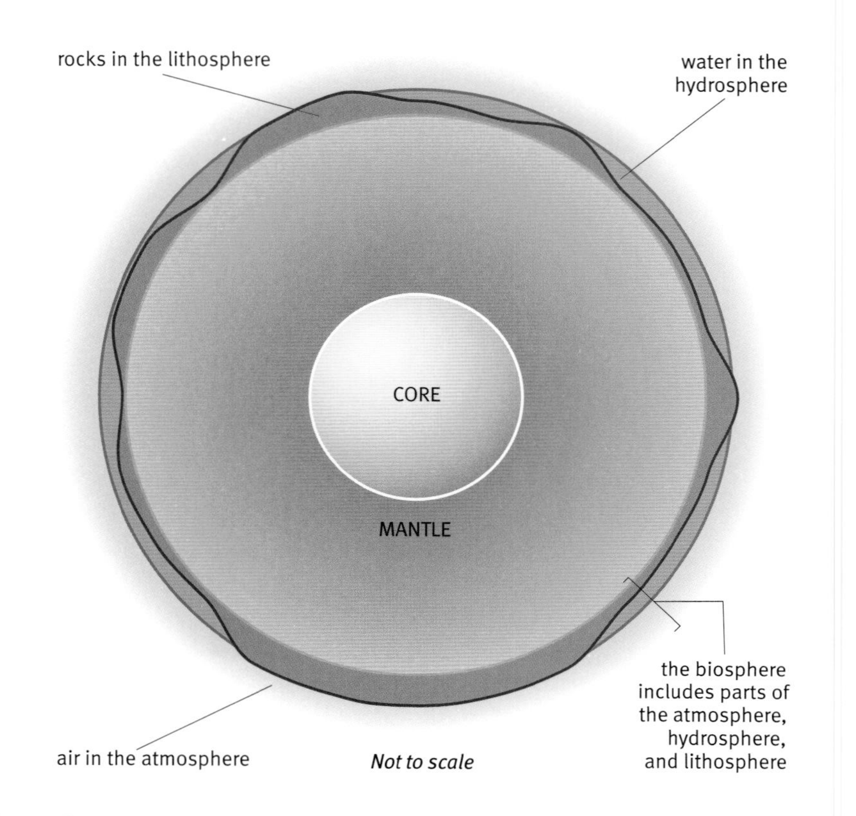

The Earth's spheres. The spheres are much smaller than this relative to the core and mantle.

The **lithosphere** is broken into giant plates that fit around the globe like puzzle pieces. These are the tectonic plates which move a little bit each year as they slide on top of the upper part of the mantle. The **crust** and upper **mantle** make up the lithosphere. They are about 80 km deep.

The oceans and rivers make up the **hydrosphere**. This is not a complete sphere, but the oceans cover two-thirds of the globe, so it almost is.

There are living things in the sea, rivers, and lakes. Large areas of land are covered with living things, and there are billions of living things in soil. So scientists like to think of a sphere of life encasing the planet; they call this the **biosphere**.

Finally, wrapped like a big fluffy duvet around the Earth, keeping it warm, is the layer of air we call the **atmosphere**.

Key words

lithosphere
crust
mantle
hydrosphere
biosphere
atmosphere

Elements in the spheres

The spheres vary greatly in their chemical composition. The air consists mainly of two free elements: nitrogen and oxygen. The hydrosphere is almost entirely the compound water.

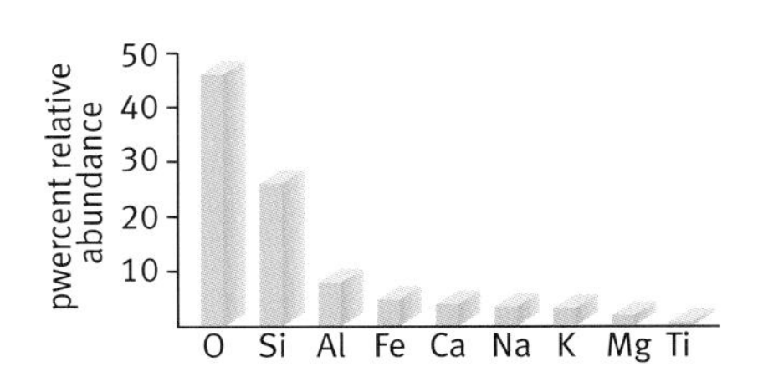

The abundant elements in the lithosphere

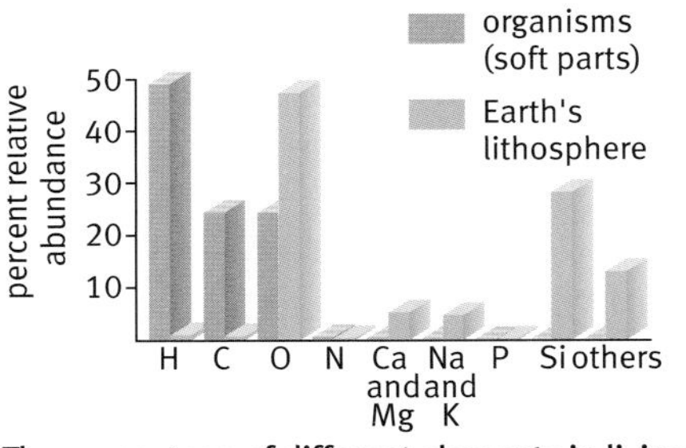

The percentage of different elements in living tissue and in the lithosphere compared

The rocks of the lithosphere are made up mainly of silicates. These are compounds of silicon and oxygen together with much smaller quantities of other elements.

The elements exist as different chemical species in the four spheres. In the lithosphere, carbon combines with hydrogen to make the hydrocarbons in crude oil. Carbon is also hidden in the calcium carbonate of chalk and limestone. In the atmosphere, carbon is present as the gas carbon dioxide. Carbon is a vitally important element in the biosphere: most of the chemicals that make up living things are compounds of carbon with hydrogen, oxygen, and a few other elements.

Flowing between the spheres

Chemicals do not always stay in one sphere. They are constantly on the move between the spheres (see the diagram on the right). Think of a carbon atom: it may start in the atmosphere; be taken into a plant in the biosphere; be washed into water in the hydrosphere; then buried in sediment of the lithosphere.

Water flows freely between the spheres. The obvious place for water is the hydrosphere. But think of clouds in the atmosphere; then rain sinking into the lithosphere, reappearing as a spring, where it gets drunk by an animal or soaked up by roots back into the biosphere.

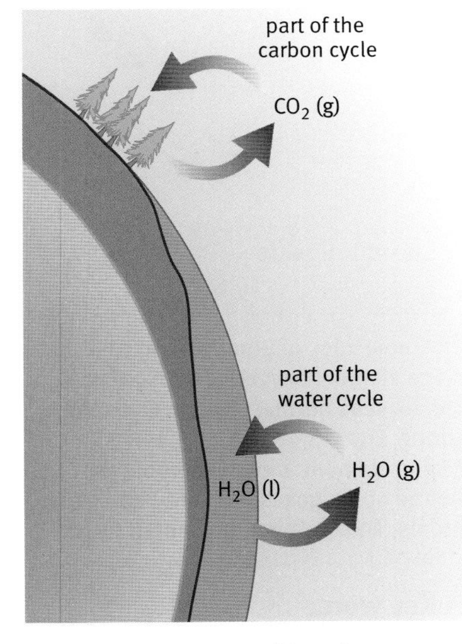

Cycles between the Earth's spheres. The most active places are the junctions between the spheres. Photosynthesis happens at the junction between the atmosphere and biosphere. Volcanoes move gases from the lithosphere to the atmosphere and the hydrosphere.

Questions

1. 'The biosphere overlaps with the atmosphere, hydrosphere, and lithosphere.' Give some examples to illustrate this statement.
2. Make a list of 10 different things you use (including what you wear and eat) during a day. Identify which sphere each item has come from.
3. Look at the bar charts on this page. What are the three most abundant elements in living tissue? What are the three most abundant elements in the lithosphere? Comment on your answers,

Find out about:

- gases in the air
- weak attractions between molecules
- strong covalent bonding

B Chemicals of the atmosphere

The Earth is just the right size for its gravity to hold on to gases. It is also just the right distance from the Sun to have the right temperature for liquid water to exist. This water, together with the carbon dioxide and oxygen in the atmosphere, means that Earth can support a great variety of plant and animal life.

The composition of the atmosphere is 78% N_2, 21% O_2, 1% Ar, 0.03% CO_2, with small amounts of water vapour.

The presence of argon in the air was not discovered until the 1890s – over a hundred years after the discovery of oxygen. Argon is a noble gas but it is not a rare gas. Every time you breathe in you take about 5 cm^3 argon into your lungs.

All the chemicals in the atmosphere are gases at normal temperatures – that is why they have ended up in the atmosphere. This means they have low melting and boiling points.

The other similarity between the chemicals in the atmosphere is that they are all either non-metallic elements (O_2, N_2, Ar), or they are compounds made from atoms of non-metallic elements. For example, CO_2 is made from carbon and oxygen.

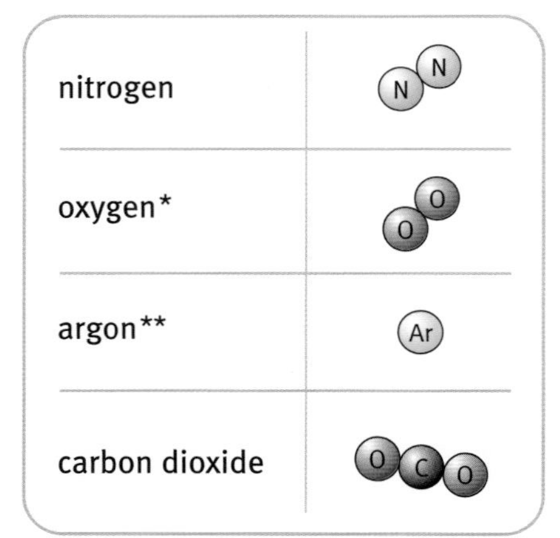

The molecules of atmospheric gases.
* The element oxygen is unusual as it can exist as normal oxygen, O_2, or as ozone, O_3.
** The element argon is a noble gas. All the noble gases are made up of single atoms, so argon is represented by Ar.

Atoms and molecules in the air

Most of the chemicals in the atmosphere are made of **small molecules**. Only the noble gases exist as single atoms.

All molecules have a slight tendency to stick together. For example, there is an attraction between one O_2 and another O_2. But these **attractive forces** are weak. This is why the chemicals that make up the atmosphere are gases with low melting and boiling points.

One way to picture this is that the molecules are moving so quickly that, when two O_2 molecules come close to each other, the attractive force between them is not strong enough to hold them together.

Strong bonds in molecules

The forces inside molecules that hold the atoms together are very strong, many times stronger than the weak attractions between molecules. Small molecules such as O_2 or H_2 do not split up into atoms except at very, very high temperatures.

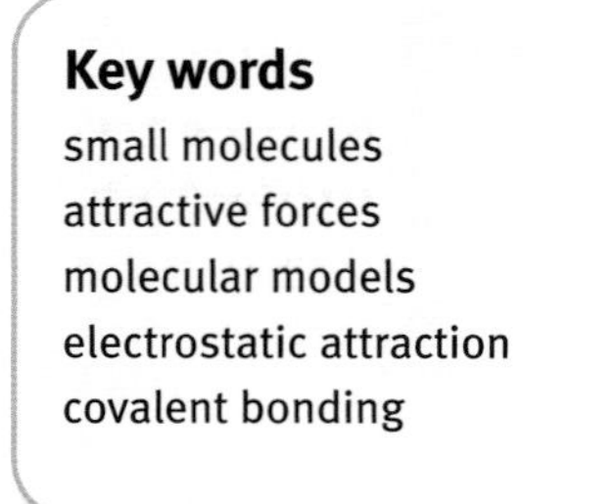

Key words
small molecules
attractive forces
molecular models
electrostatic attraction
covalent bonding

Chemists often use a single line to represent a single bond between atoms in a molecule. For example, the simple molecules in hydrogen, oxygen, and carbon dioxide can be represented by the molecular formulae H_2, O_2, and CO_2. But if you want to show their bonds, they can be represented by:

Some **molecular models** use the same idea. A coloured ball represents each atom, and a stick or a spring is used for each bond joining them together.

Electrons and bonding

A knowledge of atomic structure can help to explain the bonding in molecules (see Module C4 *Chemical patterns*, Section K). When non-metal atoms combine to form molecules, they do so by sharing electrons in their outer shells.

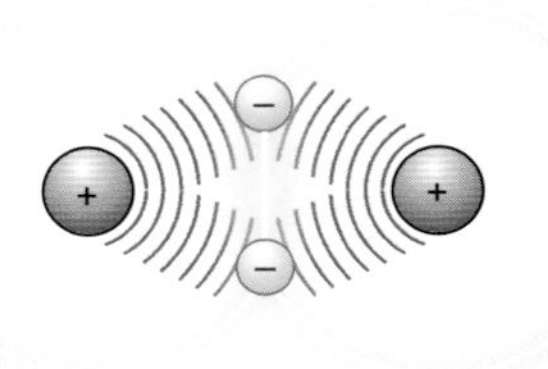

The formation of a covalent bond between two hydrogen atoms

The molecule of hydrogen is held together by the **electrostatic attraction** between the two nuclei and the shared pair of electrons. This is a single bond.

This type of strong bonding is called **covalent bonding**. 'Co' means 'together' or 'joint', while the Latin word 'valentia' means strength. So we have strength by sharing.

The number of bonds that an element can form depends on the electrons in the outer shell of its atoms. The table on the right gives the number of covalent bonds normally formed by atoms of some common non-metal elements.

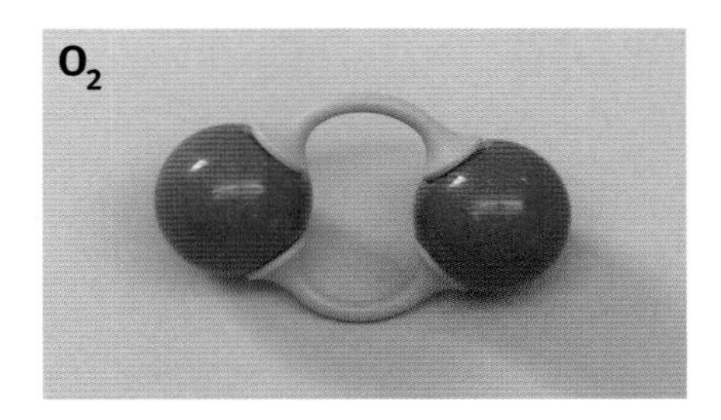

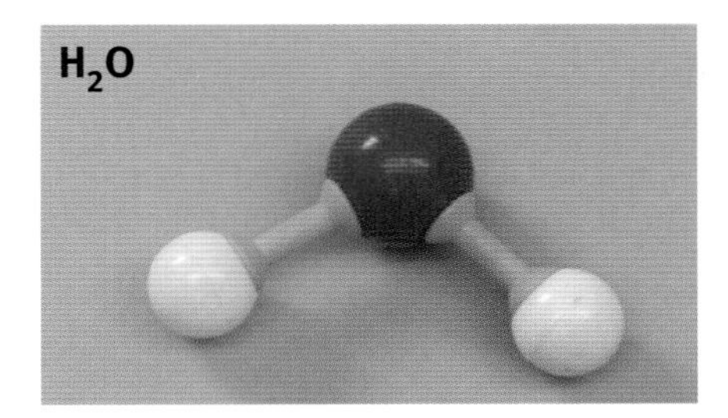

Ball-and-stick models of H_2, O_2, H_2O and CO_2. Notice that the molecules have a definite shape. There are fixed angles between the bonds in molecules

Atom	Usual number of covalent bonds
H, hydrogen	1
Cl, chlorine	1
O, oxygen	2
N, nitrogen	3
C, carbon	4
Si, silicon	4

Questions

1 Estimate the volume of argon in the room in which you are sitting.

2 Draw diagrams to show the covalent bonding in these molecules:

a hydrogen chloride, HCl
b ammonia, NH_3
c methane, CH_4
d ethene C_2H_4

Find out about:
- unusual properties of water
- bonding within and between water molecules
- ions in solution

On Earth, water exists mostly in the liquid state (the oceans and the clouds), with some in the solid state (ice) and a smaller amount as gas (water vapour).

C Chemicals of the hydrosphere

Properties of water

Water is such a familiar chemical that it not obvious that chemically it has some very special properties.

One of these special properties is that it is a liquid at room temperature. The H_2O molecule has a smaller mass than molecules of O_2, N_2, or CO_2, which are all gases at normal temperatures. So you might expect H_2O to be a gas at normal temperatures. But H_2O melts at 0 °C and boils at 100 °C. For some reason, water molecules of H_2O have a greater tendency to stick together than the molecules of a gas like oxygen.

Strange things happen when water cools from room temperature. At first, the liquid contracts as expected. This goes on until the temperature falls to 4 °C. Then, on cooling further to 0 °C, it starts to expand. This means that ice is less dense than the cold water surrounding it, so it floats. This is very important in nature. It means that winter ice does not sink to the bottom of lakes where it would be far from the warming rays of spring sunshine.

A third odd property of water is that it is a good solvent for **salts**. Most common solvents do not **dissolve** ions, but water does.

Pure water does not conduct electricity – in this respect it resembles other liquids that are made up of small molecules. This shows that it does not contain charged particles that are free to move. Even though pure water does not conduct, you should never touch electrical devices with wet hands. This is because the water on your hands is not pure, so it does conduct electricity and increases the risk of you receiving an electric shock.

Water molecules

Knowledge of the structure of water molecules can help to explain its remarkable properties. The three atoms in a water molecule are not arranged in a straight line but are at an angle.

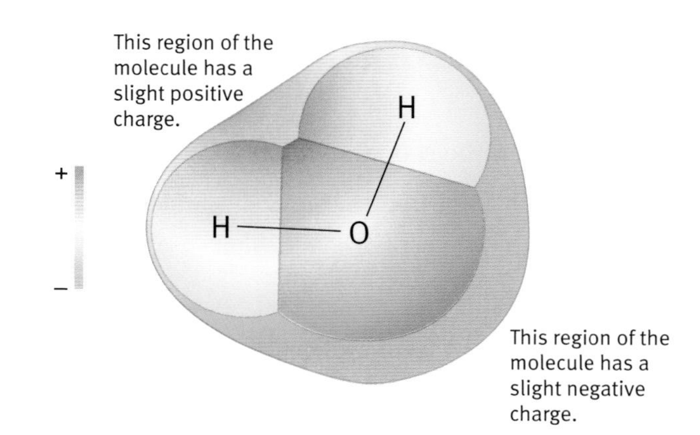

A water molecule, showing that oxygen has a slightly greater share of the pair of electrons in each bond

Key words
salts
dissolve
weathering

In the covalent bonds between the atoms, the electrons are not evenly shared. The oxygen atoms have more than their fair share. This means that there is a slight negative charge on the oxygen side of each molecule and a slight positive charge on the hydrogen side. Overall the molecules are still electrically neutral.

The small charges on opposite sides of the molecules cause slightly stronger attractive forces between the molecules. These small charges also help water dissolve ionic compounds by attracting the ions out of their crystals.

The attractions between water molecules and their angular shape mean that in ice they line up to create a very open structure. As a result, ice is less dense than liquid water.

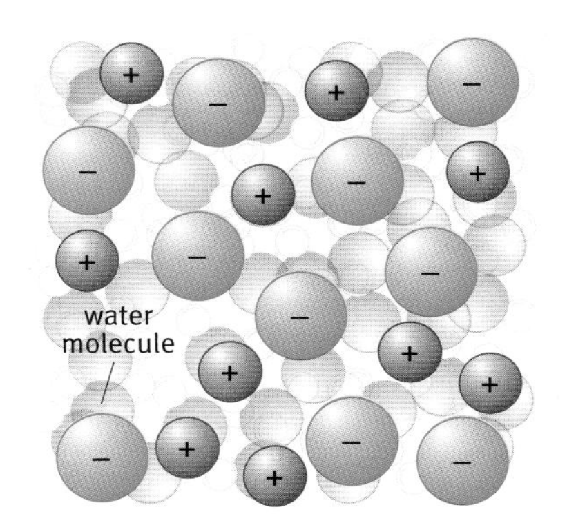

Ions separate and can move about freely when dissolved in water. In sea water there is a mixture of positive ions and negative ions.

Why is sea water salty?

The diagram below shows how soluble chemicals get carried from rocks to the sea during part of the water cycle.

The main soluble chemical carried into the sea is sodium chloride. River water does not taste salty because the concentration is so low, but the concentration of salt in the sea has built up over millions of years, and so it tastes salty.

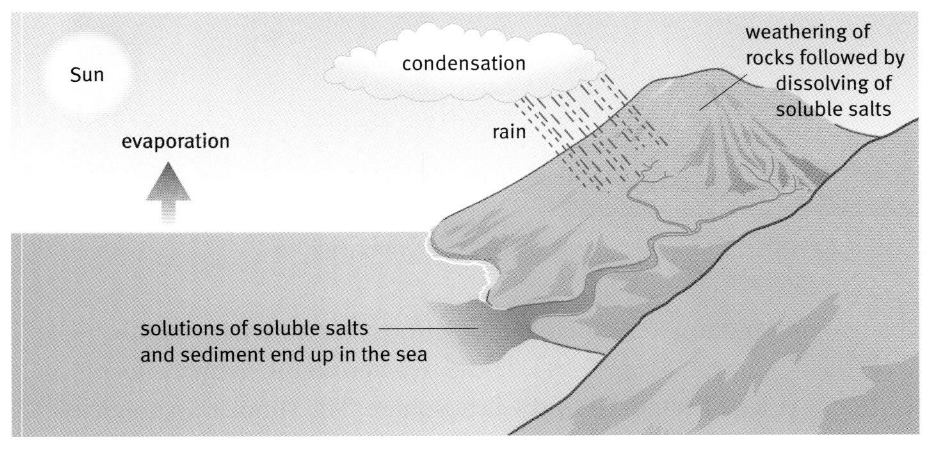

Weathering slowly breaks down rocks. This exposes the inside of the rocks to water. Soluble chemicals in the rocks dissolve in the water and get washed away.

Other chemicals that dissolve include potassium chloride, potassium bromide, potassium iodide, magnesium chloride, and magnesium sulfate. One litre of typical sea water contains about 40 g of dissolved chemicals from rocks.

Most of the compounds that are dissolved in sea water are made up of positively charged metal ions and negatively charged non-metal ions. They are salts.

Questions

1. How would the possibility of life on Earth have been affected had ice been denser than water?
2. Draw up a table to list in one column the properties of water that are typical of molecular chemicals and the properties that are unusual.
3. Write out the formulae of these salts with the help of the table showing the ion charges on page 115 in Module C4. The sulfate ion is SO_4^{2-}.
   - **a** potassium iodide
   - **b** magnesium chloride
   - **c** magnesium sulfate

Find out about:
- chemicals in the crust of the Earth
- ionic bonding
- silica and silicates
- giant covalent structures

D Chemicals of the lithosphere

Rocks and minerals

The outer rigid layer of the Earth is the lithosphere. 'Lithos' is the Greek word for stone or rock. The top part of the lithosphere – the part we live on – is the crust.

Some rocks are massive, like the cliffs on the side of a mountain. Scientists include boulders, stones, and pebbles as rocks. A stone is just a small piece of rock.

Rocks are made of one or more minerals. Sandstone, for example, is made of mainly one mineral: silicon dioxide. Limestone is also made of a single mineral: calcium carbonate. Granite is a mixture of quartz, feldspar, and mica.

Granite is made from a mixture of minerals. There are glassy grains of silica (silicon dioxide), black crystals of mica, and large crystals of feldspar, which may be pink or white.

Minerals are naturally occurring chemicals. They may be elements, like gold and silver, which are found free in rocks. More commonly, they are compounds, such as silicon dioxide, SiO_2; calcium carbonate, $CaCO_3$; rock salt, NaCl; and iron oxides such as Fe_2O_3.

Haematite, Fe_3O_4. Crystals of this mineral range from metallic black to dull red.

Calcite, $CaCO_3$

Pyrites, FeS_2

The two most common elements in the lithosphere are non-metals: oxygen (47%) and silicon (28%). These two **abundant** elements form the major types of minerals in the lithosphere. The simplest example is silicon dioxide, SiO_2, which can take various crystalline forms, including quartz.

Key words
granite
mineral
abundant

Questions

1 Write a sentence that makes clear the difference between the words *rock*, *mineral*, and *lithosphere*.

2 Look at the graphs on page 123.
  a What is the most abundant metal in the lithosphere?
  b What is the most abundant non-metal?

3 Name a mineral which is:
  a an oxide **b** a chloride **c** a sulfide
  d a carbonate

4 Name the elements combined together in:
  a quartz **b** galena **c** calcite
  d pyrite

Evaporite minerals

Sea water contains abundant dissolved chemicals. When the water evaporates, **ionic compounds** crystallize. Sodium chloride, NaCl, or rock salt, is a common example. The mineral is halite.

Minerals that are formed in this way are called evaporites. Roughly 200 million years ago, in the Triassic era, vast salt deposits were laid down as sea water evaporated. The salt was later covered with other sediments and is now under the county of Cheshire in England. People have been extracting and trading this salt since before Roman times.

Salt deposits in the Great Basin. In this hot and dry area of the western USA, the salty water evaporates so fast that the salts crystallize as salt flats.

The structure and properties of salts

Crystals of sodium chloride are cubes. They are made up of sodium and chloride ions (see Module C4 *Chemical patterns*, Section J).

In every crystal of sodium chloride, the ions are arranged in the same regular pattern. This means that all crystals of the compound have the same cubic shape.

As a sodium chloride crystal forms, millions of Na^+ ions and millions of Cl^- ions pack closely packed together. The ions are held together very strongly by the attraction between their opposite charges. This is called **ionic bonding**, and the structure is called a **giant ionic structure**. Unlike compounds such as water which are made up of individual molecules of H_2O, there is *not* an individual NaCl molecule.

Because of the very strong attractive forces, it takes a lot of energy to break down the regular arrangement of ions. So NaCl has to be heated to 808 °C before it melts. Compare this to melting ice (melting point 0 °C), where you only need to supply sufficient energy to overcome the relatively weak attractive forces between the H_2O molecules.

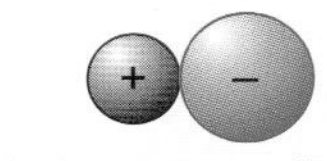

1 The Na^+ also attracts other Cl^- ions that are close to it. The Cl^- attracts other Na^+ ions that are close to it.

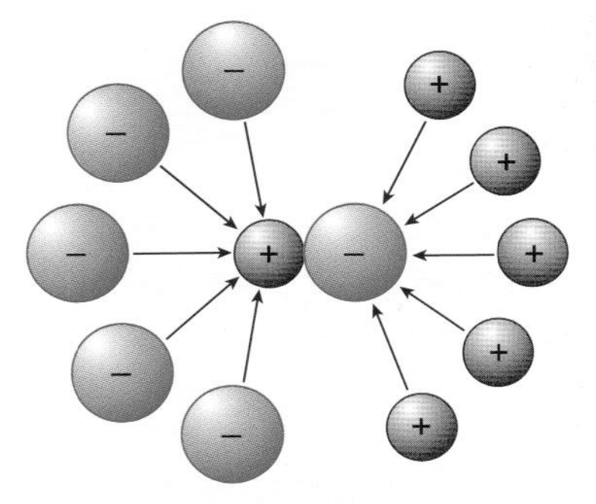

2 Another 5 Cl^- ions can fit around the Na^+, making 6 in total, and another 5 Na^+ ions can fit around the Cl^- ion, making 6 in total.

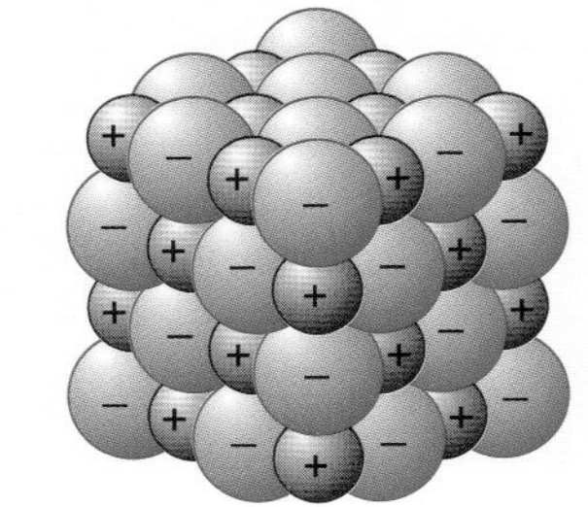

3 Each of these ions then attracts other ions of the opposite charge, and the process continues until millions and millions of oppositely charged ions are all packed closely together.

How a sodium chloride crystal forms.

Questions

5 Draw up a table to compare the properties of a molecular compound such as water and an ionic compound such as sodium chloride.

Key words

ionic compound
ionic bonding
giant ionic structure

Silica and silicates

Quartz

The mineral silica consists of silicon dioxide, SiO_2. Its commonest crystalline form is **quartz**. Crystalline silica has helped to shape human history. From the sand used for making glass, to the piezoelectric quartz crystals used in advanced communication systems, crystalline silica has been a part of human technological development.

Pure quartz crystals. Pure SiO_2 is transparent and very hard.

SiO_2 is found in many rocks. As the rocks get weathered, the SiO_2 stays behind and ends up as sand in rivers and beaches. When sand is compressed, it forms the rock called **sandstone**.

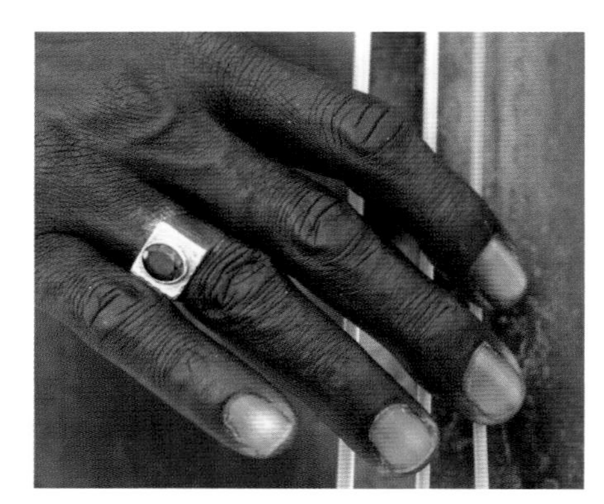

Amethyst is a form of quartz used as a **gemstone** in jewellery. It contains traces of manganese and iron oxides, which give the mineral its violet colour.

When silicon atoms bond to other atoms, they normally form four bonds per Si atom. Oxygen, on the other hand, normally forms two bonds per O atom. So, in SiO_2, each Si atom forms a covalent bond to four O atoms. Each O atom forms a covalent bond to two Si atoms.

The diagram below shows the arrangement of the Si and O atoms when they have bonded together in SiO_2. You can see that, instead of forming small molecules, they form a three-dimensional **giant covalent structure** that goes on and on. The Si—O covalent bond is very strong, so this giant structure is very strong and rigid. This is what gives SiO_2 its special properties (see the table on page 131).

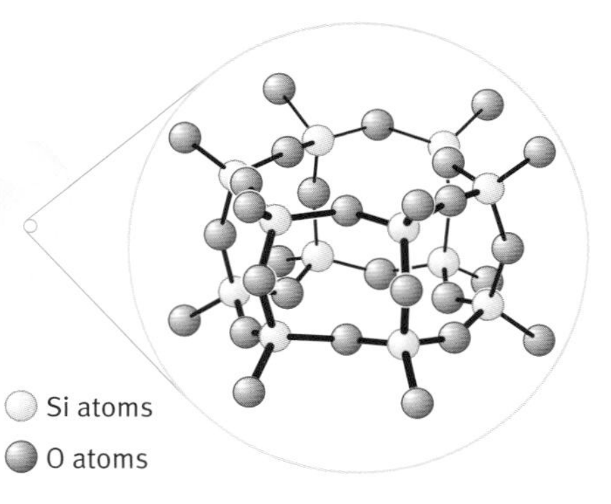

In SiO_2, each Si atom (grey) is covalently bonded to four O atoms (red). Each O atom is bonded to two Si atoms. On average there are four O atoms to every two Si atoms, so the formula simplifies to SiO_2.

Key words
- quartz
- sandstone
- gemstone
- giant covalent structure

Property of SiO_2	Comments	Uses
very hard	strong rigid structure; will scratch steel	used as abrasive in sandpaper and scouring powders
high melting (1610 °C) and boiling (2230 °C) points	strong, rigid structure difficult to break down	used to make linings for furnaces and high-temperature laboratory glassware
insoluble in water	resists weathering, ending up as sand in rivers, on beaches, and in deserts	sandstone used as building stone
electrical insulator	no free electrons or ions to carry electricity	silica glass used as an insulator in electrical devices

Some properties of SiO_2

Questions

1 What properties of silicon dioxide make it useful in sandpaper?

2 What properties of quartz as amethyst make it suitable for a gemstone?

3 Show how the properties of silicon dioxide can be explained in terms of its structure and bonding.

Silicate minerals

Over 95% of the rocks in the Earth's continental crust are formed by silica and the silicate minerals. However, the other 5% includes some very important minerals such as limestone and gold.

The Deccan Traps in India show how igneous rocks made of silicates can shape the landscape on a huge scale. The Traps cover an area of 500 000 square kilometres. They rise to over 2000 metres above sea level and consist of layer upon layer of basalt. These are the remains of lava flows which spread out about 66 million years ago. Basalt is an igneous rock which consists of tiny grains of a mixture of silicate minerals.

The structure of silicate minerals is based on giant structures of silicon and oxygen atoms. They include atoms of other elements too such as aluminium, iron, calcium, and potassium. The mica and feldspar in granite are silicate minerals.

Find out about:
- chemicals in living things
- proteins, carbohydrates, and DNA

E Chemicals of the biosphere

All living things are made from chemicals. The study of the chemicals of life is called biochemistry. Most biochemicals are based on three elements: carbon, hydrogen, and oxygen, sometimes with nitrogen, sulfur and phosphorus. The compounds consist of large molecules.

Why carbon?

The skeleton of all biochemicals is made from carbon. Carbon is the element on which all life is based. The special things about carbon that make this possible are:

- Carbon atoms can form chains by joining to themselves.
- Carbon forms four strong covalent bonds, so other atoms can join onto the chains (see below). Very often these are hydrogen atoms.

These properties of carbon mean it can make an amazing variety of compounds. This is why life itself is amazingly varied. Most biochemicals are polymers, built by joining together smaller, simpler molecules.

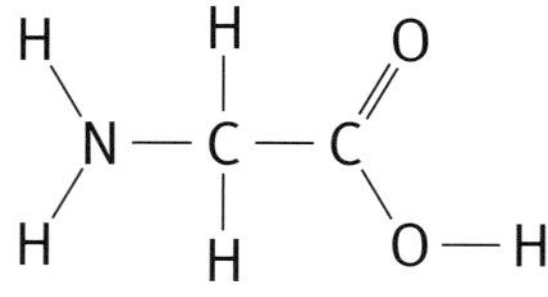

glycine

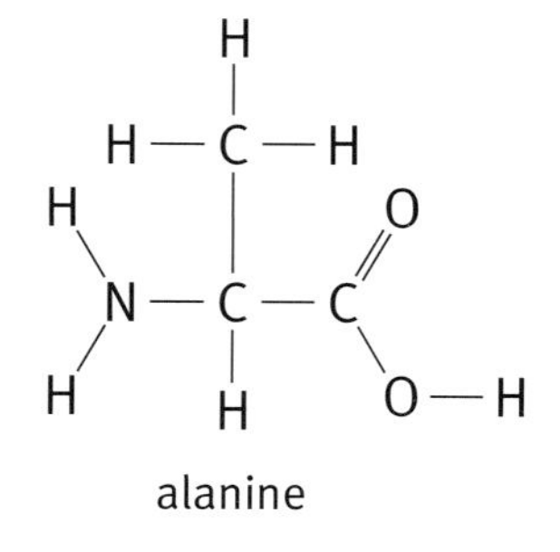

alanine

Models of two amino acid molecules and their structures shown in symbols

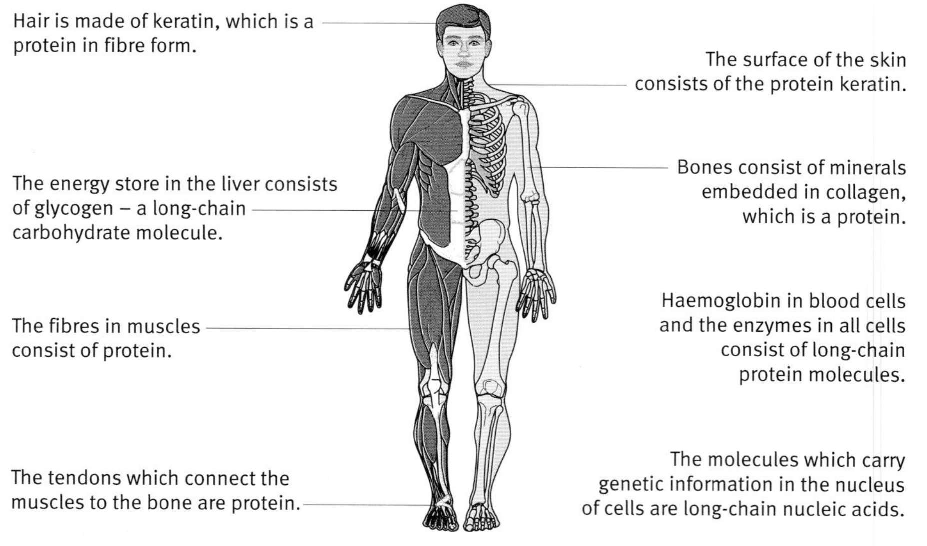

Polymers in the human body

Proteins

Hair, skin, and muscle are built from **proteins**. The enzymes that control biochemical reactions are made of proteins. Everywhere you look in your body, you will find proteins doing different kinds of jobs. Proteins are so varied because the carbon chains they are built from can be so varied.

Proteins are polymers, built by joining together monomers called amino acids. There are 20 different amino acids, and proteins have hundreds of them joined together.

Each protein has a unique sequence of amino acids in its polymer chains.

Carbohydrates

Photosynthesis in the leaves of plants turns carbon dioxide and water into glucose. Glucose is a sugar, which belongs to the family of compounds called **carbohydrates**.

There are just three elements in carbohydrates. They are compounds of carbon and the elements in water. This explains their name:

'carbo-' for carbon

'-hydrate' from the Greek word for water

Glucose is a very soluble sugar. Plants convert it to starch as an energy store. Starch is an insoluble polymer made up of long chains of glucose units.

Plants also join up glucose molecules in a different way to produce another polymer called cellulose. This is the polymer that makes up plant cell walls.

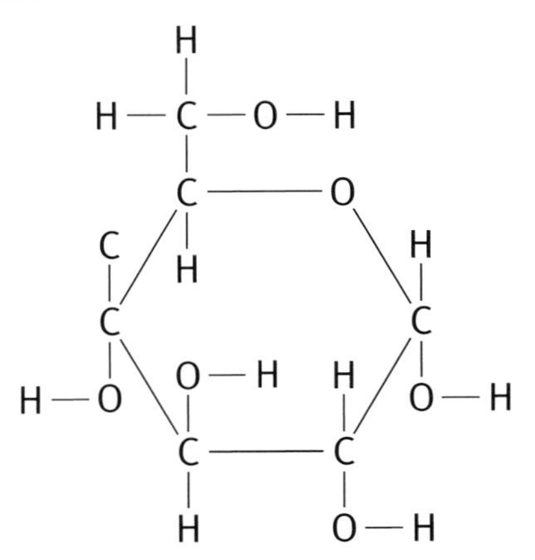

A model of a glucose molecule and its structure in symbols

Nucleic acids

DNA and RNA are nucleic acids. They are the molecules that carry the genetic code.

The backbone of a DNA molecule is a polymer with alternating sugar and phosphate groups. Attached to the backbone are four bases: adenine, cytosine, guanine, and thymine.

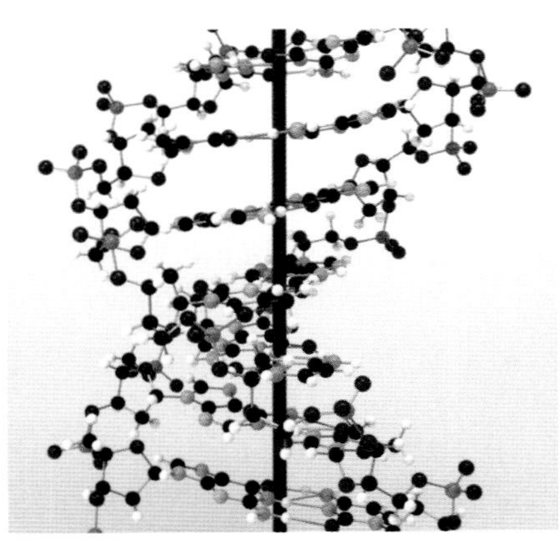

A model of a short length of a DNA molecule. The blue atoms represent nitrogen and the violet atoms represent phosphorus.

Questions

1 Look at the models of three amino acid molecules.

a Which five elements are present in the molecules?

b In these molecules, how many bonds are formed by:

i each carbon atom?
ii each oxygen atom?
iii each nitrogen atom?

c Write the molecular formulae for the three amino acids

d In what ways are the structures of the amino acids the same, and how do they differ?

2 Name the three elements which make up carbohydrates.

3 If the formula of glucose is written $C_xH_yO_z$, what are the values of x, y, and z?

4 Look at the model of a DNA molecule on the right. Which elements are present in DNA?

Key words

protein
photosynthesis
carbohydrate
DNA

Find out about:

- natural cycles of elements
- human impacts on the environment

F Human impacts on the environment

Element cycles

As living things grow, die, and decay, the elements move between the biosphere, hydrosphere, atmosphere, and lithosphere. This happens naturally on a large scale.

The carbon cycle

Some human activities can make a difference to these **natural cycles**. An important example is the effect on the **carbon cycle** of burning fossil fuels.

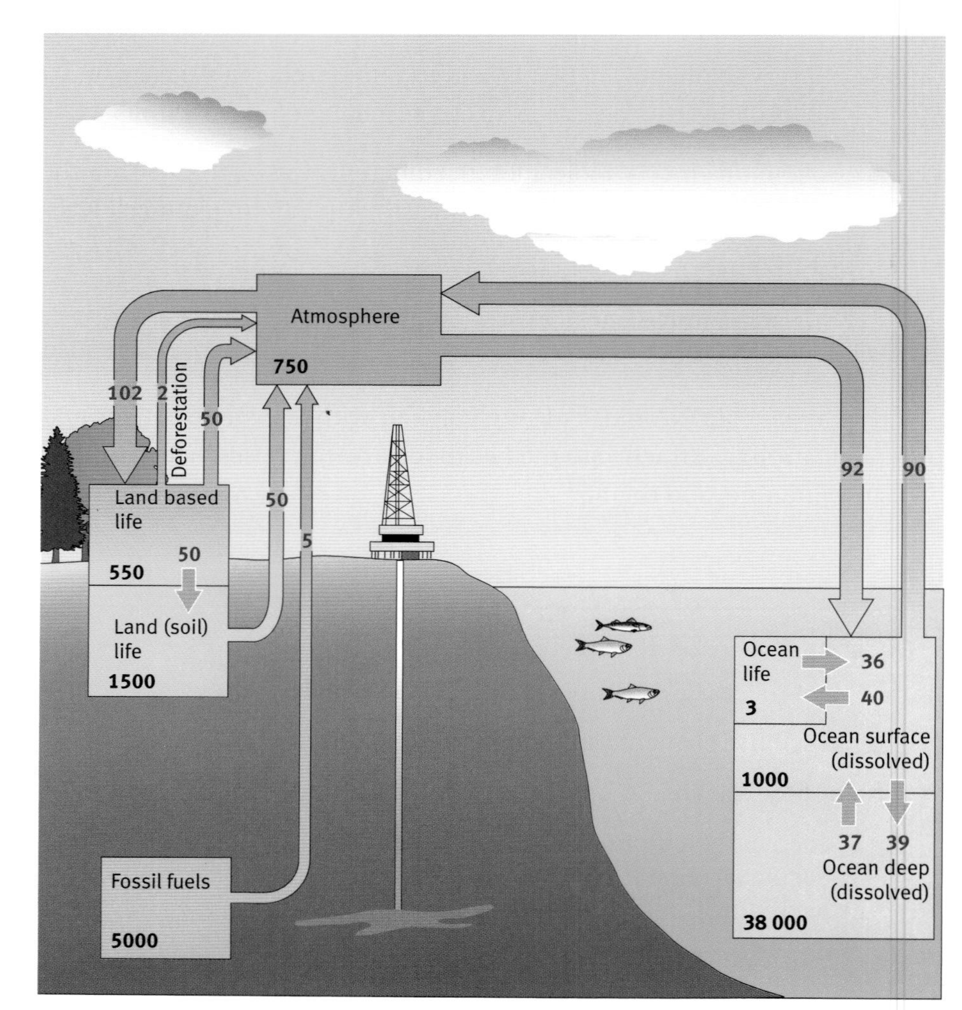

A simplified version of the carbon cycle. The figures are in 1000 million tonnes (gigatonnes). The figures in black are estimates of the total mass of carbon (worldwide) in the different spheres. The figures in red are the estimated flows of carbon between spheres.

The flow of carbon dioxide into the air from burning fuels is small compared with the natural flows due to respiration and photosynthesis. Even so, it is large enough to have raised the concentration of carbon dioxide in the air from about 277 parts per million before the Industrial Revolution to around 360 parts per million today.

Questions

1 Give an example of a chemical species containing carbon in:
 a the atmosphere
 b the hydrosphere
 c the lithosphere
 d the biosphere

2 According to the figures in the diagram of the carbon cycle, what is the total mass, in gigatonnes, of carbon in the biosphere?

3 a What is the total mass of carbon passing into the atmosphere each year?
 b What is the total mass of carbon taken out of the atmosphere each year?
 c What is the net change in the mass of carbon in the atmosphere each year?

4 a What mass of carbon is removed from the atmosphere each year by the photosynthesis of plant life on land?
 b What mass of carbon is added to the atmosphere each year by the respiration of living things on or in the land?

5 Does the diagram of the carbon cycle suggest that human activity is having a significant effect on the global carbon cycle?

The nitrogen cycle

Nitrogen is essential for making many biochemicals, especially proteins, which are the main nitrogen-containing compounds in the biosphere.

N_2 (nitrogen gas) has small molecules with weak bonds between them. It is a gas and is found in the atmosphere.

NO_3^- (nitrate) and NH_4^+ (ammonium) are ions. They are attracted to water molecules, which makes them dissolve. So they may get into the hydrosphere. They are also attracted to ions in the soil, so they get into the lithosphere. When they are absorbed by plant roots, they enter the biosphere.

Growing crops take in the nitrogen they need from the soil in the form of nitrate ions. Farmers and gardeners add fertilizers to the soil. This puts back the nitrogen removed from the soil when crops are grown and harvested.

The problem for intensive agriculture is that plants cannot use nitrogen directly from the air. Only a few specialized bacteria and algae can convert N_2 gas to ammonium ions and nitrates. This is the process called 'fixing' nitrogen. Nitrogen fixation makes the nitrogen available for use by plants.

There are three main ways of fixing nitrogen from the air:

- the action of microorganisms (bacteria or algae)
- a chemical reaction in the air during lightning flashes
- a manufacturing process called the Haber process used to make fertilizers

The manufacture of fertilizers now makes a major impact on the **nitrogen cycle**. As much nitrogen is fixed by industry as is fixed naturally by the natural processes supplying nitrogen to the soil.

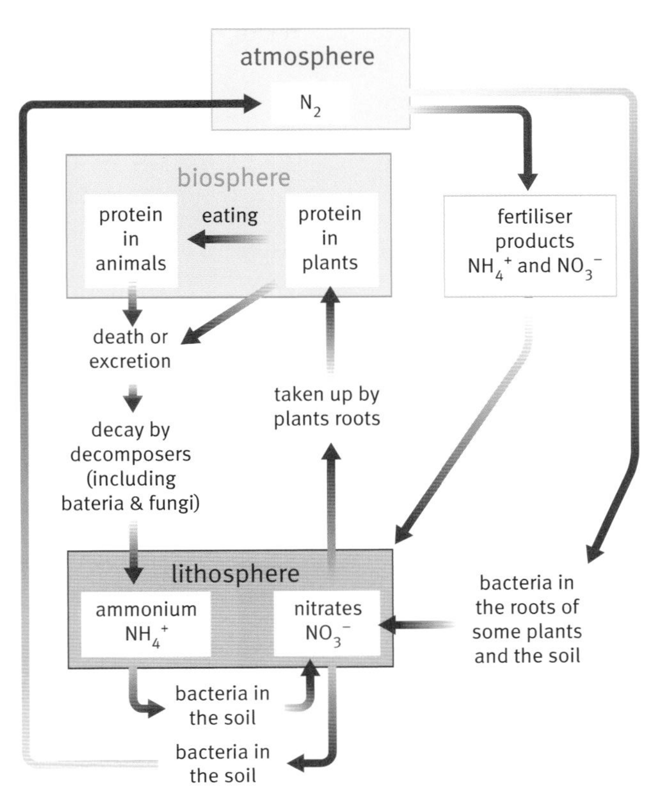

Nitrogen gets converted to different chemical species as it goes round the nitrogen cycle.

Key words
natural cycle
carbon cycle
nitrogen cycle

Questions

6 Explain why:

a there is lots of N_2 in the atmosphere, but very little in the lithosphere

b there is lots of NH_4^+ in the lithosphere, but very little in the atmosphere

7 Explain why plants can be short of nitrogen when there is so much nitrogen in the air.

8 Suggest possible consequences of the large-scale fixing of nitrogen by industry and the use of synthetic nitrogen compounds as fertilizers.

Find out about:
- metal ores
- extracting metals
- chemical quantities
- electrolysis

G Metals from the lithosphere

Metal ores

The wealth of societies has often depended on their ability to extract and use metals. Mining and quarrying for metal ores takes place on a large scale and can have a major impact on the environment.

All metals come from the lithosphere, but most metals are too reactive to exist on their own in the ground. Instead, they exist combined with other elements as compounds. Like other compounds found in the lithosphere, they are called minerals

Rocks which contain useful minerals are called **ores**. The valuable minerals are very often the **oxides** or sulfides of metals.

Metal	Name of the ore	Mineral in the ore
aluminium	bauxite	aluminium oxide, Al_2O_3
copper	copper pyrites	copper iron sulfide, $CuFeS_2$
gold	gold	gold, Au
iron	haematite	iron oxide, Fe_2O_3
sodium	rock salt	sodium chloride, NaCl

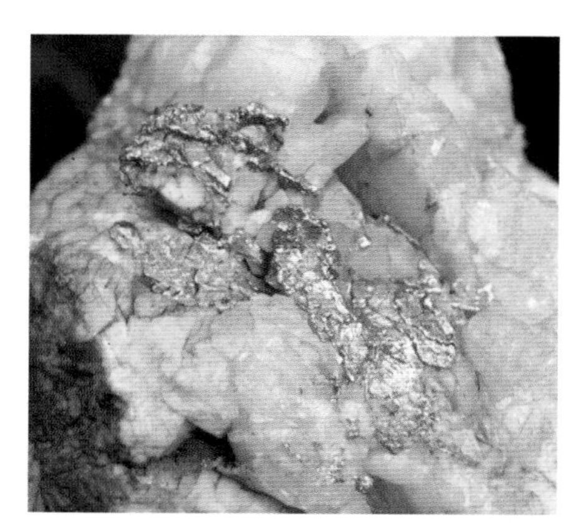

Gold is so unreactive that it occurs uncombined in the lithosphere. But most metals occur as compounds.

Because it occurs in an uncombined state, gold has been used by humans for more than 5000 years. More **reactive metals** like iron were not used by humans until methods for extracting them had been developed.

Mineral processing

Over hundreds of millions of years, rich deposits of ores have built up in certain parts of the Earth's crust. But even the richest deposits do not contain pure mineral. The valuable mineral is mixed with lots of useless dirt and rock, which have to be separated off as much as possible. This is called concentrating the ore.

Some ores are already fairly concentrated when they are dug up – iron ore is often over 85% pure Fe_2O_3. But other ores are much less concentrated – copper ore usually contains less than 1% of the pure copper mineral.

An open-pit copper mine in Utah, USA. Mining on this scale makes a big impact on the environment

Extracting metals: some of the issues

There is a range of factors to weigh up when thinking about the method for **extracting** a metal.

1 How can the ore be reduced?

The more reactive the metal, the harder it is to reduce its ore. The table on the right compares the methods used to reduce different ores.

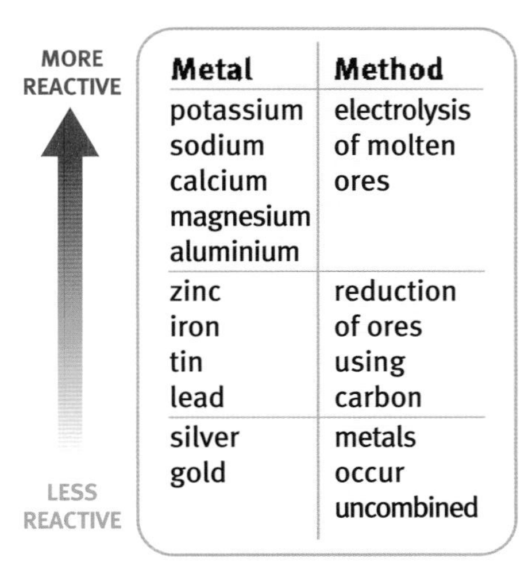

Metal	Method
potassium sodium calcium magnesium aluminium	electrolysis of molten ores
zinc iron tin lead	reduction of ores using carbon
silver gold	metals occur uncombined

2 Is there a good supply of ore?

Metals ores are mined in different parts of the world. If ore is not very pure, it may not be worth using – the cost of concentrating the ore may be too great. The more valuable the metal, the lower the quality of ore which can be used.

3 What are the energy costs?

It takes energy to extract metals, as well as a good supply of ore. This is especially true if the metal is extracted by electrolysis. For example, a quarter of the cost of making aluminium is the cost of electricity.

4 What is the impact on the environment?

Metals like iron and aluminium are produced on a huge scale. Millions of tonnes of ore are needed. Mining this ore can have a big environmental impact. This is why it is important to recycle metals. It takes about 250 kg of copper ore to make 1 kg of copper. So recycling 1 kg of copper means that 250 kg of ore need not be dug up.

Hot metal can be poured.

Questions

1. Suggest explanations for these facts:
   - **a** The Romans used copper, iron, and gold, but not aluminium.
   - **b** Iron is cheap compared with many other metals.
   - **c** Gold is expensive, even though it is found uncombined in nature.
   - **d** About half the iron we use is recycled, but nearly all the gold is recycled.
   - **e** The tin mines in Cornwall have closed, even though there is still some tin ore left in the ground.

Key words

ores
oxides
reactive metals
extracting (a metal)

Extracting metals from ores

Zinc is a metal that can be extracted from its oxide. Zinc is found in the lithosphere as ZnS, called zinc blende. This can be easily turned to ZnO by heating it in air.

The task is to remove the oxygen from the zinc, to convert ZnO to Zn. Removing oxygen in this way is called **reduction**. The process needs a **reducing agent** which will remove oxygen. In removing oxygen, the reducing agent is **oxidized**.

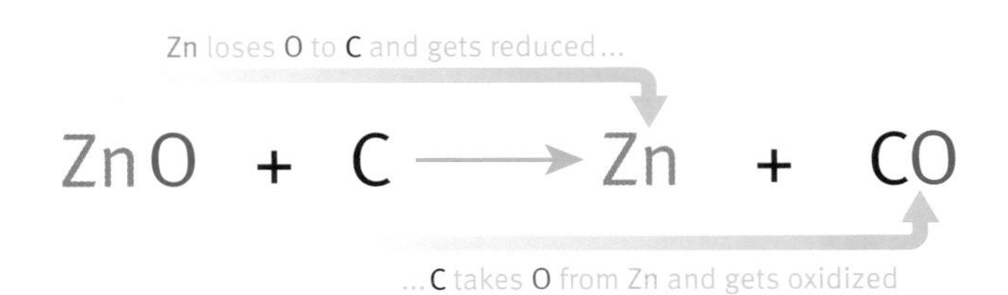

Reducing zinc oxide to zinc using carbon

Carbon is often used as a reducing agent to extract metals. Carbon, in the form of coke, can be made cheaply from coal. At high temperatures, carbon has a strong tendency to react with oxygen, so it is a good reducing agent. What is more, the carbon monoxide formed is a gas, so it is not left behind to make the zinc impure.

Key words

reduction
reducing agent
oxidized

Questions

1 a Write an equation for the reaction of zinc sulfide with oxygen to make zinc oxide and sulfur dioxide.

b What problems might arise from the formation of the sulfur dioxide on a large scale.

c What might be done to deal with this problem?

2 Why do oxidation and reduction always go together when carbon extracts a metal from a metal oxide?

How much metal?

Chemists often ask the 'How much?' question. It is useful to know, say, how much iron it is possible to get from 100 kg of pure iron ore, Fe_2O_3.

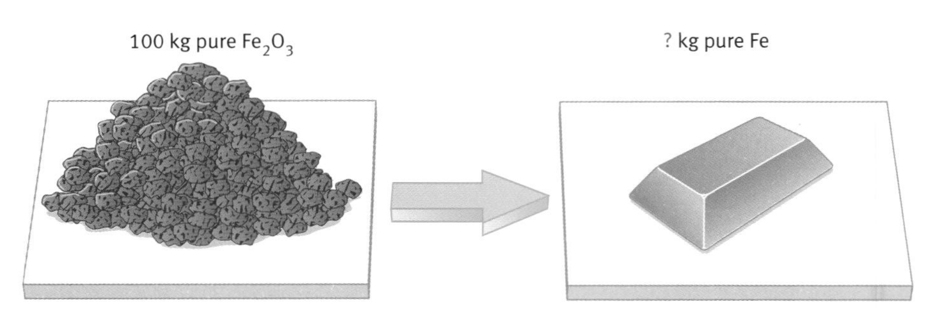

Relative atomic masses

Chemists need to know the relative masses of the atoms involved to answer questions such as 'How much Fe could you get from 100 kg of Fe_2O_3?'

Atoms are far too small to weigh directly. For example, it would take getting on for a million million million million hydrogen atoms to make one gram.

Instead of working in grams, chemists find the mass of atoms relative to one another. Chemists can do this using an instrument called a mass spectrometer.

Values for the **relative atomic masses** of elements are shown in the periodic table on page 95. The relative mass of the lightest atom, hydrogen, is 1.

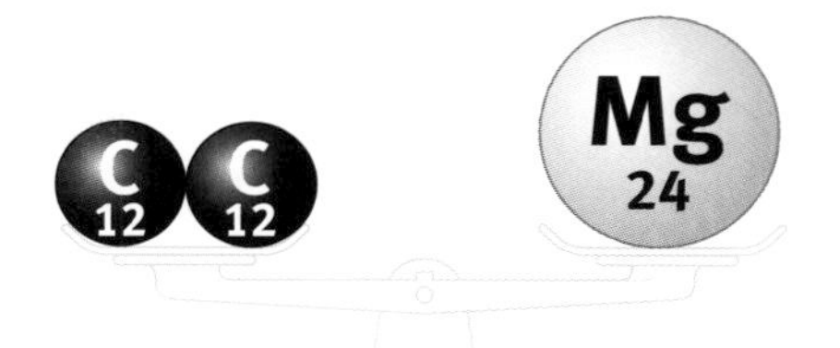

Mg atoms weigh twice as much as C atoms.

Formula masses

If you know the formula of a compound, you can work out its **relative formula mass** by adding up the relative atomic masses of all the atoms in the formula:

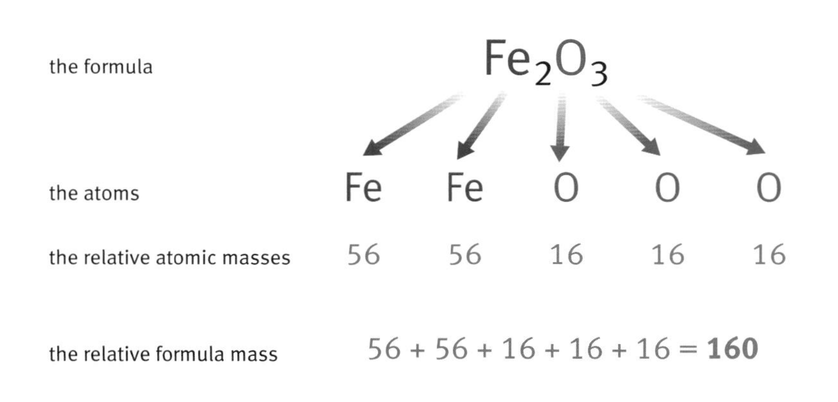

Finding the formula mass of Fe_2O_3

Example

How much Fe could you get from 100 kg of Fe_2O_3?

Fe_2O_3 has a relative formula mass of **160**

In this formula, there are 2 atoms of Fe
2Fe has relative mass 2 ¥ 56 = **122**

This means that, in 160 kg of Fe_2O_3, there must be 112 kg of Fe.

So 1 kg of Fe_2O_3 would contain 112/160 kg of Fe.

So 100 kg of Fe_2O_3 would contain 100 kg ¥ 112/160 of Fe = 70 kg.

Another way of saying this is that the percentage of Fe in Fe_2O_3 is 70%.

Key words

relative atomic mass
relative formula mass

Questions

Look up relative atomic masses in the periodic table on page 95.

1 What is the relative formula mass of carbon dioxide?

2 What mass of Al could be made from 1 tonne of Al_2O_3?

3 What mass of Na could be made from 2 tonnes of NaCl?

4 The main ore of chromium is $FeCr_2O_4$. What is the mass percentage of Cr in $FeCr_2O_4$?

5 1000 tonnes of copper ore are dug out of the ground. Only 1% of this is the pure mineral, $CuFeS_2$. What mass of the Cu could be made from 1000 tonnes of the ore?

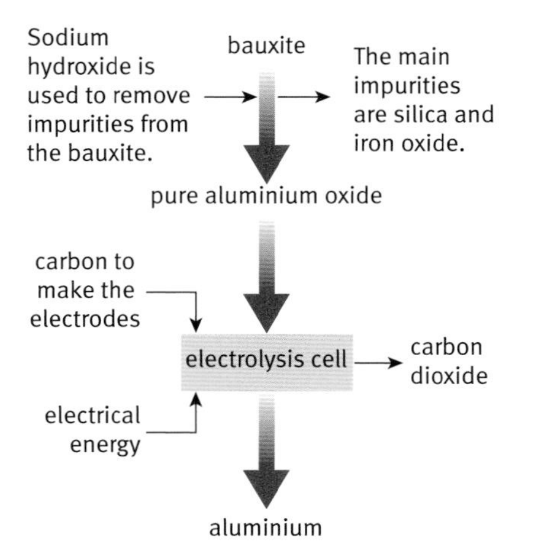

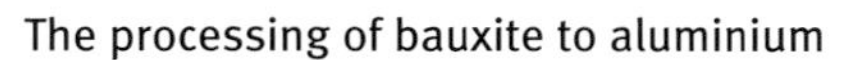
The processing of bauxite to aluminium

Extracting aluminium

Some reactive metals, such as aluminium, hold on to oxygen so strongly that they cannot be extracted using carbon as a reducing agent. To extract these metals, the industry has to use **electrolysis**.

Aluminium is the most abundant metal in the lithosphere. Much of the metal is in aluminosilicates. It is very hard to separate the aluminium from these minerals.

The main ore of aluminium is bauxite. This consists mainly of aluminium oxide, Al_2O_3. There is some iron in bauxite which has to be removed before extraction of aluminium.

The diagram below shows the equipment used to extract aluminium by **electrolysis**. The process takes place in steel tanks lined with carbon. The carbon lining is the negative **electrode**.

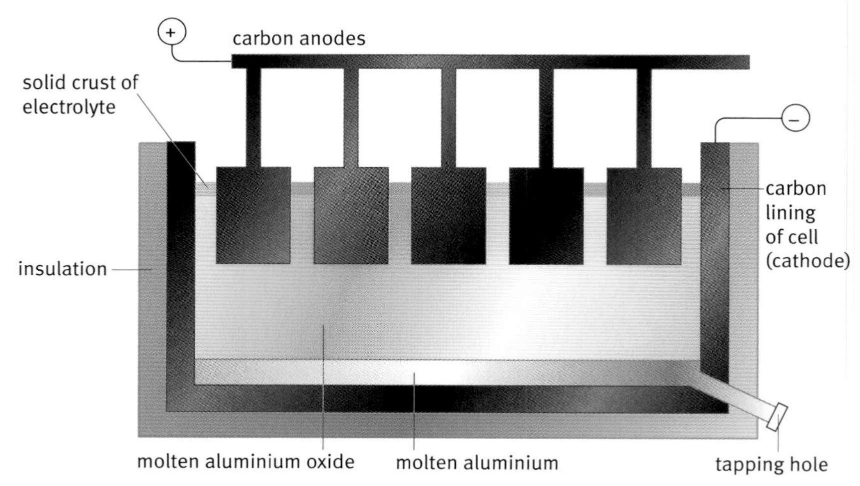

Equipment for extracting aluminium from its oxide by electrolysis

The **electrolyte** is hot, molten Al_2O_3, which contains Al^{3+} and O^{2-} ions. Aluminium forms at the negative electrode. Aluminium is a liquid in the hot furnace and forms a pool of molten metal at the bottom of the tank.

The positive electrodes are blocks of carbon dipping into the molten oxide. Oxygen forms at the positive electrodes.

Ions into atoms and molecules

Electrolysis turns ions back into atoms. Metal ions are positively charged, so they are attracted to the negative electrode. It is a flow of electrons from the power supply into this electrode that makes it negative.

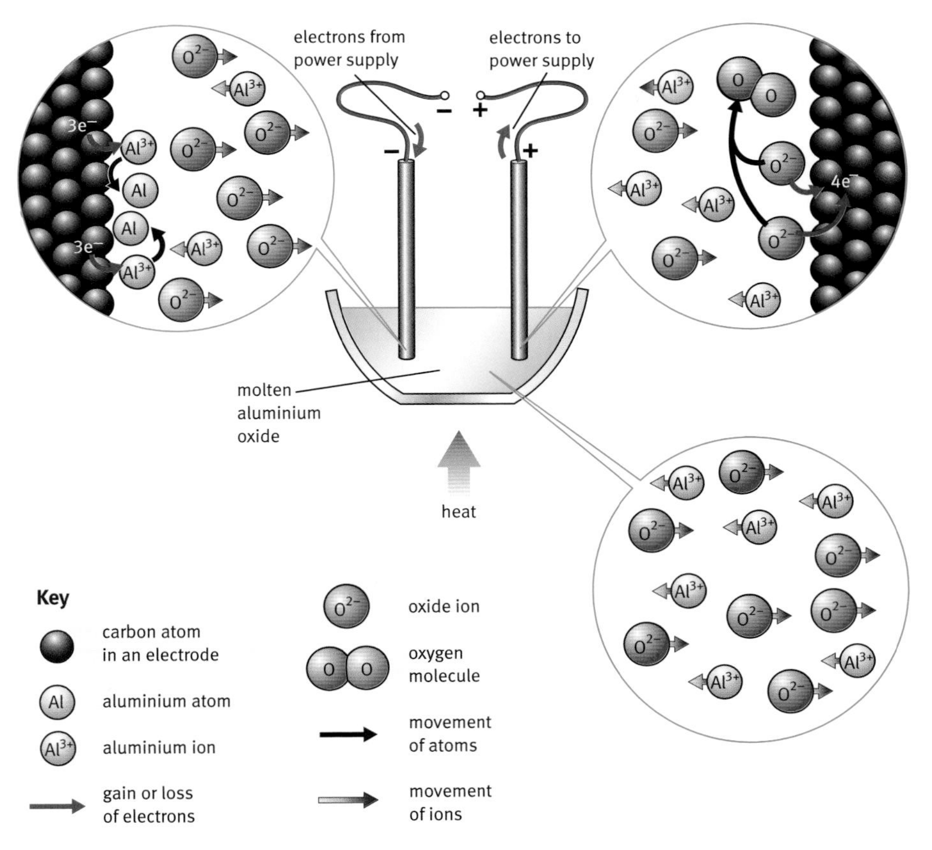

Changes to ions during the electrolysis of molten aluminium oxide

Positive metal ions gain electrons from the negative electrode and turn back into atoms. During the electrolysis of molten aluminium oxide, the aluminium ions turn into aluminium atoms:

Al^{3+}	+	$3e^-$	$\longrightarrow$	Al
ion		electrons supplied by the negative electrode		atom

Non-metal ions are negatively charged, so they are attracted to the positive electrode. This electrode is positive because electrons flow out of it to the power supply.

Negative ions give up electrons to the positive electrode and turn back into atoms. During the electrolysis of molten aluminium oxide, the oxide ions turn into oxygen atoms, which then pair up to make oxygen molecules:

O_2^-	$\longrightarrow$	O	+	$2e^-$
ion		atom		electrons removed by the positive electrode

O	+	O	$\longrightarrow$	O_2
atom		atom		molecule

Key words
electrolysis
electrode
electrolyte

Questions

1 Draw a diagram to show the electron arrangements in:
- **a** an aluminium atom
- **b** an aluminium ion

2 Sodium is extracted by the electrolysis of molten sodium chloride.
- **a** What are the two products of the process?
- **b** Use words and symbols to describe the changes at the electrodes during this process.

Find out about:
- the properties of metals
- the structure of metals
- bonding in metals

H Structure and bonding in metals

Metal properties

Metals have been part of human history for thousands of years. Our lives still depend on metals, despite the development of new materials, including all the different plastics. The varied uses of metals reflect their properties.

Technologists have learnt to use new metals and alloys so that, as well as steel and aluminium, other metals such as titanium and magnesium can now be used in engineering.

Many metals are strong. The titanium hull of this research submarine is strong enough to withstand the pressure at a depth of 6 km. Titanium is also used to make hip joints and racing cars.

Metals can be bent or pressed into shape. They bend without breaking. They are malleable. Aluminium sheet can be moulded under pressure to make cans.

Most metals have high melting points.

Metals conduct electricity. Copper and aluminium are commonly used as conductors.

Metallic structures

Materials scientists use a model for the structure of metals which imagines that metal atoms are:

- tiny spheres
- arranged in a regular pattern
- packed close together in a crystal as a giant structure

Key words
metallic bonding

The diagram on the right shows the arrangement of atoms in copper, a typical metal. You can see how closely together the atoms of copper are packed. In fact, they are packed as close together as it is possible to be. Every atom has 12 other atoms touching it – the maximum number possible. The atoms are held together by strong metallic bonds. Because the bonds are strong, copper is strong and difficult to melt.

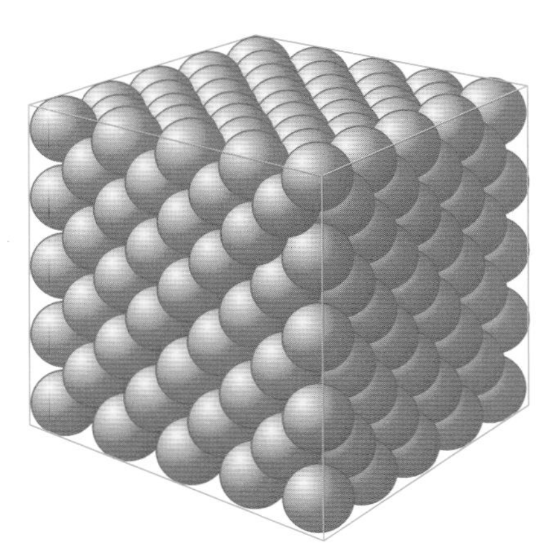

The arrangement of atoms in copper. Because the metallic bonds are strong, copper is strong and difficult to melt. Because the bonds are flexible, copper is malleable – the atoms can be moved around without shattering the structure.

Metallic bonding

Metals have a special kind of bonding – not ionic, nor covalent, but metallic. **Metallic bonds** are strong, but flexible, so they allow the atoms to move to a new position.

Metal atoms tend to lose the electrons in their outer shell easily. In the solid metal, the atoms lose these electrons and become positive ions. The electrons, no longer held by the atoms, drift freely between the metal atoms, which are now positively charged. The attraction between the 'sea' of negative electrons and the positively charged metal atoms holds the structure together.

Overall, a metal crystal is not charged. This is because the total negative charge on the electrons balances the total positive charge on the metal ions.

The electrons can move freely between the ions, which explains why metals conduct electricity well. When an electric current flows through a metal wire, the free electrons drift from one end of the wire towards the other. Although the electrons are free, the metal ions themselves are packed closely together in a regular lattice.

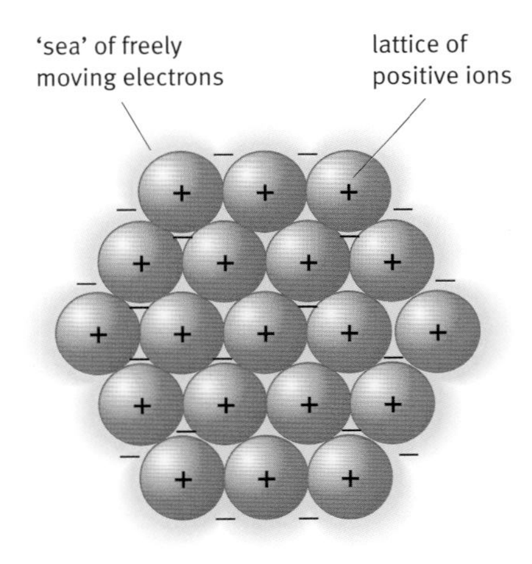

A model of metallic bonding

Questions

1. Give five examples of metals used for their strength. For each metal, give an example of a use that depends on the strength of the metal.
2. Someone looking at a model showing the arrangement of atoms in a copper crystal might think that the following statements are true. Which of these ideas are correct? Which ideas are false? What would you say to put someone right who believed the false ideas?
   - **a** Copper crystals are shaped like cubes because the atoms are packed in a cubic pattern.
   - **b** There is air between the atoms in a crystal of copper.
   - **c** Copper is dense because the atoms are closely packed.
   - **d** The atoms in a copper crystal are not moving at room temperature.
   - **e** Copper has a high melting point because the atoms are strongly bonded in a giant structure.
   - **f** Copper melts when strongly heated because the atoms melt
3. There are positive metal ions in a metal crystal, but a metal is not an ionic compound. Explain.

Find out about:
- impacts of extracting, using, and disposing of metals

1 The life cycle of metals

Mining, mineral processing, and metal extraction produce many valuable metal products, but these activities can also have a serious impact on the environment. There can be a conflict between those that want to build up profitable industries and those whose aim is to protect the natural world.

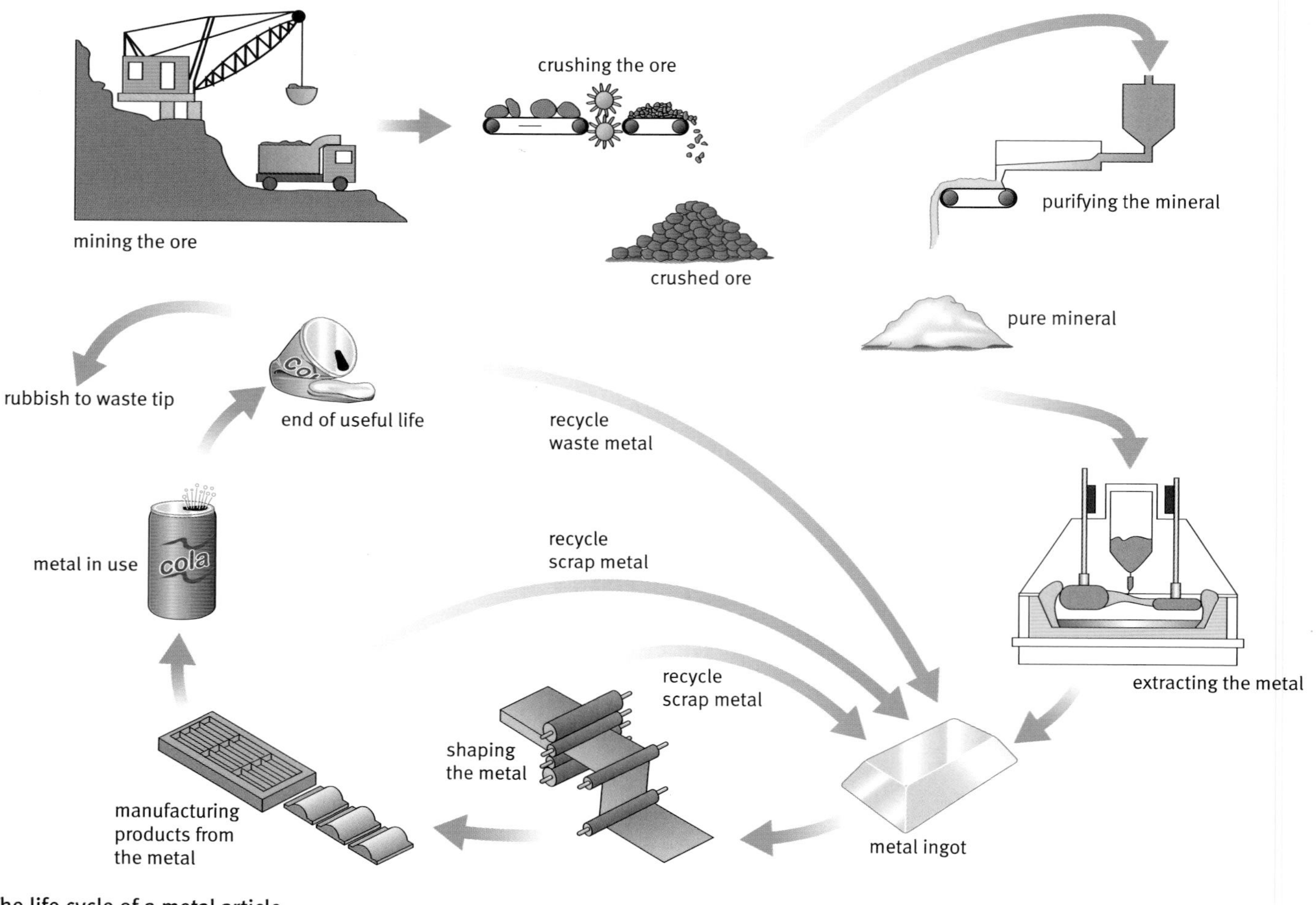

The life cycle of a metal article

Mining

Mining produces large volumes of waste rock and can leave very large holes in the ground. Miners use explosives to blast the rock. This is noisy and produces dust.

Open-cast mining for bauxite takes place in tropical countries such as Jamaica, Brazil, and Surinam. Separating the pure aluminium oxide leaves behind large volumes of red mud made up mostly of iron oxide. This has to be stored in large ponds where the very fine material can settle out.

Processing ores

Many metal ores are high value but low grade. The ore in an open-pit copper mine may contain as little as 0.4% of the metal and still be profitable. This means that 99.6% of the rock dug from the ground becomes waste. Near any mine there are waste tips. These can be a hazard if they contain traces of toxic metals such as lead or mercury.

Metal extraction

All the stages of metal extraction and metal fabrication need energy, use large volumes of water, and give off air pollutants. Higher expectations from society and tighter regulation mean that industries have to do more to prevent harmful chemicals escaping into the environment. Economic pressures favour the development of equipment and procedures that minimize the use of energy, water, and other resources.

Metals in use

Careful choice of metals can reduce the environmental impact of our life style. In transport, for example, lighter cars, trucks, and trains mean less fuel consumption and emissions, as well as less wear and tear on roads and tracks. Vehicles can be designed to be lighter by replacing steel with lighter metals such as aluminium and with plastics.

Recycling

Recycling is well established in the metal industries. Scrap metal from all stages of production is routinely recycled. Much metal is also recycled at the end of the useful life of metal products.

Recycled steel can be as good as new after reprocessing. The scrap is fed to a furnace and melted with fresh metal to make new steel. For every tonne of steel recycled, there is a saving of 1.5 tonnes of iron ore and half a tonne of coal. There is also a big reduction in the total volume of water needed, since large quantities of water are used in mineral processing.

Recycling aluminium is particularly cost-effective because so much energy is needed to extract the metal from its oxide by electrolysis. Recycling also reduces the impact on the environment by cutting the use of raw materials – and the associated mining and processing.

A large pond in Jamaica used to contain the waste from a mine to extract bauxite. Bauxite is impure aluminium oxide. The main impurity is iron oxide, which ends up in the rusty-looking waste.

Questions

1 Why are recycling rates for metal waste from manufacturing higher than recycling of metals after use?

(2) According to an international company: 'Steel and aluminium products play a role in everything we do in modern-day life. So, although the production of steel and aluminium consumes resources and energy, both materials make a major contribution to our quality of life.' Draw up a table to list the benefits and costs of our use of these two metals. Do you agree with the claim made by this company?

(3) The aluminium industry argues that aluminium is a sustainable material, because known bauxite reserves will last for hundreds of years at the current rates of production. Do you agree?

C5 Chemicals of the natural environment

Summary

You have learnt how the theories of structure and bonding can help to explain the properties of the chemicals we find in the atmosphere, hydrosphere, biosphere and lithosphere. You have also learnt about some of the methods used to extract metals from their ores.

Molecules

- Most non-metal elements and most compounds between non-metal elements are molecular.
- The atoms in molecules are held together by strong covalent bonds.
- The attractive forces between molecules are weak, so that small molecules are often gases, such as the gases in the atmosphere.
- Water is a compound with small molecules that is a liquid which makes up most of the hydrosphere.
- Molecular compounds do not conduct electricity because their molecules are not charged.
- Living things are mainly made up from molecular compounds containing the elements carbon, hydrogen, oxygen, and nitrogen with small amounts of other elements.
- Carbohydrates, proteins, and DNA consist of long-chain molecules.

Giant ionic structures

- Compounds made of metals and non-metals have giant ionic structures.
- Ionic compounds have high melting points because of the strong attraction between the ions.
- Ionic compounds conduct electricity when molten or dissolved in water, when the ions are free to move.
- Salts are ionic compounds: some occur as minerals in the lithosphere; some dissolve in water and make the sea salty.

Giant covalent structures

- Silicon dioxide and diamond have giant covalent structures with atoms held together in a regular network with strong bonds.
- Chemicals with giant covalent structures have high melting points and do not dissolve in water.
- Giant structures do not conduct electricity because there are no free electrons or ions.
- Much of the lithosphere is made of giant covalent structures based on silicon, oxygen, and other elements, including aluminium.

Metallic structures

- All metal structures have a giant structure of metal atoms.
- The metallic bonding between the atoms is strong.
- Metals conduct electricity when solid and when molten because the bonding electrons are free to move.
- In a metal crystal there are positively charged ions held closely together by a sea of electrons that are free to move.

Ions into atoms

- Electrolysis turns ions back into atoms.
- Electrolysis splits an ionic compound and turns it back into its elements so it can be used to extract metals from ores.
- At a negative electrode, positive metal ions gain electrons and become metal atoms. H
- At a negative electrode, negative ions lose electrons and turn back into non-metal atoms (which may then join up to make molecules). H

Questions

1 A concept map is a web-like diagram for summarizing ideas and the links between them. Every concept map has 'nodes' which are boxes, with 'links' that are the lines between the boxes. Each node contains a concept. Each link includes a few words to show the relationships between the concepts. Arrow heads on the lines show the direction of relationship.

Create your own concept map for the topic of structure and bonding in chemicals of the environment. Include as many of the key words in this chapter as you can.

2 **a** Give two properties of a mineral that make it valued as a gemstone.

b Diamond is a popular and valued gemstone.

i Name the element in diamond.

ii Why is diamond such a hard material?

c An amethyst is a purple variety of quartz. It is a crystal of silicon dioxide with small amounts of iron impurity.

i Why is amethyst cheaper than diamond?

ii Why does amethyst have a very high melting point?

3 Passing an electric current through molten calcium bromide produces a metallic bead at one electrode and a red-brown gas at the other. Calcium bromide consists of calcium ions, Ca^{2+}, and bromide ions, Br^{-} The metallic bead reacts with water to produce a gas that burns with a squeaky pop.

a What is the name for the process of splitting a compound with an electric current?

b **i** Identify the metal formed and the red-brown gas.

ii Which of the products forms at the negative electrode?

c Draw and label a diagram of an apparatus that could be used to pass an electric current through molten calcium bromide.

d **i** Why does solid calcium bromide not conduct electricity?

ii Write the chemical formula of calcium bromide.

e Explain the changes when the current flows at the electrode that produces the metallic bead.

4 This is one of the reactions for extraction iron from its ore in a furnace:

$$Fe_2O_3(s) + 3CO(g) \rightarrow 2Fe(s) + 3CO_2(g)$$

a In this reaction which chemical is:

i oxidized?

ii reduced?

iii the reducing agent?

b What mass of iron can be obtained from 16 tonnes of iron oxide? (See page 39 for the relative atomic masses.)

c The carbon monoxide comes from coke (C) in the furnace. What mass of coke is needed to form the carbon monoxide required to extract the metal from 16 tonnes of iron oxide?

4 Aluminium is a more reactive metal than iron. It can be extracted from its oxide, Al_2O_3, only by electrolysis.

a Name one other metal that can only be extracted by electrolysis.

b Write an equation to show what happens at the negative electrode during the electrolysis of molten aluminium oxide.

c Calculate the mass of aluminium that can be extracted from 1000 kg of aluminium oxide.

5 Make a list of five metals in common use.

a For each metal:

i give an example of how it is used

ii state a property of the metal that it makes it more suitable for that use than the other metals in your list

b Explain why all metals conduct electricity.

Why study chemical synthesis?

We use chemicals to preserve food, treat disease, and decorate our homes. Many of these chemicals are synthetic. Developing new products, such as drugs to treat disease, depends on the chemists who synthesize and test new chemicals.

The science

Chemists who synthesize new chemicals need knowledge of science explanations combined with practical skills. It is important to understand how to control reactions so that they are neither too slow nor dangerously fast. Planning a synthesis involves calculating how much of the reactants to mix together to make the amount of product required.

Acids are important reagents in synthesis. Ionic theory can explain the characteristic behaviours of these chemicals.

Chemistry in action

Synthesis provides many of the chemicals that we need for food processing, health care, cleaning and decorating, modern sporting materials, and many other products. The chemical industry today is developing new processes for manufacturing these chemicals more efficiently and with less impact on the environment.

Chemical synthesis

Find out about:

- the importance of the chemical industry
- the steps involved in the synthesis of a new chemical
- techniques for controlling the rate of chemical change
- a theory to explain acids and alkalis
- ways to measure the efficiency of chemical synthesis

Find out about:

- the chemical industry
- bulk and fine chemicals
- the importance of chemical synthesis

A The chemical industry

The chemical industry converts raw materials, such as crude oil, natural gas, minerals, air and water, into useful products. The products include chemicals for use as dyes, food additives, fertilizers, dyestuffs, paints, and pharmaceutical drugs.

The industry makes **bulk chemicals** on a scale of thousands or even millions of tonnes per year. Examples are ammonia, sulfuric acid, sodium hydroxide, chlorine, and ethene.

On a much smaller scale, the industry makes **fine chemicals** such as drugs and pesticides. It also makes small quantities of speciality chemicals needed by other manufacturers for particular purposes. These include such things as flame retardants, food additives, and the liquid crystals for flat-screen televisions and computer displays.

The chemical industry converts raw materials into pure chemicals which are then used in synthesis to make a wide range of products.

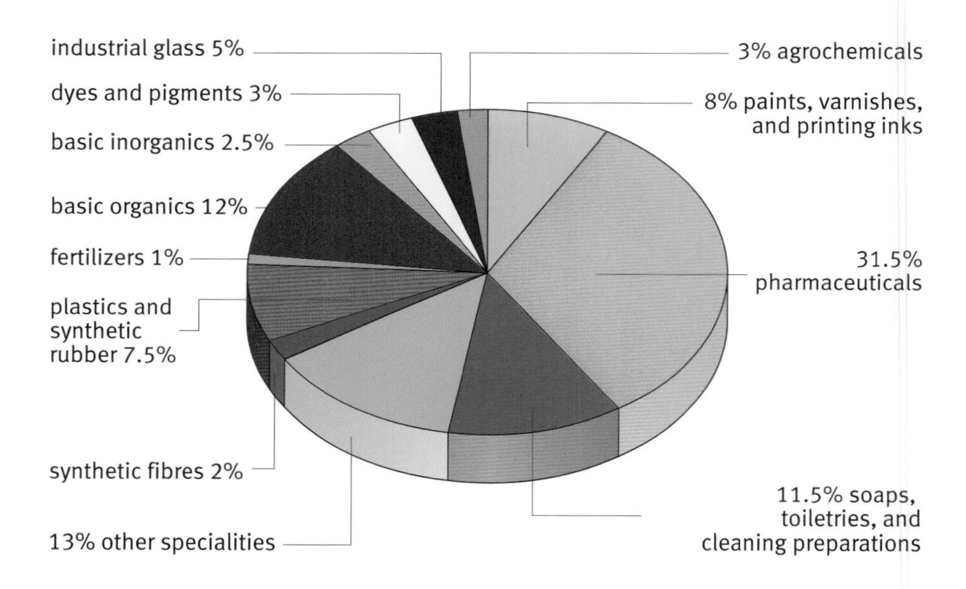

The pie chart shows the range of products made by the chemical industry in Britain and their relative value.

The part of a chemical works which produces a chemical is called a plant. Some of the chemical reactions occur at a high temperature, so that a source of energy is needed. Also, a lot of electric power is need for pumps to move reactants and products from one part of the plant to another. Sensors monitor the conditions at all the key points in the plant. The data is fed to computers in the control centre, where the technical team controls the plant.

Key words

bulk chemicals
fine chemicals
scale up

People in the chemical industry

People with many different skills are needed in the industry. Research chemists work in laboratories to find new processes and develop new products.

The industry needs new processes so that it can be more competitive and more sustainable. The aim is to use smaller amounts of raw materials and energy while creating less waste.

People devising new products have to work closely with people in the marketing and sales department. They are able to say if the novel product is wanted. If the new product is promising it may first be tried out by making it in a pilot plant.

As part of the market research, possible new products are given to customers for trial. At the same time, financial experts estimate the value of the new product in the market. They then compare this with the cost of making the product to check that the new process will be profitable.

Chemical engineers have to **scale up** the process and design a full-scale plant. This can cost hundreds of millions of pounds.

Some chemicals from the industry go directly on sale to the public, but most of them are used to make other products. Transport workers carry the chemicals to the industry's customers.

Every chemical plant needs managers and administrators to control the whole operation. There are also people in service departments look after the needs of the people working there which includes medical and catering staff, training, and safety officers.

Questions

1. Give the name and chemical formula of a bulk chemical.
2. What percentage value of products of the chemical industry in Britain are used:
   - **a** in agriculture and horticulture?
   - **b** to make polymers?
   - **c** for medical diagnosis and treatment?
3. List these chemicals under two headings: 'bulk chemical' and 'fine chemical'.
   - the drug aspirin
   - the hydrocarbon ethene
   - the perfume chemical citral
   - the acid sulfuric acid
   - the herbicide glyphosate
   - the alkali sodium hydroxide
   - the food dye carotene
   - the pigment titanium dioxide

Plant operators monitor the processes from a control room.

Maintenance workers help to keep the plant running.

Find out about:
- acids and alkalis
- the pH scale
- reactions of acids

B Acids and alkalis

Acids

The word **acid** sounds dangerous. Nitric, sulfuric, and hydrochloric acids are very dangerous when they are concentrated. You must handle them with great care. These acids are less of a hazard when diluted with water. Dilute hydrochloric acid, for example, does not hurt the skin if you wash it away quickly, but it stings in a cut and rots clothing.

Not all acids are dangerous to life. Many acids are part of life itself. Biochemists have discovered the citric acid cycle. This is a series of reactions in all cells. The cycle harnesses the energy from respiration for movement and growth in living things.

Organic acids

Organic acids are molecular. They are made of groups of atoms. Their molecules consist of carbon hydrogen and oxygen atoms. The acidity of these acids arises from the hydrogen in the —COOH group of atoms.

Acetic acid (chemical name: ethanoic acid) is the acid in vinegar. Most white vinegar is just a dilute solution of acetic acid. Brown vinegars have other chemicals in the solution that give the vinegar its colour and flavour. Most micro-organisms cannot survive in acid, so vinegar is used as a preservative in pickles (E260). The pure acid is a liquid.

Citric acid is found in citrus fruits like oranges and lemons. The human body turns over about 2 kg of citric acid a day during respiration. The acid is manufactured on a large scale for the food industry. Citric acid (E330) and its salts (E331 and E332) are added to food to prevent them reacting with oxygen in the air – they are antioxidants. They also give a tart taste to drinks and sweets.

Mineral acids

Sulfuric, hydrochloric, and nitric acids come from inorganic or mineral sources. The pure acids are all molecular. Sulfuric and nitric acids are liquids at room temperature. Hydrogen chloride is a gas which becomes hydrochloric acid when it dissolves in water.

Alkalis

Pharmacists sell antacids in tablets to control heartburn and indigestion. The chemicals in these medicines are the chemical opposites of acids. They are designed to neutralize excess acid produced in the stomach – hence the name 'antacids'.

Some chemical antacids are soluble in water to give a solution with a pH above 7. Chemists call them **alkalis**. Common alkalis are sodium hydroxide, NaOH; potassium hydroxide, KOH; and calcium hydroxide, $Ca(OH)_2$.

The traditional name for sodium hydroxide is 'caustic soda'. The word caustic means that the chemical attacks living tissue including skin. Alkalis can do more damage to delicate tissues than dilute acids. Caustic alkalis are used in the strongest oven and drain cleaners. They clearly have to be used with great care.

Sulfuric acid, H_2SO_4, is manufactured from sulfur, oxygen, and water. The pure, concentrated acid is an oily liquid. The chemical industry in the UK makes about 2 million tonnes of the acid each year. The acid is essential for the manufacture of other chemicals, including detergents, pigments, dyes, plastics, and fertilizers.

The old name for hydrogen chloride was 'spirits of salt'. It forms when concentrated sulfuric acid is added to salt crystals. Hydrogen chloride, HCl, is a gas which fumes in moist air and is very soluble in water. Today, the chemical industry makes most of the hydrogen chloride it needs as a by-product of the production of other chemicals, such as the polymer PVC.

Oven cleaners often contain caustic alkalis.

Key words
acid
alkalis

Questions

1 Work out from the pictures of molecules the formulae of:
- **a** acetic acid
- **b** tartaric acid

2 What are the formulae of these antacids?
- **a** Magnesium hydroxide made up of magnesium ions, Mg^{2+}, and hydroxide ions, OH^-.
- **b** Aluminium hydroxide made up of aluminium ions, Al^{3+}, and hydroxide ions, OH^-.

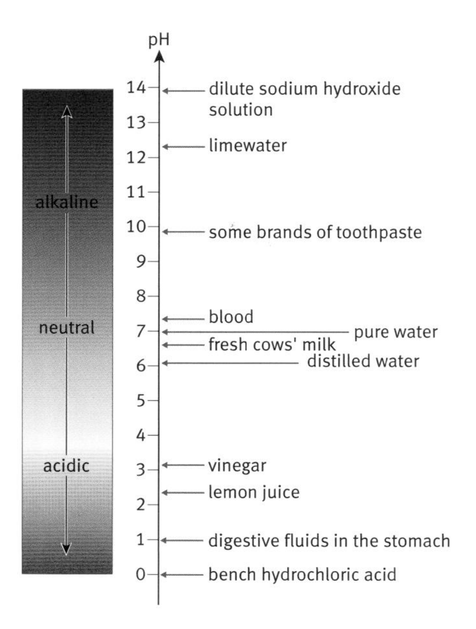

The pH scale

Reactions of acids

The reactions of acids are important not just to chemists but to everyone living in a consumer society.

Acids with indicators

The term pH appears on many cosmetic, shampoo and food labels. It is a measure of acidity. The **pH scale** is a number scale which shows the acidity or alkalinity of a solution in water. Most laboratory solutions have a pH in the range 1–14.

Indicators change colour to show whether a solution is acidic. Litmus turns red in acid solution. Special mixed indicators, such as universal indicator, show a range of colours and can be used to estimate pH values.

Acids with metals

Acids react with **metals** to produce **salts**. The other product is hydrogen gas.

$$\text{acid} + \text{metal} \longrightarrow \text{salt} + \text{hydrogen}$$

Not all metals will react in this way. You may remember the list of metals in order of reactivity in Module C5 *Chemical of the natural environment*, page 137. Metals below lead in the list do not react with acids, and even with lead it is hard to detect any change in a short time.

Etching metal

One method of chemical etching is based on the reaction of acids with metals. Etching is a way of producing multiple copies of printed pictures that have the quality of original drawings.

- First a metal plate of zinc or steel is covered with wax.
- Then the artist scrapes the wax with a stylus to make the drawing.
- The plate is then dipped in an acid bath. The acid reacts only with the metal that has been exposed. The metal still coated with wax is protected and so does not react.
- The artist uses a feather to brush away hydrogen bubbles that stick to the plate. If they were left on the plate the acid would not 'bite' into that spot.
- When the acid has removed enough metal, the plate is taken from the acid and rinsed. The remaining wax is removed.
- The plate is now ready to be covered with ink, which flows into the grooves etched by the acid.

Using a feather to brush away hydrogen bubbles while etching a metal plate with acid

Acids with metal oxides or hydroxides

An acid reacts with a **metal oxide** or **hydroxide** to form a salt with water. No gas forms.

$$\text{acid} + \text{metal oxide (or hydroxide)} \longrightarrow \text{salt} + \text{water}$$

The reaction between an acid and a metal oxide is often a vital step in making useful chemicals from ores.

Acids with carbonates

Acids react with **carbonates** to form a salt, water, and bubbles of carbon dioxide gas. Geologists can test for carbonates by dripping hydrochloric acid onto rocks. If they see any fizzing, the rocks contain a carbonate. This is likely to be calcium carbonate or magnesium carbonate.

$$2HCl(aq) + CaCO_3(aq) \longrightarrow CaCl_2(aq) + H_2O(l) + CO_2(g)$$

This is a foolproof test for the carbonate ion. So the term 'the acid test' has come to be used to describe any way of providing definite proof.

Testing for limestone using hydrochloric acid

Key words

pH scale
indicators
metals
salts
metal oxide
metal hydroxide
carbonates

Questions

3 Write a balanced equation for the reaction which takes place when a steel plate is etched with hydrochloric acid. The salt formed is iron(II) chloride.

4 Magnesium hydroxide, $Mg(OH)_2$, is an antacid used to neutralize excess stomach acid, HCl. Write a balanced equation for the reaction.

5 There is a volcano in Tanzania, Africa, whose lava contains sodium carbonate, Na_2CO_3. The cooled lava fizzes with hydrochloric acid. Write a balanced equation for the reaction.

6 Limescale forms in kettles, boilers, and pipes where hard water is heated. Limescale consists of calcium carbonate. Three acids are often used to remove limescale: citric acid, acetic acid (in vinegar), and dilute hydrochloric acid. Which acid would you use to de-scale an electric kettle and why?

Find out about:
- an ionic explanation for neutralization reactions
- salts and their formulae
- uses of salts

C Salts from acids

What makes an acid an acid?

Chemists have a theory to explain why all the different compounds that are acids behave in a similar way when they react with indicators, metals, carbonates, and metal oxides.

It turns out that acids do not simply mix with water when they dissolve. They react, and when they react with water they produce hydrogen ions. For example, hydrochloric acid is a solution of hydrogen chloride in water. The HCl molecules react with the water to produce **hydrogen ions** and chloride ions.

$$HCl(g) + water \longrightarrow H^+(aq) + Cl^-(aq)$$

The theory of acids is an ionic theory. Any compound is an acid if it produces hydrogen ions when it dissolves in water.

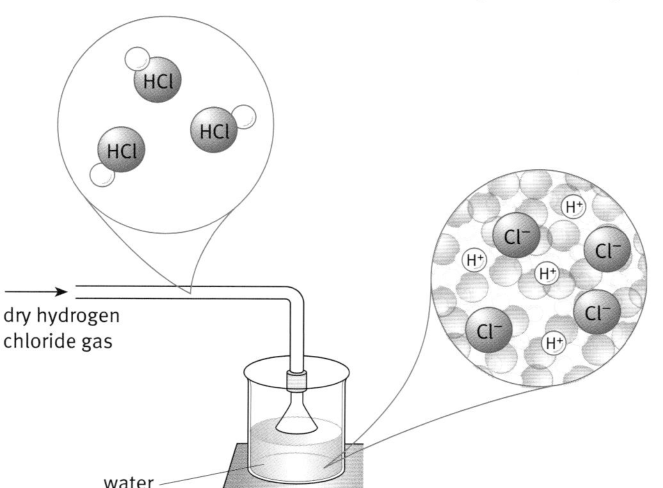

Hydrogen chloride dissolves in water to make hydrochloric acid. The HCl molecules react with water to form ions

All acids contain hydrogen in their formula. Nitric acid, HNO_3, and phosphoric acid, H_3PO_3 both contain hydrogen. But not all chemicals that contain hydrogen are acids. Ethane, C_2H_6, and ethanol, C_2H_5OH, are not acids.

In organic acids it is only the hydrogen atoms in the —COOH groups that can ionize when the acid dissolves in water.

Strong and weak acids

Some acids ionize completely when they dissolve in water. Chemists call them strong acids. Hydrochloric, sulfuric and nitric acids are strong acids.

Organic acids only ionize slightly when they dissolve. They are weak acids. In vinegar, for example, only about one in a hundred of the ethanoic acid molecules are ionized. This helps to explain why vinegar is pH 3 but dilute hydrochloric acid is pH 1.

What makes a solution alkaline?

Alkalis such as the soluble metal hydroxides are ionic compounds. They consist of metal ions and **hydroxide ions**. When they dissolve, they add hydroxide ions to water. It is these ions which make the solution alkaline.

$$NaOH(aq) + water \longrightarrow Na^+(aq) + OH^-(aq)$$

Neutralization

Sodium hydroxide and hydrochloric acid react to produce a salt (sodium chloride) and water.

$$Na^+(aq) + OH^-(aq) + H^+(aq) + Cl^-(aq) \longrightarrow Na^+(aq) + Cl^-(aq) + H_2O(l)$$

Key words
hydrogen ions
hydroxide ions
neutralization reaction

During a **neutralization reaction** the hydrogen ions from an acid react with hydroxide ions from the alkali to make water.

$$H^+(aq) + OH^-(aq) \longrightarrow H_2O(l)$$

The remaining ions in the solution make a salt.

Salts

Salts form when a metal oxide, or hydroxide, neutralizes an acid. So every salt can be thought of as having two parents. Salts are related to a parent metal oxide or hydroxide and to a parent acid.

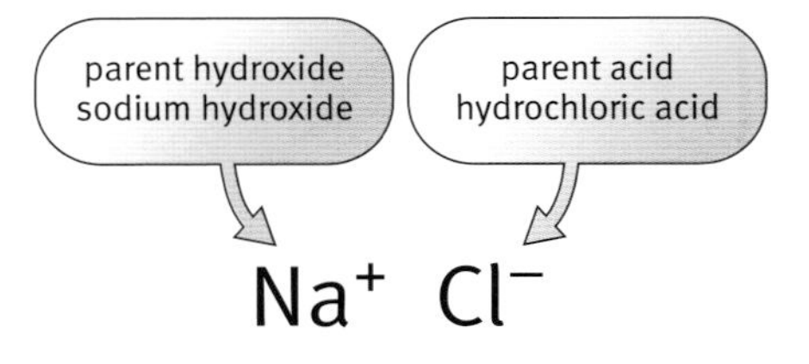

Salts are ionic (see Module C4 *Chemical patterns*, Section J). Most salts consist of a positive metal ion combined with a negative non-metal ion. The metal ion comes from the parent metal oxide or hydroxide. The non-metal ion comes from the parent acid.

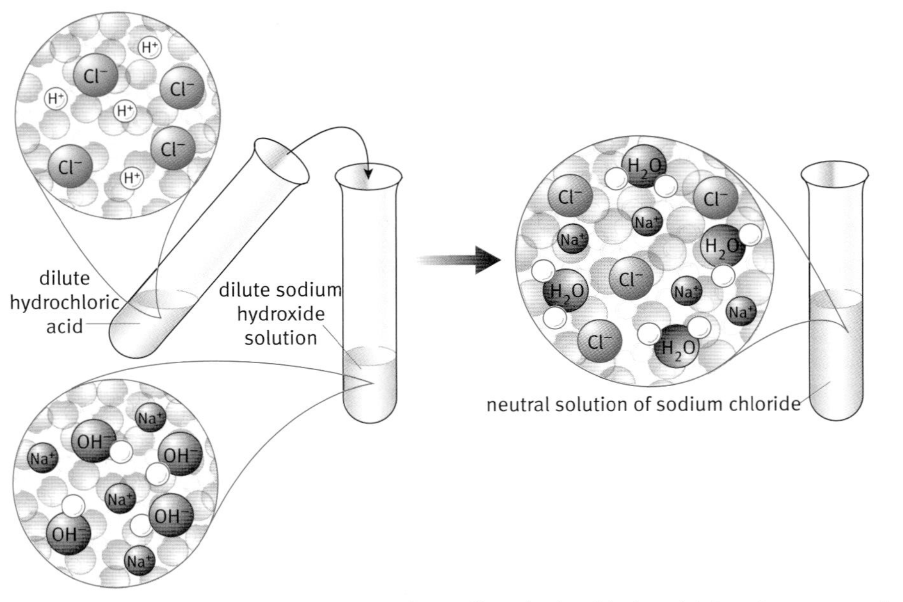

Dilute sodium hydroxide solution neutralizes dilute hydrochloric acid, forming a neutral solution of sodium chloride.

It is possible to work out the formulae of salts knowing the charges on the ions (see Module C4 *Chemical patterns*, page 115). Some non-metal ions consist of more than one atom. The table below includes some examples. In the formula for magnesium nitrate, $Mg(NO_3)_2$, the brackets around the NO_3 show that two complete nitrate ions appear in the formula.

Non-metal ions that consist of more than one atom	Symbols
carbonate	CO_3^{2-}
hydroxide	OH^-
nitrate	NO_3^-
sulfate	SO_4^{2-}

Questions

1 Write equations to show what happens when these compounds dissolve in water:
- **a** nitric acid
- **b** sulfuric acid
- **c** calcium hydroxide

2 Identify the parent acid and a possible parent metal oxide or hydroxide that can react to form:
- **a** lithium chloride
- **b** calcium nitrate
- **c** magnesium sulfate

3 Use the tables of ions on page 115 and on this page to write down the formulae of these salts:
- **a** potassium nitrate
- **b** magnesium carbonate
- **c** sodium sulfate
- **d** calcium nitrate

Find out about:

- everyday uses of salts
- salts used in dialysis

D Salts in our lives

Soluble salts find their way into many areas of our lives. It is the job of chemists to make them for us. In Module C5 *Chemicals of the natural environment*, you learned about how most of the starting materials for the stuff we use in our daily lives comes from the lithosphere. These raw materials have to be transformed into the right chemicals.

The reactions of acids with metals, oxides, hydroxides, and carbonates can all be used to make valuable salts. For uses such as food or medicines, these salts have to be made pure.

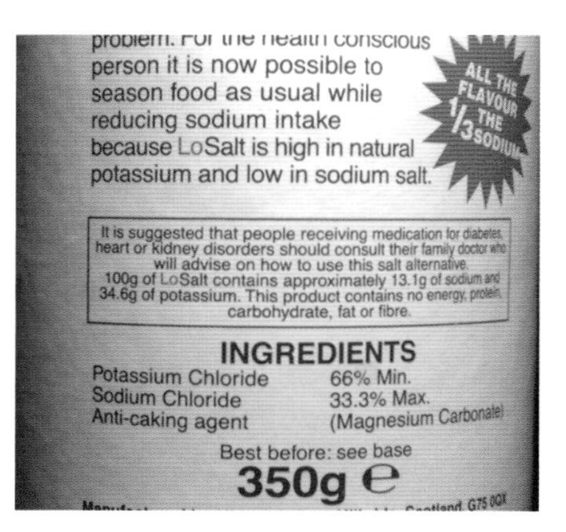

Potassium chloride, KCl, is used by people who are trying to cut down the amount of sodium in their diet, very often to help reduce high blood pressure. It is also added to fertilizers to provide the potassium ions that plants need for growth.

Potassium nitrate, KNO_3, is used in curing meat to make things like bacon and pastrami. It is also used as fertilizer and as an important constituent of gunpowder and fireworks.

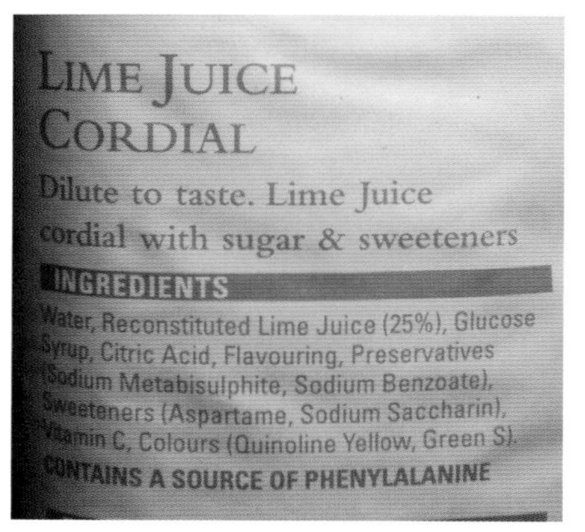

Sodium benzoate, $C_7H_5O_2Na$, is widely used as a food preservative (E211). It is used to prevent bacteria spoiling food. The salt works best to preserve food if the pH is low, and so it is added to foods such jams, salad dressing, fruit juices, pickles, and carbonated drinks.

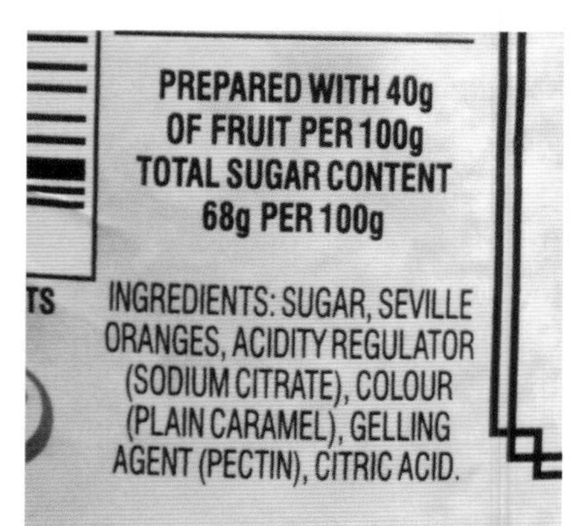

Monosodium citrate, $C_6H_7O_7Na$, is a food additive (E331) which helps to control the pH of foods and to make an antioxidant more effective. So it prevents food going off in the presence of air.

Making calcium chloride for dialysis

The kidneys remove toxic chemicals from blood. In cases of kidney failure, patients are put on dialysis machines that do the job outside the body. Blood passes out of the body through a tube into the dialysis machine.

Inside the machine the blood goes past a special membrane. On the other side of the membrane is a solution containing a mixture of salts at the same concentrations as the same salts in the blood. The toxic chemicals pass from the blood through the membrane into the solution. It is also possible for salts to pass back into the blood.

One of the salts in the dialysis solution is calcium chloride. It is a particularly important salt as the level of calcium in the blood has to be maintained at a particular level. Just a little bit too much or too little and the patient could become very ill indeed. The calcium chloride therefore has to be very pure, and the quantity added to the solution has to be measured accurately.

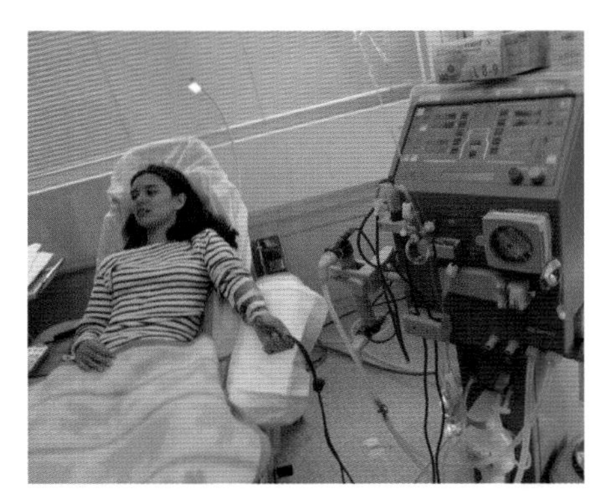

Kidney dialysis

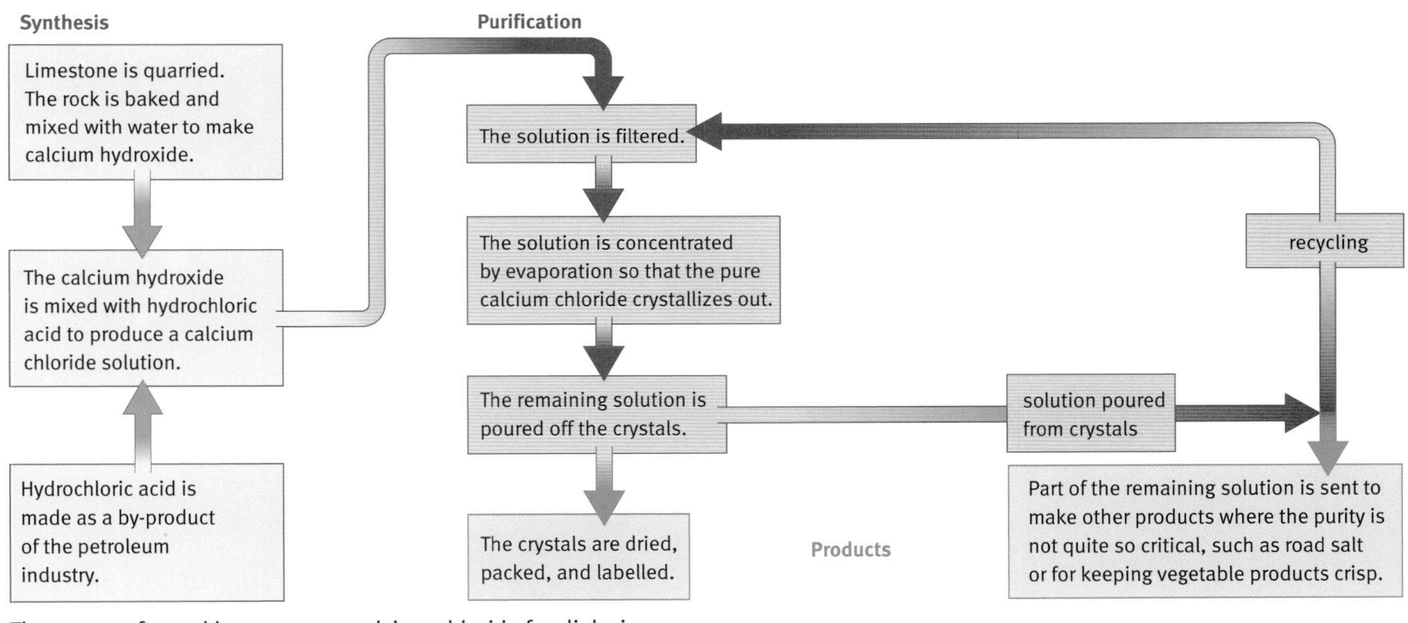

The process for making very pure calcium chloride for dialysis

Questions

1 Identify the parent metal oxide or hydroxide and a possible parent acid that can react to form:
- **a** potassium chloride
- **b** potassium nitrate
- **c** sodium citrate

2 Refer to the flow diagram for the process to make calcium chloride.
- **a** Write the balanced equation for the reaction used to make the salt.
- **b** Identify steps taken to make the yield of the pure salt as large as possible.

3 Chemical food additives are given E numbers. E331 can be monosodium citrate, disodium citrate, or trisodium citrate. With the help of the model shown on page 152 suggest an explanation for the fact that citric can form three sodium salts.

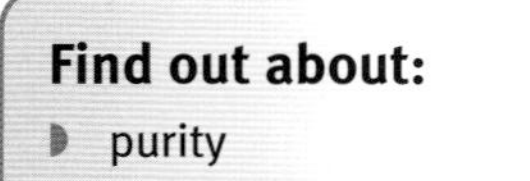

Find out about:

- purity
- titrations for testing purity

E Purity of chemicals

Grades of purity

Suppliers of chemicals offer a range of grades of chemicals. In a school laboratory you might use one of these grades: technical, general laboratory, and analytical. The purest grade is the analytical grade.

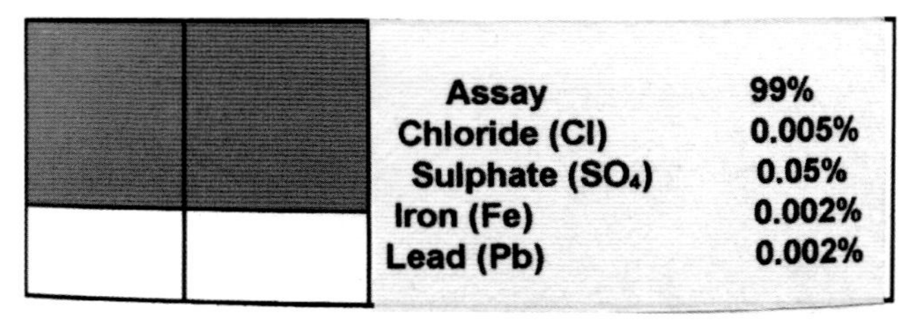

Label on a bottle of laboratory grade calcium carbonate

Calcium carbonate, for example, is used in a blast furnace to extract iron from its ores. It is also an ingredient of indigestion tablets. The iron industry can use limestone straight from a quarry. Limestone has some impurities but they do not stop it from doing its job in a blast furnace.

The calcium carbonate in an indigestion tablet must be safe to swallow. It must be very pure.

Purifying a chemical is done in stages. Each stage takes time and money, and becomes more difficult. So the higher the purity, the more expensive the chemical. Manufacturers therefore buy the quality most suitable for their purpose.

When deciding what grade of chemical to use, it is important to know:

- the amount of impurities
- what the impurities are
- how they can affect the process
- whether they will end up in the product, and whether it matters if they do

Testing purity

Medicines contain an active ingredient. Other ingredients are included to make the medicine pleasant to taste and easy to take. This means that the pharmaceutical companies that make medicines need sweeteners, food flavours, and other additives.

The companies buy in many of their ingredients. Technical chemists working for the companies have to make sure that the suppliers are delivering the right grade of chemical. The aim is to make sure that all the chemicals are between 99% and 100% pure.

Citric acid is widely used to control the pH in syrups such as cough medicines. Technicians can check the purity of the acid using a procedure called a **titration**. It is possible to calculate the purity of a sample of citric acid by measuring the volume of alkali that it can exactly neutralize. The analyst has to know the precise concentration of the alkali.

Key words

titration
burette
end point

Steps involved in a titration

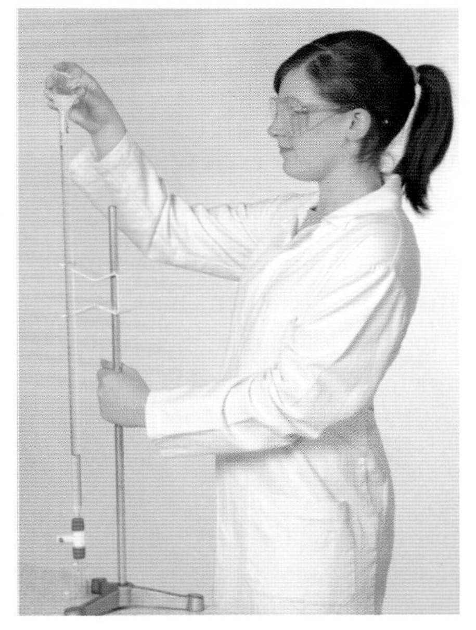

1 The technician fills a **burette** with a solution of sodium hydroxide. She knows the concentration of the alkali.

2 The technician weighs out accurately a sample of citric acid and dissolves it in water.

3 The technician dissolves the acid in pure water. Then she adds a few drops of phenolphthalein indicator. The indicator is colourless in the acid solution.

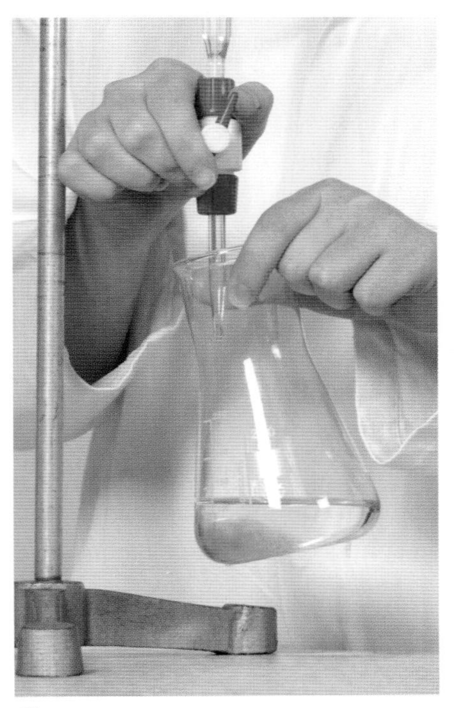

4 The technician adds alkali from the burette. She swirls the contents of the flask as the alkali runs in. Near the end she adds the alkali drop by drop. At the **end point** all the citric acid is just neutralized. The indicator is now permanently pink.

Questions

1. Make a list of some uses of sodium chloride (salt). Beside each use write which grade of salt it would be best to use.
2. Epsom salts consist of magnesium sulfate. Magnesium sulfate is soluble in water. Produce a flow diagram to show how you could remove impurities that are insoluble in water from a sample of Epsom salts. Processes you might use include: crystallization, dissolving, drying, evaporation, filtration.

Find out about:

- measuring rates of reaction
- factors affecting rates of reaction
- catalysts in industry
- collision theory

F Rates of reaction

Controlling reaction rates

Some chemical reactions seem to happen in an instant. An explosion is an example of a very fast reaction.

Other reactions take time – seconds, minutes, hours or even years. Rusting is a slow reaction and so is the rotting of food.

Chemists synthesizing chemicals must work out a procedure that is as efficient as possible. It is important that the chosen reactions happen at a convenient speed. A reaction that occurs too quickly can be hazardous. A reaction that takes several days to complete is not practical because it ties up equipment and people's time for too long.

An explosion is an example of a very fast chemical reaction

Measuring rates of reaction

Your pulse rate is the number of times your heart beats every minute. The production rate in a factory is a measure of how many articles are made in a particular time. Similar ideas apply to chemical reactions.

Chemists measure the **rate of a reaction** by finding the quantity of product produced or the quantity of reactant used up in a fixed time.

For the reaction

$$Mg(s) + 2HCl(aq) \longrightarrow MgCl_2(aq) + H_2(g)$$

the rate can be measured quite easily by collecting and measuring the hydrogen gas.

$$\text{average rate} = \frac{\text{change in the volume of hydrogen}}{\text{time for the change to happen}}$$

In most chemical reactions the rate changes with time. The graph on the left is a plot of the volume of hydrogen formed against time for the reaction of magnesium with acid. The graph is steepest at the start showing that the rate of reaction was greatest at that point. As the reaction continues the rate decreases until the reaction finally stops. The steepness of the line is a measure of the rate of reaction.

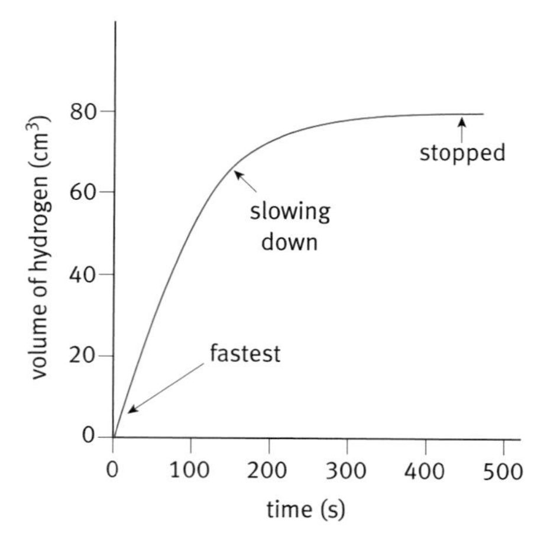

A plot of the volume of hydrogen formed against time for a reaction of magnesium with hydrochloric acid

Questions

1 For each of these reactions, pick from page 163 a method that could be used to measure the rate of reaction:

a $CaCO_3(aq) + 2HCl(aq) \longrightarrow CaCl_2(aq) + CO_2(g) + H_2O(l)$

b $Zn(s) + H_2SO_4(aq) \longrightarrow ZnSO_4(aq) + H_2(g)$

c $Na_2S_2O_3(aq) + 2HCl(aq) \longrightarrow 2NaCl(aq) + SO_2(aq) + S(s) + H_2O(l)$

Key words

rate of reaction

Methods of measuring rates of reaction

Collecting and measuring a gas product

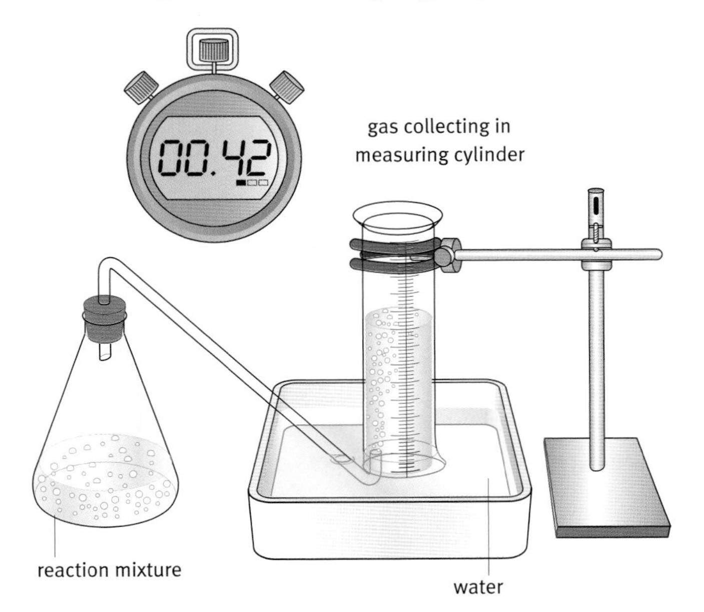

Record the volume at regular intervals, such as every 30 or 60 seconds.

Measuring the loss of mass as a gas forms

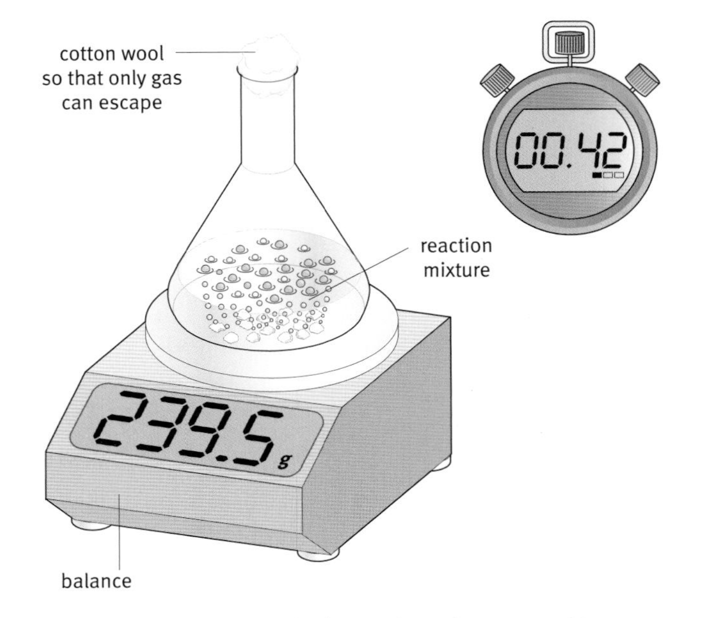

Record the mass at regular intervals such as every 30 or 60 seconds.

Timing how long it takes for a small amount of solid reactant to disappear

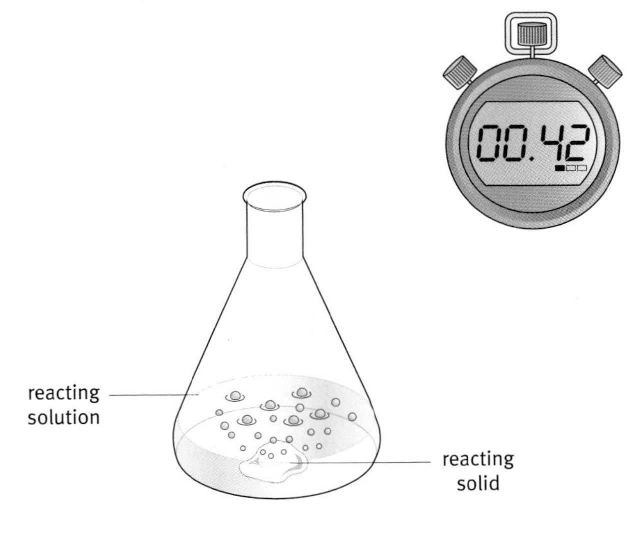

Mix the solid and liquid in the flask and start the timer. Stop it when you can no longer see any solid.

Timing how long it takes for a solution to turn cloudy

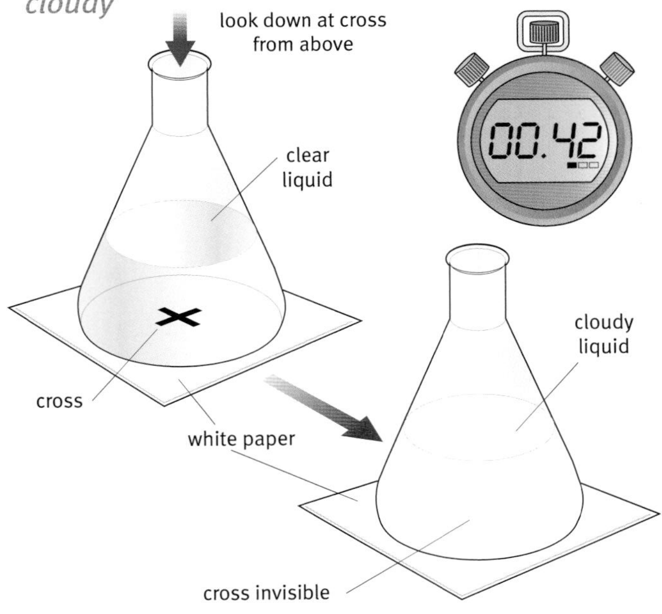

This is for reactions that produce an insoluble solid. Mix the liquids in the flask and start the timer. Stop it when you can no longer see the cross on the paper through the solution.

Factors affecting reaction rates

A sliced loaf goes stale faster than an unsliced loaf. Milk standing in a warm kitchen goes sour more quickly than milk kept in a refrigerator. Changing the conditions alters the rate of these processes and may others.

Factors which affect the rate of chemical reactions are:

- The *concentration* of reactants in solution – the higher the concentration the faster the reaction.
- The *surface area* of solids – powdering a solid increases the surface area in contact with a liquid or solid and so speed up the reaction.
- The *temperature* – typically a 10 °C rise in temperature can roughly double the rate of reaction.
- *Catalysts* – catalysts are chemicals which speed up a chemical reaction without being used up in the process.

The factors in action

The apparatus in the diagram was used in an investigation into the effect of changing the conditions on the reaction of zinc metal with sulfuric acid. The graph below shows the results.

$$Zn(s) + H_2SO_4(aq) \longrightarrow ZnSO_4(aq) + H_2(g)$$

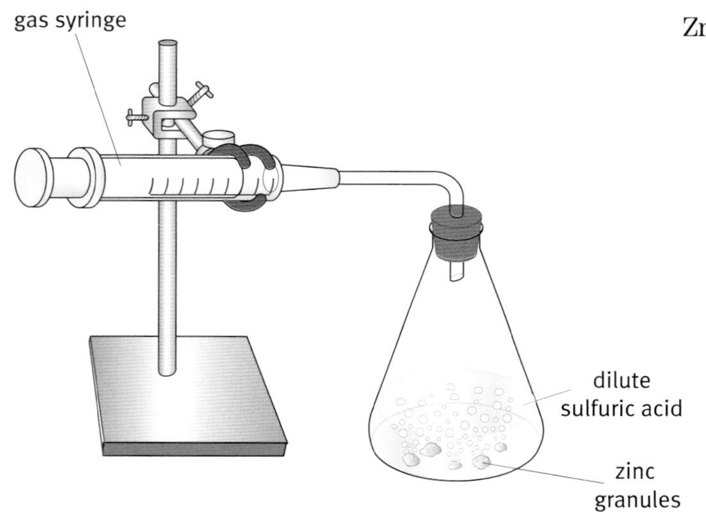

Apparatus used to investigate the factors affecting the rate of reaction of zinc with sulfuric acid.

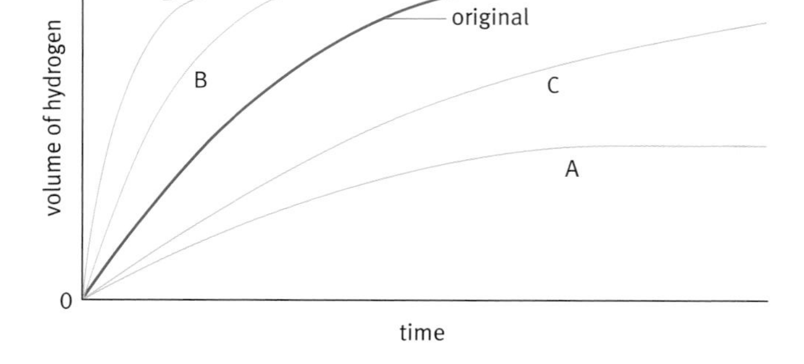

Plots showing the volume of hydrogen formed over time during an investigation of the factors affecting the rate of reaction of zinc with sulfuric acid. The investigator used the same mass of zinc each time. There was more than enough metal to react with all the acid.

The red line on the graph plots the volume of hydrogen gas against time using zinc granules and 50 cm^3 of dilute sulfuric acid at 20 °C. The reaction gradually slows down and stops because the acid concentration falls to zero. There is more than enough metal to react with all the acid. The zinc is in excess.

The effect of concentration

Line A on the graph shows the results of repeating the procedure using acid that was half as concentrated while leaving all the other conditions the same.

The investigator added 50 cm^3 of this more dilute acid. Halving the acid **concentration** lowers the rate at the start. Again the reaction slows down because the concentration falls as the acid reacts with the metal. The final volume of gas is cut by half because there was only half as much acid to start with.

The effect of temperature

Line B shows the result of carrying out the reaction at 30 °C while leaving all the other conditions the same as in the original set up. This speeds up the reaction and more or less doubles the rate at the start. The quantities of chemicals are the same so the final volume of gas collected at room temperature is the same as it was originally.

The effect of surface area

Line C shows the result of keeping to all the original conditions but using the same excess of zinc metal in larger pieces. Fewer larger lumps of metal have a smaller **surface area** so the reaction starts more slowly. The amount of acid is unchanged and the metal is still in excess so that the final volume of hydrogen is the same.

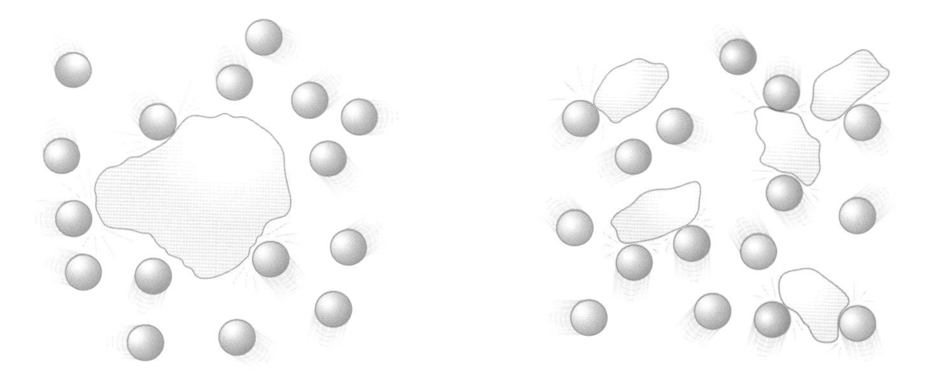

Breaking up a solid into smaller pieces increases the surface area. This increases the amount of contact between the solid and the solution, making it possible for the reaction to go faster.

The effect of adding a catalyst

Line D shows what happens when the investigation is repeated with everything the same as in the original set up but with a **catalyst** added. Adding a few drops of copper(II) sulfate solution produces this effect. The reaction starts more quickly and the graph is steeper. Catalysts do not change the final amount of product, so the volume of gas at the end is the same as before.

Key words

concentration
surface area
catalyst

Questions

2 How would you account for the fact that:
- **a** sliced bread goes stale more quickly than unsliced bread?
- **b** there is a danger of explosions in flour mills?

3 How is it possible to control conditions to slow down or stop these changes:
- **a** the rusting of iron?
- **b** a chip-pan fire?
- **c** milk going sour?

(4) How is it possible to control conditions to speed up these changes:
- **a** the setting of an epoxy glue?
- **b** the cooking of an egg?
- **c** the conversion of oxides of nitrogen in car exhausts to nitrogen?
- **d** the conversion of sugar to alcohol and carbon dioxide?

Catalysts in industry

What is a catalyst?

A catalyst is a chemical that speeds up a chemical reaction. It takes part in the reaction, but is not used up.

Modern catalysts can be highly selective. This is important when reactants can undergo more than one chemical reactions to give a mixture of products. With a suitable catalyst it can be possible to speed up the reaction that gives the required product, but not speed up other possible reactions that create unwanted by-products.

Better catalysts

Catalysts are essential in many industrial processes. They make many processes possible economically. This means that chemical products can be made at a reasonable cost and sold at affordable prices.

Research into new catalysts is an important area of scientific work. This is shown by the industrial manufacture of ethanoic acid (see page 152) from methanol and carbon monoxide. This process was first developed by the company BASF in 1960 using a cobalt compound as the catalyst at 300 °C and at a pressure 700 times atmospheric pressure.

About six years later the company Monsanto developed a process using the same reaction, but a new catalyst system based on rhodium compounds. This ran under much milder conditions: 200 °C and 30–60 times atmospheric pressure.

In 1986, the petrochemical company BP bought the technology for making ethanoic acid from Monsanto. They have since devised a new catalyst based on compounds of iridium. This process is faster and more efficient. Iridium is cheaper, and less of the catalyst is needed. Iridium is even more selective so the yield of methanol is greater and there are fewer by-products. This makes it easier to make pure methanol and there is less waste.

The manufacture of ethanoic acid from methanol and carbon monoxide is only possible in the presence of a catalyst.

methanol + carbon monoxide → ethanoic acid

$CH_3OH(g)$ + $CO(g)$ → $CH_3COOH(g)$

Reaction conditions:
pressure: 30 atmospheres
temperature: 200 °C
catalyst: iridium.

Collision theory

Chemists have a theory to explain how the various factors affect reaction rates. The basic idea is that molecules can only react if they bump into each other. Imagining molecules colliding with each other leads to a theory which can account for the effects of concentration, temperature and catalysts on reaction rates.

According to **collision theory**, when molecules collide some bonds between atoms can break while new bonds form. This creates new molecules.

Molecules are in constant motion in gases and liquid and there are millions upon millions of collisions every second. Most reactions would be explosive if every collision led to reaction. It turns out that only a very small fraction of all the collisions are 'successful' and actually lead to reaction. These are the collisions in which the molecules are moving with enough energy to break bonds between atoms.

Any change that increases the number of 'successful' collisions per second has the effect of increasing the rate of reaction.

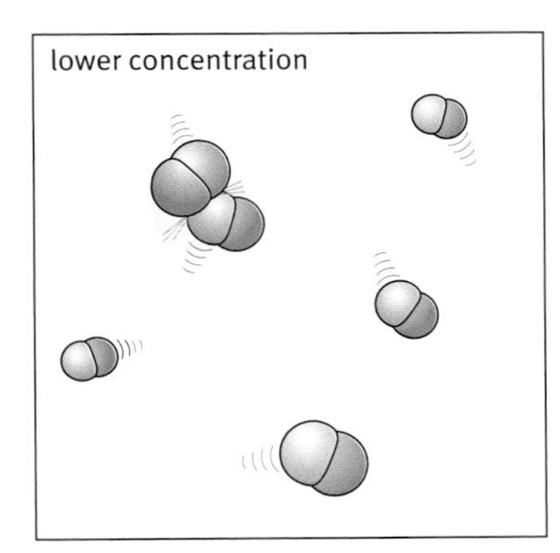

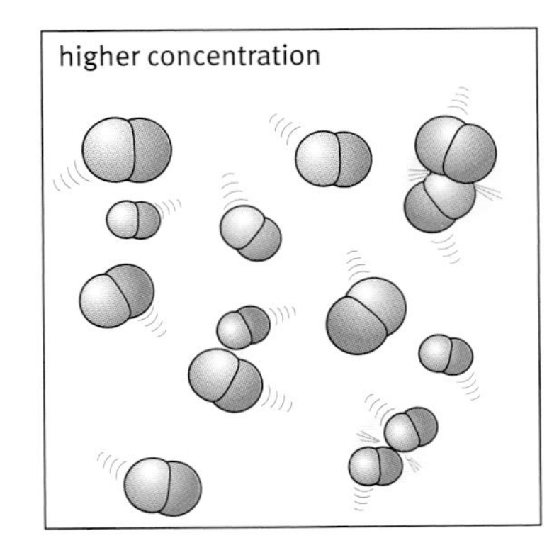

Molecules have a greater chance of colliding in a more concentrated solution. More collisions means more reaction. Reactions get faster if the reactants are more concentrated.

Questions

1 a Where do cobalt, rhodium and iridium appear in the periodic table (see Module C4, page 95).

b Why is not surprising that these three metals can be used to make catalysts for the same process?

2 Suggest reasons why it is important to develop industrial processes which:

a run at lower temperatures and pressures

b produce less waste

Key words

collision theory

Find out about:
- reacting masses
- yields from chemical reactions

G Chemical quantities

Chemists wanting to make a certain quantity of product need to work out how much of the starting materials to order. Getting the sums right matters, especially in industry, where a higher yield for a lower price can mean better profits.

The trick is to turn the symbols in the balanced chemical equation into masses in grams or tonnes. This is possible given the relative masses of the atoms in the periodic table (see Module C4 *Chemical patterns*, page 39).

Key words
relative formula mass
reacting mass
actual yield
theoretical yield
percentage yield

Reacting masses

Adding up the relative atomic masses for all the atoms in the formula of a compound gives the **relative formula mass** of chemicals (see Module C5 *Chemicals of the natural environment*, page 139). Given the relative formula masses it is then possible to work out the masses of reactants and products in a balanced equation. These are the **reacting masses**.

Questions

1 What mass of:
  a copper(II) oxide reacts with 98 g H_2SO_4 in dilute sulfuric acid
  b HCl in hydrochloric acid reacts with 100 g calcium carbonate
  c HNO_3 in dilute nitric acid neutralizes 56 g of potassium hydroxide?

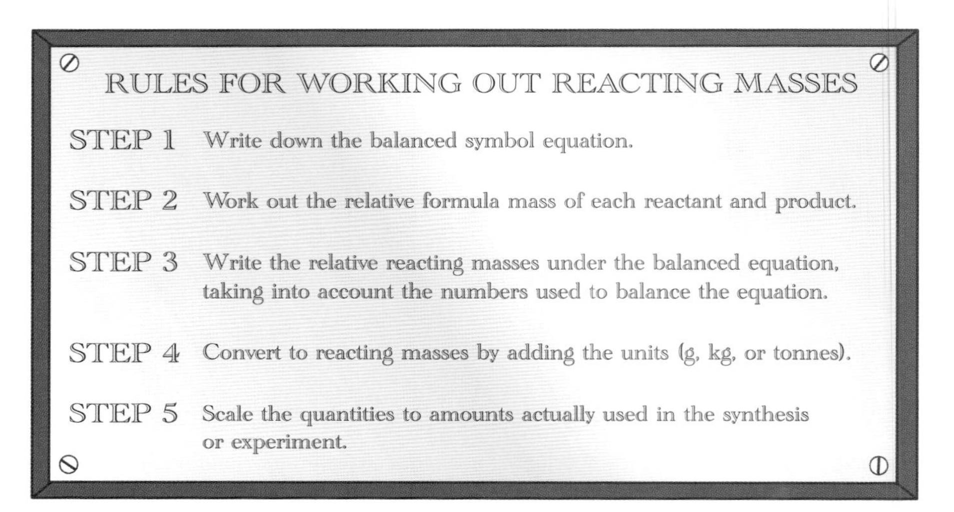

Example

What are the masses of reactants and products when sulfuric acid reacts with sodium hydroxide?

Step 1 $2NaOH + H_2SO_4 \longrightarrow Na_2SO_4 + 2H_2O$

Step 2 relative formula mass of NaOH = 23 + 16 + 1 = 40

relative formula mass of $H_2SO_4 = (2 \times 1) + 32 + (4 \times 16) = 98$

relative formula mass of $Na_2SO_4 = (2 \times 23) + 32 + (4 \times 16) = 142$

relative formula mass of $H_2O = (2 \times 1) + 16 = 18$

Steps 3 & 4

2NaOH	+	H_2SO_4	⟶	Na_2SO_4	+	$2H_2O$
2 × 40 = 80		98		142		2 × 18 = 36
80 g		98 g		142 g		36 g

Yields

The yield of any synthesis is the quantity of product obtained from known amounts of starting materials. The **actual yield** is the mass of product after it is separated from the mixture and purified and dried.

Theoretical yield

The **theoretical yield** is the mass of product expected if the reaction goes exactly as shown in the balanced equation. This is what could be obtained in theory if there are no by-products and no losses while chemicals are transferred from one container to another. The actual yield is always less than the theoretical yield.

Example

What is the theoretical yield of ethanoic acid made from 8 tonnes methanol? (See page 167.)

Step 1 *Write down the balanced equation*

methanol + carbon monoxide ⟶ ethanoic acid

$CH_3OH(g)$ + $CO(g)$ ⟶ $CH_3COOH(g)$

32 60

Step 2 *Work out the relative formula masses*

methanol: 12 + 4 + 16 = 32

ethanoic acid: 24 + 4 + 32 = 60

Steps 3 & 4 *Write down the relative reacting masses and convert to reacting masses by adding the units*

Theoretically, 32 tonnes of methanol should give 60 tonnes of ethanoic acid.

Step 5 *Scale to the quantities actually used*

If the theoretical yield of ethanoic acid = x tonnes, then

$$\frac{\text{mass of ethanoic acid}}{\text{mass of methanol}} = \frac{60 \text{ tonnes}}{32 \text{ tonnes}} = \frac{x \text{ tonnes}}{8 \text{ tonnes}}$$

So, the yield of ethanoic acid from 8 tonnes of methanol should be

$$8 \text{ tonnes} \times \frac{60 \text{ tonnes}}{32 \text{ tonnes}} = 15 \text{ tonnes}$$

Percentage yield

The **percentage yield** is the percentage of the theoretical yield that is actually obtained. It is always less than 100%.

Questions

1 What is the mass of salt which forms in solution when:
  a hydrochloric acid neutralizes 4 g sodium hydroxide,
  b 12.5 g zinc carbonate reacts with excess sulfuric acid?

2 A preparation of sodium sulfate began with 8.0 g of sodium hydroxide.
  a Calculate the theoretical yield of sodium sulfate from 8.0 g sodium hydroxide.
  b Calculate the percentage yield given that the actual yield was 12.0 g.

Example

What is the percentage yield if 8 tonnes of methanol produces 14.7 tonnes of ethanoic acid?

From the previous example:

theoretical yield = 15 tonnes

actual yield = 14.7 tonnes

$$\text{percentage yield} = \frac{\text{actual yield}}{\text{theoretical yield}} \times 100$$

$$= \frac{14.7 \text{ tonnes}}{15 \text{ tonnes}} \times 100$$

$$= 98\%$$

Find out about:
- steps in synthesis
- making a soluble salt

An operator emptying magnesium sulfate into the tank of a sprayer on a farm. As well as being a micronutrient needed for healthy plant growth, the salt is needed as a:
- raw material in soaps and detergents
- laxative in medicine
- refreshing additive in bath water
- raw material in the manufacture of other magnesium compounds
- supplement in feed for poultry and cattle
- coagulant in the manufacture of some plastics

H Stages in chemical synthesis

Chemical synthesis is a way of making new compounds. Synthesis puts things together to make something new. It is the opposite of analysis which takes things apart to see what they are made of.

The process of making a soluble salt on a laboratory scale illustrates the stages in a chemical synthesis.

In the following method an excess of solid is added to make sure that all the acid is used up. This method is only suitable if the solid added to the acid is either insoluble in water or does not react with water.

Making a sample of magnesium sulfate

Choosing the reaction

Any of the characteristic reactions of acids can all be used to make salts:

- acid + metal ⟶ salt + hydrogen
- acid + metal oxide or hydroxide ⟶ salt + water
- acid + metal carbonate ⟶ salt + carbon dioxide + water

Magnesium metal is relatively expensive because it has to be extracted from one of its compounds. So it makes sense to use either magnesium oxide or carbonate as the starting point for making magnesium sulfate from sulfuric acid.

Carrying out a risk assessment

It is always important to minimize exposures to risk. You should take care to identify hazardous chemicals. You should also look for hazards arising from equipment or procedures. This is a **risk assessment**.

In this preparation the magnesium compounds are not hazardous. The dilute sulfuric acid is an irritant, which means that you should keep it off your skin and especially protect your eyes. You should always wear eye protection when handling chemicals, for example.

Working out the quantities to use

In this procedure the solid is added in excess. This means that the amount of product is determined by the volume and concentration of the sulfuric acid. The concentration of dilute sulfuric acid is 98 g/litre. It turns out that a volume of 50 cm^3 dilute sulfuric acid is suitable. This contains 4.9 g of the acid.

Carrying out the reaction in suitable apparatus under the right conditions

The reaction is fast enough at room temperature, especially if the magnesium carbonate is supplied as a fine power.

This reaction can be safely carried out in a beaker. Stirring with a glass rod makes sure that the magnesium carbonate and acid mix well. Stirring also helps to prevent the mixture frothing up and out of the beaker.

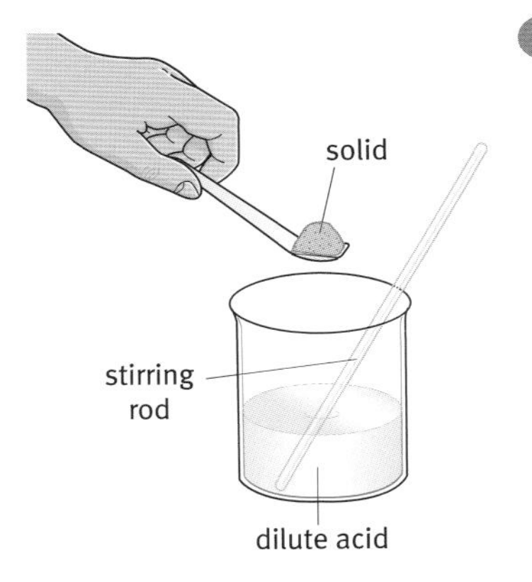

1 Measure the required volume of acid into a beaker. Add the metal or the insoluble oxide, hydroxide or carbonate bit by bit until no more dissolves in the acid. Warm when most of the acid has been used up. Make sure that there is a slight excess of solid before moving on to the next stage.

Separating the product from the reaction mixture

Filtering is a quick and easy way of separating the solution of the product from the excess solid. The mixture filters faster if the mixture is warm.

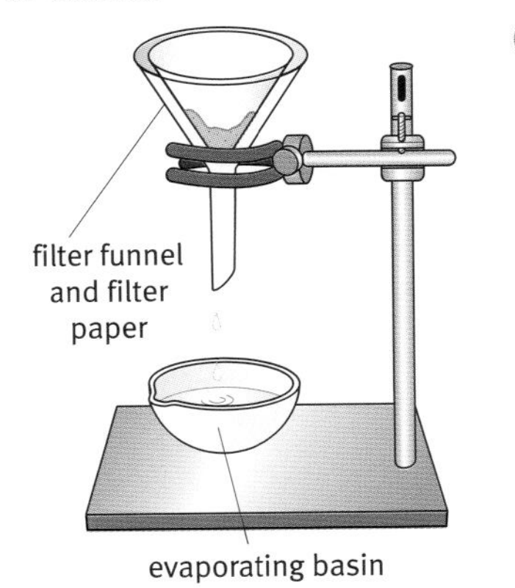

2 Filter off the excess solid collecting the solution of the salt in an evaporating basin. The residue on the filter paper is the excess solid.

Questions

1 Write the balanced equation for the reaction of magnesium carbonate with sulfuric acid.

2 Why does the mixture of magnesium carbonate and sulfuric acid froth up?

3 What is the advantage of:
- **a** using powdered magnesium carbonate?
- **b** warming when most of the acid has been used up?
- **c** adding a slight excess of the solid to the acid?

4 Why is it impossible to use this method to make a pure metal sulfate by the reaction of dilute sulfuric acid with:
- **a** lithium metal?
- **b** sodium hydroxide?
- **c** potassium carbonate?

5 Look at the procedure on pages 171–172 described for making magnesium sulfate and identify risks that might arise from:
- **a** chemicals that react vigorously and spill over
- **b** chemicals that might spit or splash on heating
- **c** hot apparatus that might cause burns
- **d** apparatus that might crack and form sharp edges

Key words

risk assessment

Purifying the product

After the mixture has been filtered, the filtrate contains the pure salt dissolved in water. Evaporating much of the water speeds up crystallization. This is conveniently carried out in an evaporating basin. The concentrated solution can then be left to cool and crystallize.

Once the crystals are nearly dry they can be transferred to a desiccator. This is a closed container which contains a solid that absorbs water strongly.

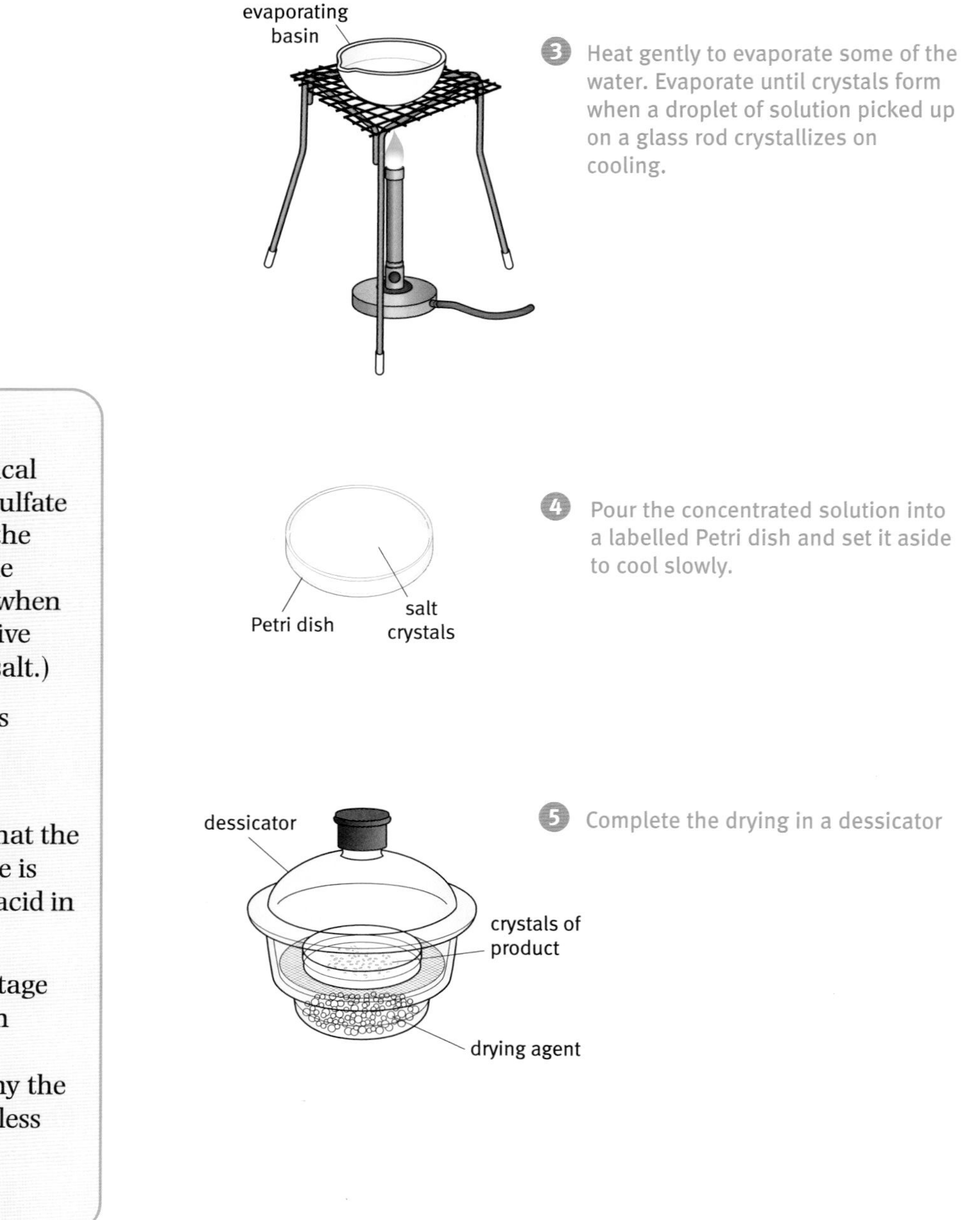

Questions

6 Calculate the theoretical yield of magnesium sulfate crystals produced in the synthesis. (Include the water in the crystals when working out the relative formula mass of the salt.)

7 Identify the impurities removed during the purification stages.

8 Why is it important that the magnesium carbonate is added to the sulfuric acid in excess?

9 **a** What is the percentage yield of magnesium sulfate?

b Suggest reasons why the percentage yield is less than 100% in this preparation.

Measuring the yield and checking the purity of the product

The final step is to transfer the dry crystals to a weighed sample tube and reweigh it to find the actual yield of crystals. Often it is important to carry out tests to check that the product is pure.

The appearance of the crystals can give a clue to the quality of the product. A microscope can help if the crystals are small. The crystals of a pure product are often well-formed and even in shape.

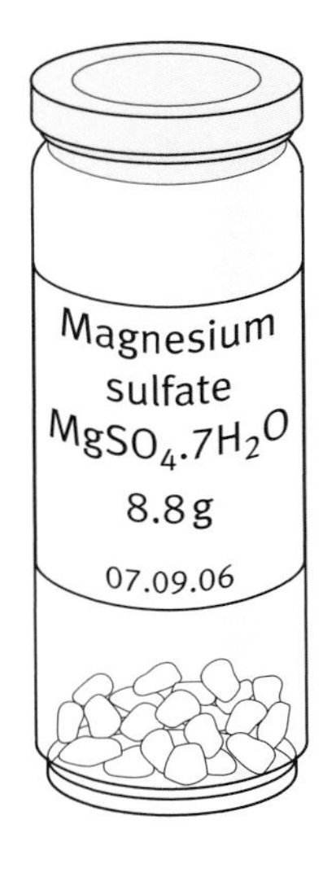

6 The weighed sample of product showing the name and formula of the chemical, the mass of product, and the date it was made.

Crystals of pure magnesium sulfate seen through a Polaroid filter (×60)

C6 Chemical synthesis

Summary

You have learnt how chemists can use their knowledge of chemical reactions to plan and carry out the synthesis of new compounds.

The chemical industry

- The chemical industry is an important part of the country's economy.
- Industry makes bulk chemicals on a large scale and fine chemicals for more specialized purposed on a smaller scale.
- Chemical synthesis provides useful products such as food additives, fertilizers, dyestuffs, paints, pigments, and pharmaceuticals.

The ionic theory of acids and alkalis

- Acids react in characteristic ways with metals, metal oxides, and metal carbonate.
- Alkalis neutralize acids to form salts.
- All acids have similar properties because they produce hydrogen ions, $H^+(aq)$, in water.
- Alkalis produce aqueous hydroxide ions, $OH^-(aq)$, when they dissolve in water.
- During a neutralization reaction, the hydrogen ions from an acid react with hydroxide ions from an alkali to make water.

Controlling the rate of change

- Chemists follow the rate of a change by measuring the disappearance of a reactant or the formation of a product.
- Factors which affect the rate of change include the concentration of reactants, the particle size of solid reactants, the temperature, and the presence of catalysts.
- Collision theory can be used to explain why changing the concentration of reactants affects the rate of reaction. H

Synthesis

- A chemical synthesis involves a number of stages, including:
  - choosing the reaction or series of reactions to make the required product
  - carrying out a risk assessment
  - working out the quantities of reactants to use H
  - carrying out the reaction in suitable apparatus in the right conditions
  - separating the product from the reaction mixture
  - purifying the product
  - measuring the yield and checking the purity of the product
- The balanced equation for the reaction is used to work out the quantities of chemicals to use and to calculate the theoretical yield.
- A titration is a technique that can be used to check the purity of chemicals used in synthesis.

Questions

1 Give examples to show why each of these products of chemical synthesis are useful and valuable:

a food additives

b fertilizers,

c dyestuffs

d paints

e pharmaceuticals

2 Use the ionic theory of acids and alkalis to explain why:

a Solutions of acids in water conduct electricity with hydrogen forming at the negative electrode.

b Solutions of the hydroxides of lithium, sodium, and potassium are alkaline.

c Water is one of the products of a neutralization reaction.

3 A small piece of metal is cut from a stick of lithium taken from the bottle of oil in which it is stored. the mass of the metal sample is 0.1 g. The piece is put into an apparatus containing excess water. After the metal and water have been allowed to mix, the volume of hydrogen produced is measured at intervals.

Time (minutes)	Volume of gas (cm^3)
1	8
2	24
3	72
4	138
5	172
6	172

a Why is lithium stored in oil?

b Draw a diagram of an apparatus that could be used to collect and measure the hydrogen produced by the reaction. Show how it could be arranged so that the lithium and water do not mix until the experimenter is ready to start the reaction.

c Plot the results on graph paper with labelled axes. Plot time on the horizontal axis.

d What is the average rate of reaction between 3 and 4 minutes. The rate can be measured by the rate of formation of gas in cm^3 per minute.

e Explain the change in rate between 4 and 5 minutes.

f It is suspected that that the results in the first 4 minutes are affected by some oil on the surface of the metal. Draw a dotted line on the graph to show the results that might be expected if the measurements are repeated using 0.1 g of lithium from which all traces of oil are removed.

4 Zinc sulfate is a soluble salt used in some dietary supplements. It can be made by reacting zinc oxide with dilute sulfuric acid. Zinc oxide is an insoluble white solid.

These are the first steps in the synthesis of zinc sulfate:

- Add small portions of powdered zinc oxide to some warm sulfuric acid. Stir the mixture after each addition. Keep doing this until the mixture is slightly cloudy.
- Filter the mixture once all the acid has been used up.

a Why is zinc oxide added until the mixture stays slightly cloudy?

b Suggest reasons for using:

i powdered zinc oxide

ii warm sulfuric acid

c What is the purpose of filtering the mixture?

d Write a word equation for the reaction and then a balanced symbol equation.

e Describe the next steps needed to obtain dry crystals of zinc sulfate from the liquid that passes through the filter paper.

5 Pure calcium chloride is used in kidney dialysis. One way of making the salt is to add an excess of powdered limestone (calcium carbonate) to dilute hydrochloric acid.

a Calculate the theoretical yield of calcium chloride that can be made from 10 kg of calcium carbonate.

b What is the percentage yield if the actual yield of calcium chloride is 9.9 kg from 10 kg calcium carbonate?

Why study chemistry?

Chemistry is the science which helps us to understand matter on an atomic scale. It is the central science. Knowledge of chemistry informs materials science and engineering as well as biochemistry, genetics, and environmental sciences.

The science

Understanding carbon chemistry helps to explain the chemistry of life. There are so many carbon compounds that chemists have had to find ways to organize their knowledge to study the compounds in families such as the alcohols, carboxylic acids, and esters.

Chemists have theories to help answer key questions about chemical reactions: How much?, How fast?, and How far?

Chemistry in action

Analytical chemists help to protect us by checking that food and water are safe. Analysis of blood helps to diagnose disease. Forensic scientists use analysis to solve crimes.

Chemists synthesize new chemicals to meet our needs. Medicinal chemists synthesize thousands of new compounds in the search for new drugs to treat or cure disease. The chemical industry synthesizes products on a large scale. The industry is changing so that it can be more sustainable.

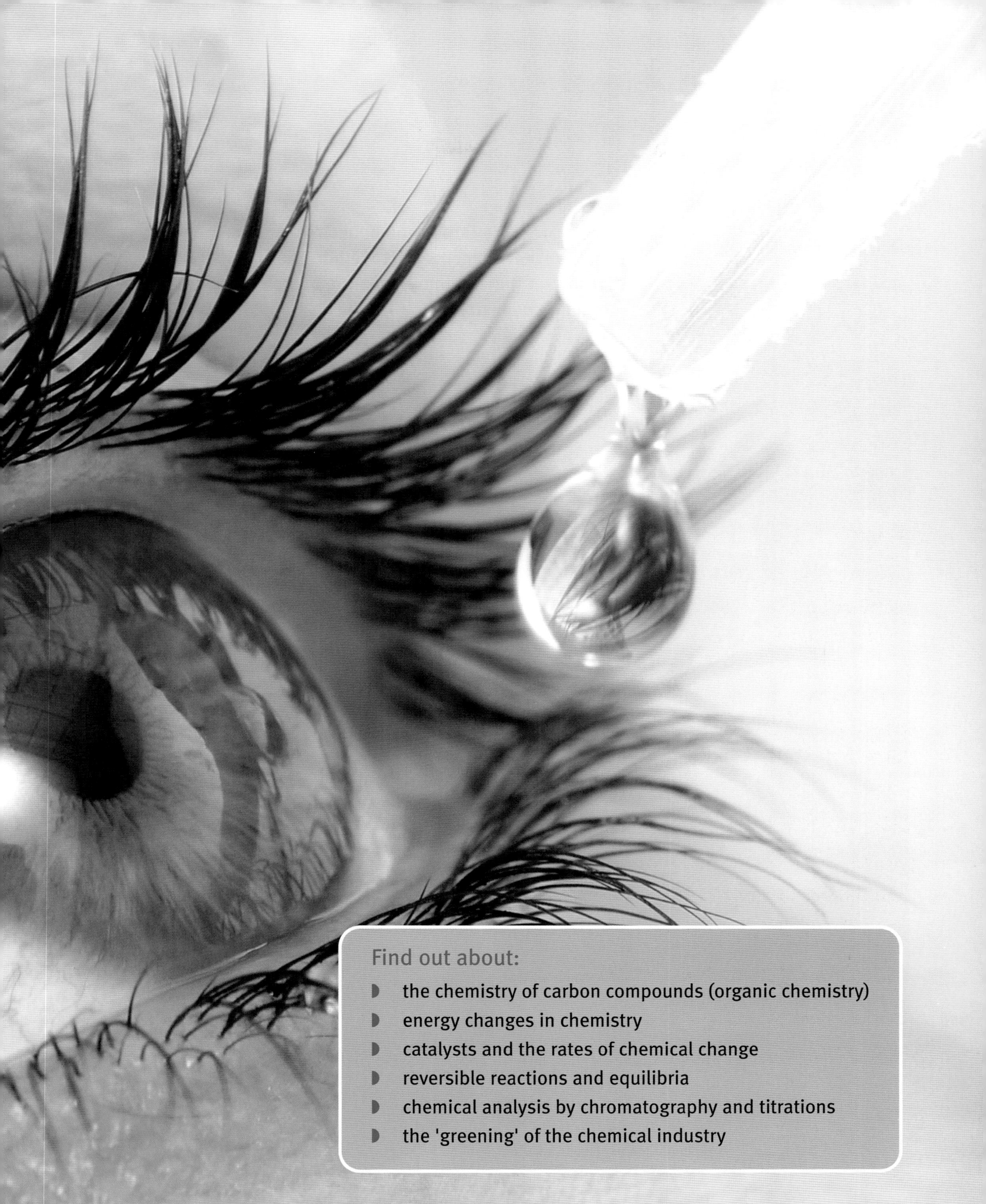

Find out about:

- the chemistry of carbon compounds (organic chemistry)
- energy changes in chemistry
- catalysts and the rates of chemical change
- reversible reactions and equilibria
- chemical analysis by chromatography and titrations
- the 'greening' of the chemical industry

Topic 1

The chemistry of carbon compounds

There are more carbon compounds than there are compounds of all the other elements put together. The chemistry of carbon is so important that it forms a separate branch of the subject called **organic chemistry**.

Synthetic polymers make good nets to catch fish for food.

Spiders spin a web with a natural organic polymer to catch food.

Organic chemistry

The word 'organic' means 'living'. At first, organic chemistry was the study of compounds from plants and animals. Now we know that all the complex variety of compounds can be made artificially. Organic chemistry includes the study of synthetic compounds, including polymers, drugs, and dyes.

Key words
organic chemistry

Chains and rings

It helps to think of organic compounds being made up of a skeleton of carbon atoms supporting other atoms. Some of the other atoms may be reactive, while others are less so. In organic compounds, carbon is often linked to hydrogen, oxygen, nitrogen, and halogen atoms.

Carbon forms so many compounds because carbon atoms can join up in many ways, forming chains, branched chains, and rings. The chains can be very long, as in the polymer polythene. A typical polythene molecule may have 10 000 or more carbon atoms linked together (see Module C2 *Material choices*, Section E). A polythene molecule is still very tiny, but much bigger than a methane molecule.

To make sense of the huge variety of carbon compounds, chemists think in terms of families, or series, of organic compounds.

Bonding in carbon compounds

The bonding in organic compounds is covalent (see Module C4 *Chemical patterns*, Section B). The structures are molecular. The structures can be worked out from knowing how many covalent bonds each type of atom can form. Carbon atoms form 4 bonds, hydrogen atoms form 1 bond, while oxygen atoms form 2 bonds.

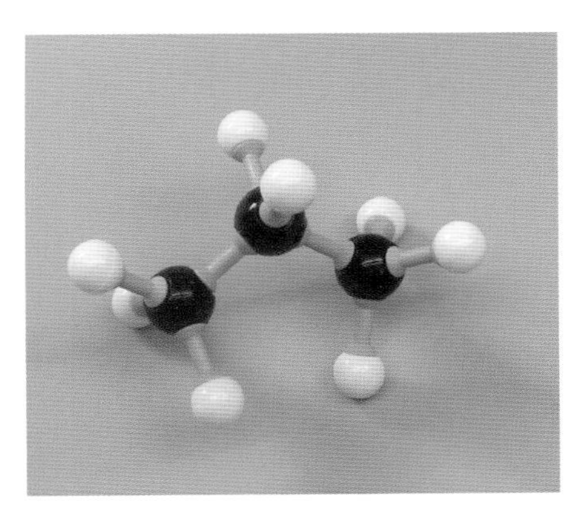

Propane: a hydrocarbon with three carbon atoms in a chain.

Methylbutane: a hydrocarbon with a branched chain.

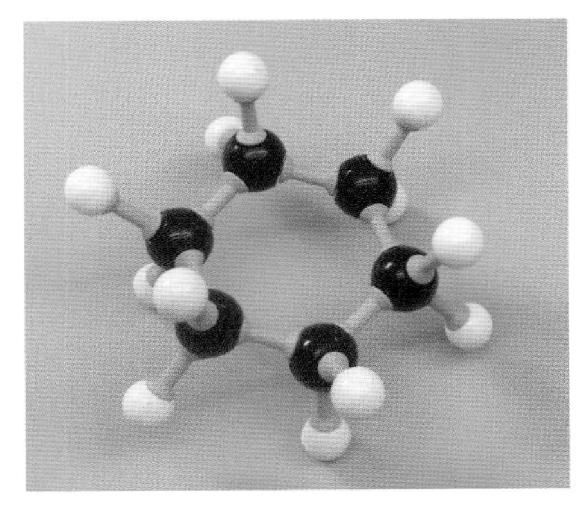

Cyclohexane: a hydrocarbon with a ring of carbon atoms.

Find out about:
- the alkane series of hydrocarbons
- physical properties of alkanes
- chemical reactions of alkanes

1A The alkanes

The **alkanes** make up an important series of **hydrocarbons**. They are well known because they are the compounds in fuels such as natural gas, liquid petroleum gas (LPG), and petrol. The simplest alkane is methane. This is the main gas in natural gas.

The table belowshows five alkanes.

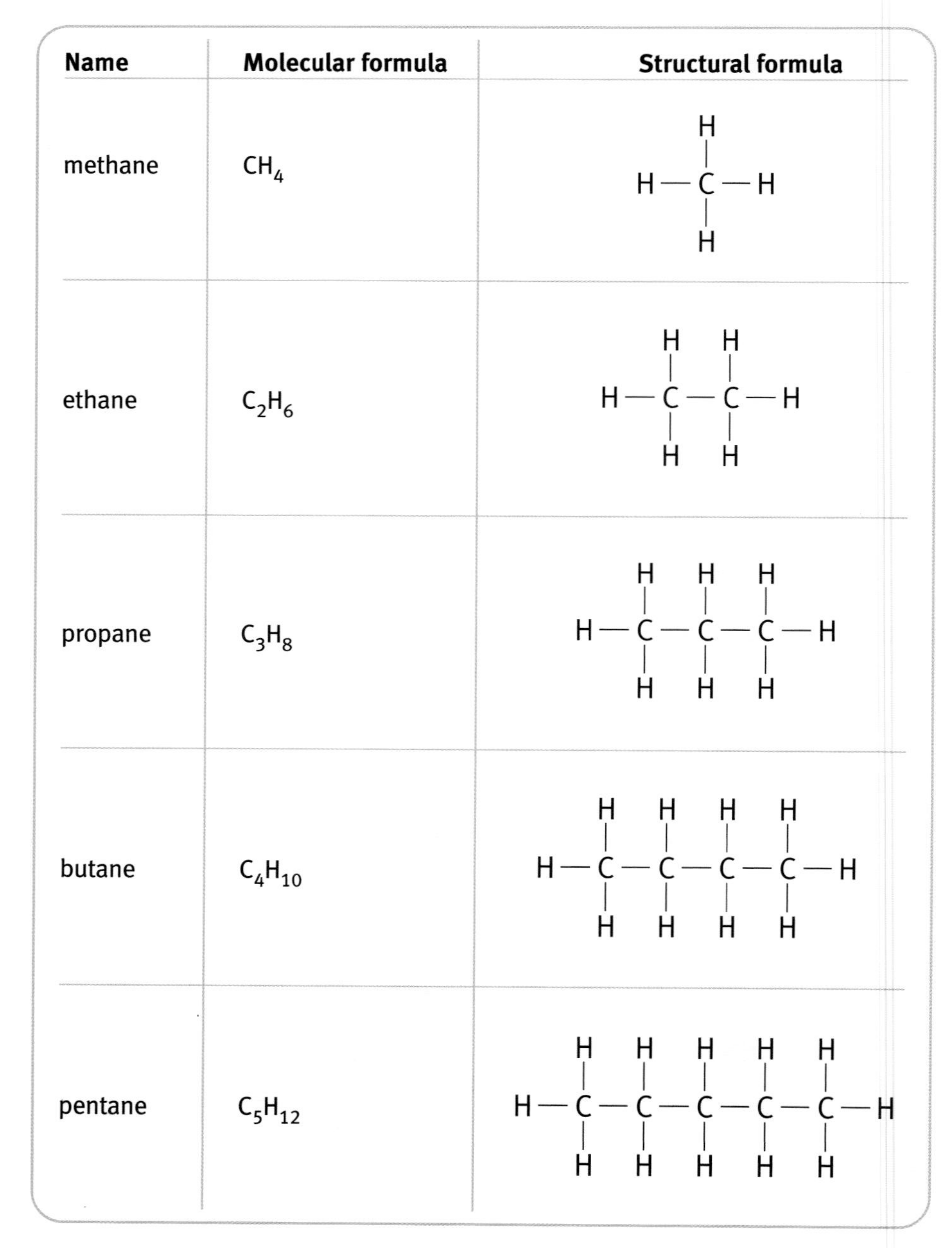

Name	Molecular formula	Structural formula
methane	CH_4	H H — C — H H
ethane	C_2H_6	H H H — C — C — H H H
propane	C_3H_8	H H H H — C — C — C — H H H H
butane	C_4H_{10}	H H H H H — C — C — C — C — H H H H H
pentane	C_5H_{12}	H H H H H H — C — C — C — C — C — H H H H H H

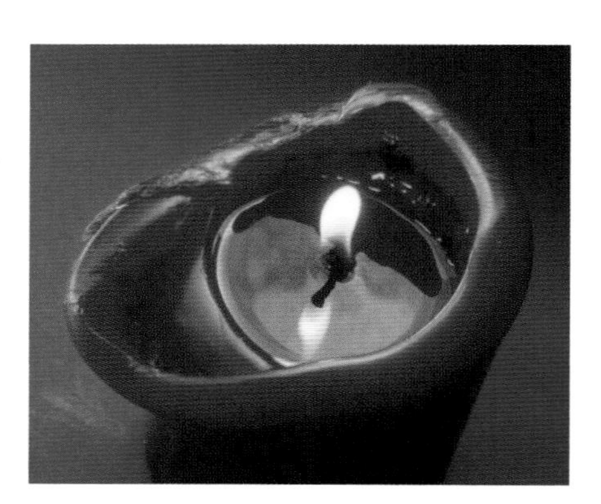

At room temperature the alkanes in candle wax are solid. The flame first melts them and then turns them to gas in the hot wick. The hot gases burn in air.

CH_4 — the molecular formula

H
H — C — H
H — the structural formula showing the chemical bonds

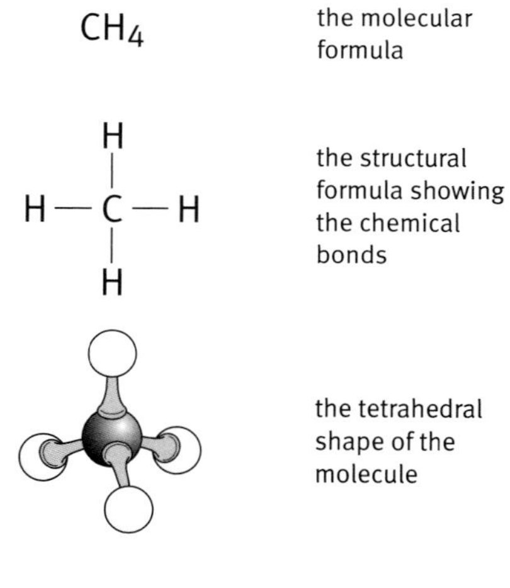

the tetrahedral shape of the molecule

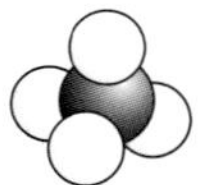

the space filled by the molecule

Ways of representing a molecule of methane

Physical properties of alkanes

The alkanes are oily. They do not dissolve in water or mix with it.

The alkanes with small molecules (up to four carbon atoms) are gases at room temperature. Those with 4–17 carbon atoms are liquids. The alkanes in candle wax have molecules with more than 17 carbon atoms, and these are solid at room temperature.

Liquid alkanes with longer molecules are sticky liquids. Their viscosity makes them suitable as ingredients of lubricants.

Key words
alkane
hydrocarbon

Chemical properties of alkanes

Burning

All alkanes burn. Many common fuels consist mainly of alkanes. The hydrocarbons burn in air, forming carbon dioxide and water.

If the air is in short supply, the products may include particles of soot (carbon) and the toxic gas carbon monoxide.

Reactions with aqueous acids and alkalis

Alkanes do not react with common laboratory reagents such as acids or alkalis. The hydrocarbons do not react because the C—C and C—H bonds in the molecules are unreactive.

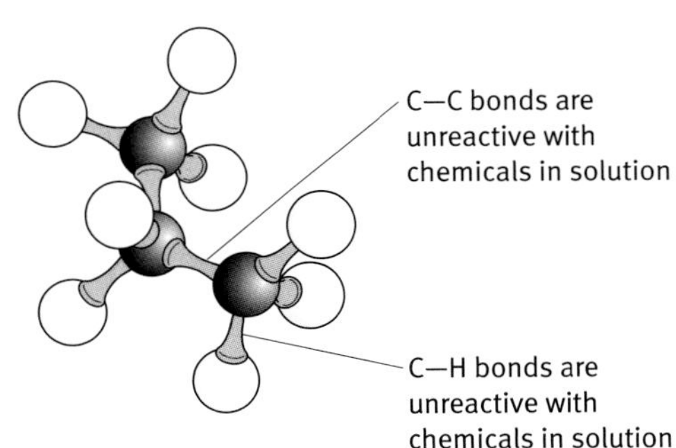

Alkanes are generally unreactive because the C—C and C—H bonds do not react with common aqueous reagents

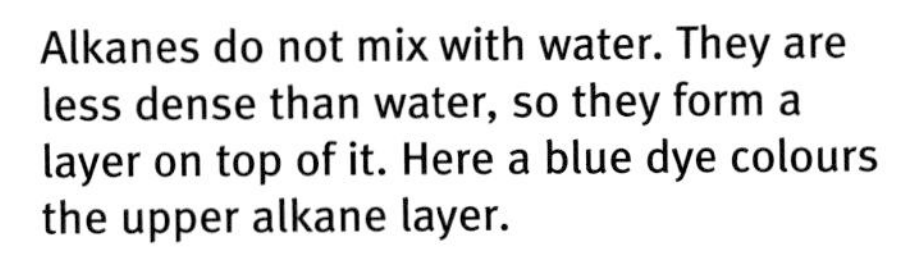

Alkanes do not mix with water. They are less dense than water, so they form a layer on top of it. Here a blue dye colours the upper alkane layer.

Questions

1 a In which group of the periodic table does carbon belong?

b How many electrons are there in the outer shell of a carbon atom?

c Which groups in the periodic table include elements that form simple ions?

d Is carbon likely to form simple ions?

2 Give two examples which show that alkanes, or mixtures of alkanes, do not mix with water?

3 Are the alkanes in petrol more or less dense than water?

4 Write a balanced equation for propane burning in plenty of air.

5 Write a balanced equation for methane burning in a limited supply of air to form carbon monoxide and steam.

Find out about:
- physical properties of alcohols
- chemical reactions of alcohols

1B The alcohols

Uses of alcohols

Ethanol is the best-known member of the series of **alcohols**. It is the alcohol in beer, wine, and spirits. Ethanol is also a very useful solvent. It is a liquid which evaporates quickly, and for this reason it is used in cosmetic lotions and perfumes. Ethanol easily catches fire and burns with a clean flame, so it can be used as a fuel.

The simplest alcohol, methanol, can be made in two steps from methane (natural gas) and steam. This alcohol is important as a chemical feedstock. The chemical industry converts methanol to a wide range of chemical products needed to manufacture products such as adhesives, foams, solvents, and windscreen washer fluid.

Structures of alcohols

The first three members of the alcohol series are methanol, ethanol, and propanol.

Ethanol: a simple alcohol. Chemists name alcohols by changing the name of the corresponding alkane to '-ol'. Ethanol is the two-carbon alcohol related to ethane.

There are two ways of looking at alcohol molecules that can help to understand their properties. On the one hand, an alcohol can be seen as an alkane with one of its hydrogen atoms replaced by an —OH group. On the other hand, the same molecule can be regarded as a water molecule with one of its hydrogen atoms replaced by a hydrocarbon chain.

Physical properties

Methanol and ethanol are liquids at room temperature. Alkanes with comparable relative molecular masses are gases. This shows that the attractive forces between molecules of alcohols are stronger than they are in alkanes. The presence of an —OH group of atoms gives the molecules this greater tendency to cling together like water.

Even so, the boiling point of ethanol at 78 °C is below that of water (100 °C). Ethanol molecules have a greater mass than water molecules, but the attractions between the hydrocarbon parts are very weak, as in alkanes.

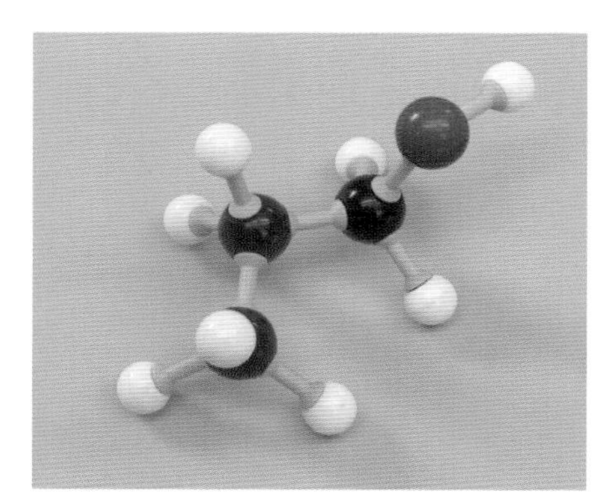

Propanol: a three-carbon alcohol related to propane.

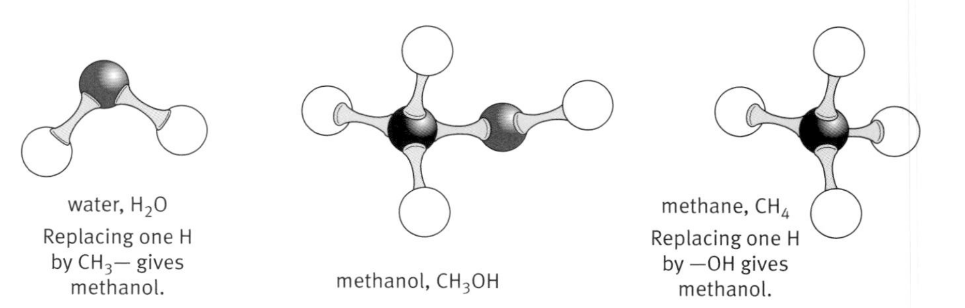

Two ways of looking at an alcohol molecule

Overall, ethanol molecules have less tendency to stick together than water molecules.

Similarly the —OH in the molecules of methanol and ethanol freely mix with water, unlike alkanes. However, alcohols with longer hydrocarbon chains, such as hexanol ($C_6H_{13}OH$), do not mix with water because the oiliness of the hydrocarbon part of the molecules dominates.

Key words
alcohols
functional group

Chemical properties

The —OH group is the reactive part of an alcohol molecule. Chemists call it the **functional group** for alcohols.

Burning

All alcohols burn. Methanol and ethanol are highly flammable and are used as fuels. These compounds can burn because of the hydrocarbon parts of their molecules.

Reaction with sodium

Alcohols react with sodium in a similar way to water. This is because both water molecules and alcohol molecules include the —OH group of atoms. With water, the products are sodium hydroxide and hydrogen. With ethanol, the products are sodium ethoxide and hydrogen.

$$2\,CH_3CH_2OH + 2Na \longrightarrow 2\,CH_3CH_2O^-Na^+ + H_2$$

sodium ethoxide

Only the hydrogen atom attached to the oxygen atom is involved in this reaction. The hydrogen atoms linked to carbon are inert.

The product has an ionic bond between the oxygen and sodium atoms. Sodium ethoxide, like sodium hydroxide, is an ionic compound and a solid at room temperature.

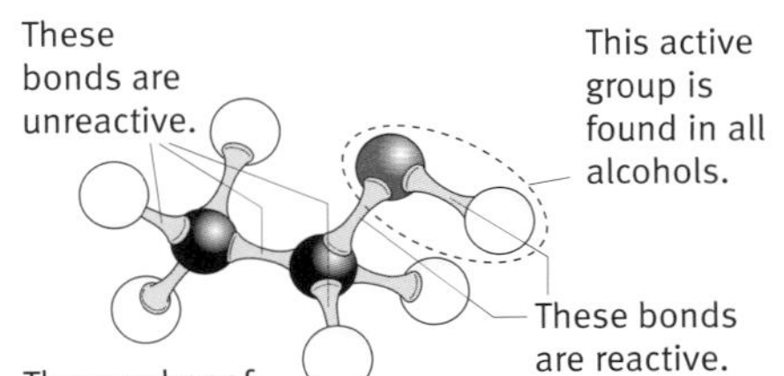

Some bonds in ethanol are more reactive than others. Alcohols are more reactive than alkanes because C—O and O—H bonds are more reactive than C—C and C—H bonds. The alcohols share similar chemical properties because they all have the —OH group in their molecules.

Questions

1 Produce a table for three alcohols similar the table of alkanes on page 180.

2 a Use values of relative atomic masses from the periodic table in Module C4, page 95, to show that propane and ethanol have the same relative mass.

b Propane boils at –42 °C, but ethanol boils at 78 °C. Suggest an explanation for the difference.

3 Write a balanced equation for propanol burning.

4 Write balanced equations for the reactions of sodium with

a water

b methanol

Find out about

- structures and properties of organic acids
- acids in vinegar and other foods
- carboxylic acids as weak acids

1C Carboxylic acids

Acids from animals and plants

Many acids are part of life itself. These are the organic acids, many of which appear in lists of ingredients on food labels (see Module C3 *Food matters*, page 217).

Acetic acid (which chemists call ethanoic acid) is the main acid in vinegar. Acetic acid helps to preserve and flavour a range of foods, including pickles and chutneys as well as vinegars.

Citric acid gives oranges and lemons their sharp taste. Lactic acid forms in muscle cells during exercise and in milk as it turns sour.

The sting of a red ant contains methanoic acid. The traditional name for this acid is formic acid, from the Latin word for ant, *formica*.

Some of the acids with more carbon atoms have unpleasant smells. The horrible odour of rancid butter is caused by the breakdown of fats to produce butyric acid. Butyric acid gets this, its traditional name, from its origins in butter. It is also the main cause of the revolting smell of vomit. The modern systematic name is butanoic acid.

Human sweat includes a wide range of chemicals, including fats. Enzymes in bacteria on the surface of skin can quickly break down these compounds into a mixture of organic acids, including butyric, caproic, and caprylic acids (among others). These compounds are largely responsible for the unpleasant, rancid smell of sweaty socks.

Goats' milk contains fats with more caproic acid and caprylic acid than cows' milk. These acids get their names from the Latin word for goat. When the fats break down to the free acids, the result is a strong goaty smell.

Structures and names of organic acids

The functional group in the molecules of organic acids is

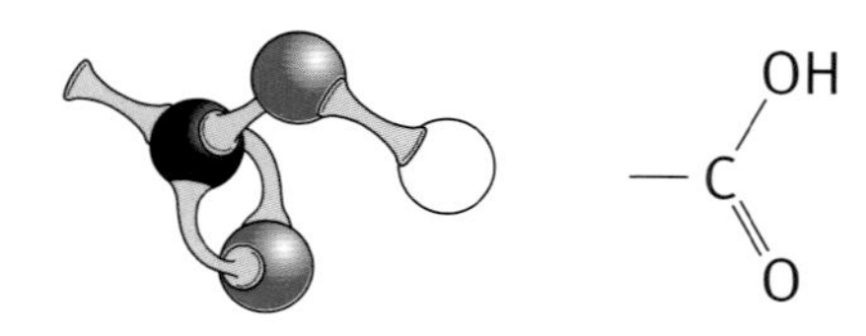

The series of compounds with this reactive group are the **carboxylic acids**. The chemical names of the compounds are related to the alkane with the same number of carbon atoms. The ending 'ane' becomes 'anoic acid'. So the systematic name for acetic acid, the two-carbon acid, is ethanoic acid.

Questions

1. Write the formula of butanoic acid. (Use the table of alkane names on page 180 to help you.)
2. The formula of caproic acid is $CH_3CH_2CH_2CH_2CH_2COOH$. What is the formula of caprylic acid, which has 8 carbon atoms in its molecules?

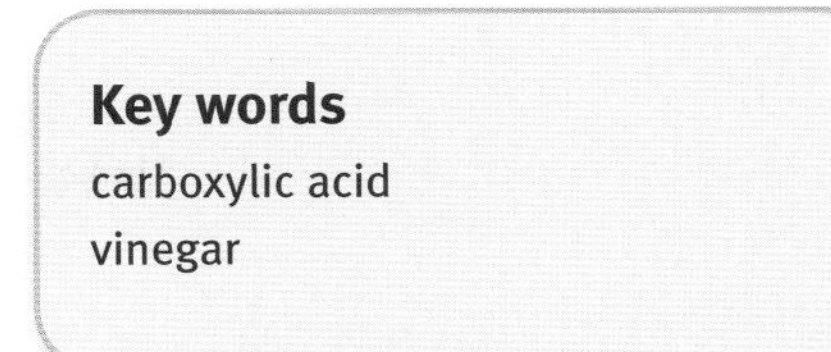

Formation of vinegar

Oxidation of ethanol produces ethanoic acid. Vinegar is manufactured by allowing solutions of alcohol to oxidize. Bacteria in the solutions help this process.

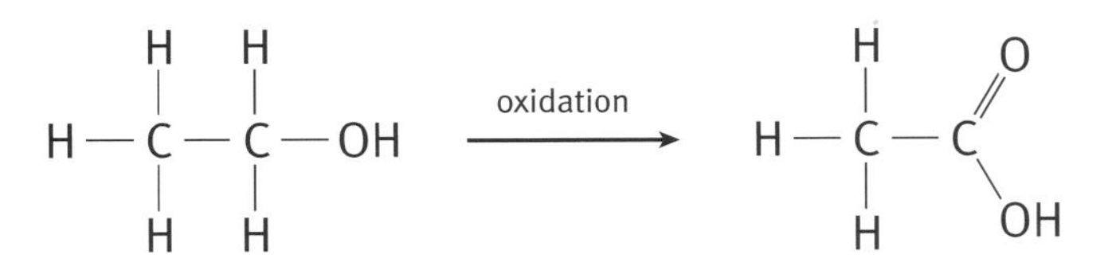

Oxidation converts beer to malt vinegar. Cider oxidizes to cider vinegar and wine to wine vinegar.

Acidity of carboxylic acids

Carboxylic acids ionize to produce hydrogen ions when dissolved in water. They only ionize to a light extent, which means that they are weak acids (see Module C6 *Chemical synthesis*, page156).

In a molecule of ethanoic acid, there are four hydrogen atoms. Three are attached to a carbon atom and one to an oxygen atom. Only the hydrogen atom attached to oxygen is reactive. This is the hydrogen atom that ionizes in aqueous solution.

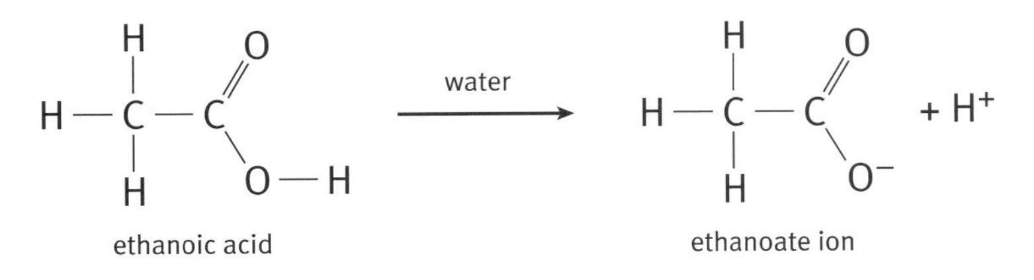

Ethanoic acid and the other carboxylic acids show the characteristic reactions of acids with metals, alkalis, and metal carbonates:

- acid + metal ⟶ salt + hydrogen
- acid + soluble hydroxide ⟶ salt + water
- acid + metal carbonate ⟶ salt + carbon dioxide + water

When ethanoic acid reacts with sodium hydroxide, the salt formed is sodium ethanoate.

There is an ionic bond between the sodium ion and the ethanoate ion:

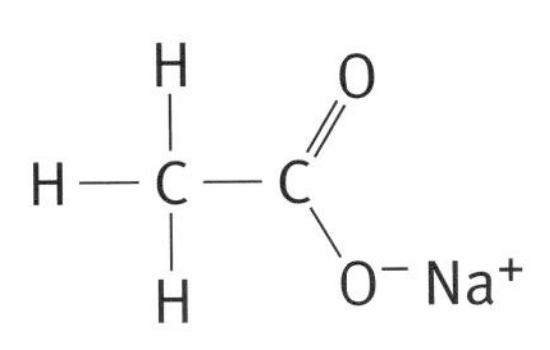

Key words
carboxylic acid
vinegar

Methanoic acid, HCOOH

Ethanoic acid, CH_3COOH

Questions

3 Write word equations and balanced symbol equations for the reactions of methanoic acid with:

a magnesium

b potassium hydroxide solution

c copper(II) carbonate

4 A good way of removing the disgusting smell of butanoic acid from vomit on a carpet or inside a car is to sprinkle it with sodium hydrogencarbonate powder. Write a word equation for the reaction that takes place. Can you explain why the smell might disappear after this reaction?

Organic acids and skin care

The story goes that Cleopatra bathed in asses' milk. This is not as daft as it sounds. All milk, including asses' milk, contains lactic acid. The acid loosens and removes dead skin cells, leaving new, smooth skin underneath.

Dermatologists sometimes use a modern, and more drastic, version of this treatment. They paint the skin with glycolic acid (hydroxyethanoic acid). This peels off the top layer of the skin, removing dead skin cells to reveal the smoother, less wrinkly skin below. But the top layer of the skin is there for a purpose – to act as a barrier to protect what is underneath. So the patient may be left with very red skin that is much more sensitive to UV radiation.

Four steps to smoother skin

Benedicte de Villeneuve is a cosmetic chemist working for L'Oreal. The company makes a milder cosmetic version for people to use at home to help remove fine wrinkles and imperfections.

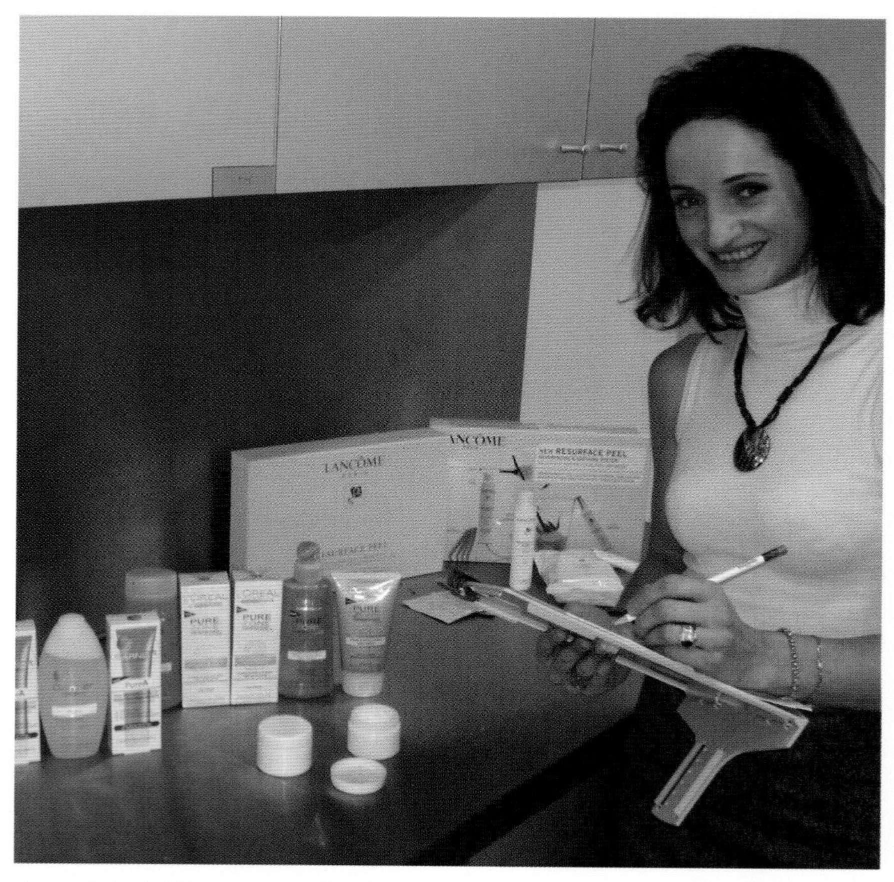

Benedicte de Villeneuve at L'Oréal

'One of our products requires a four-stage process. Users must follow the instructions exactly:

1. 'The skin is thoroughly cleansed with a cleansing wipe, which contains mainly ethanol and water.
2. 'An emulsion including esters, water, and 4% glycolic acid is applied to the skin for a few minutes only.
3. 'The solution is washed off with another wipe that contains an alkaline solution.
4. 'A moisturizer containing emollients is applied. Some products have a UV filter too.'

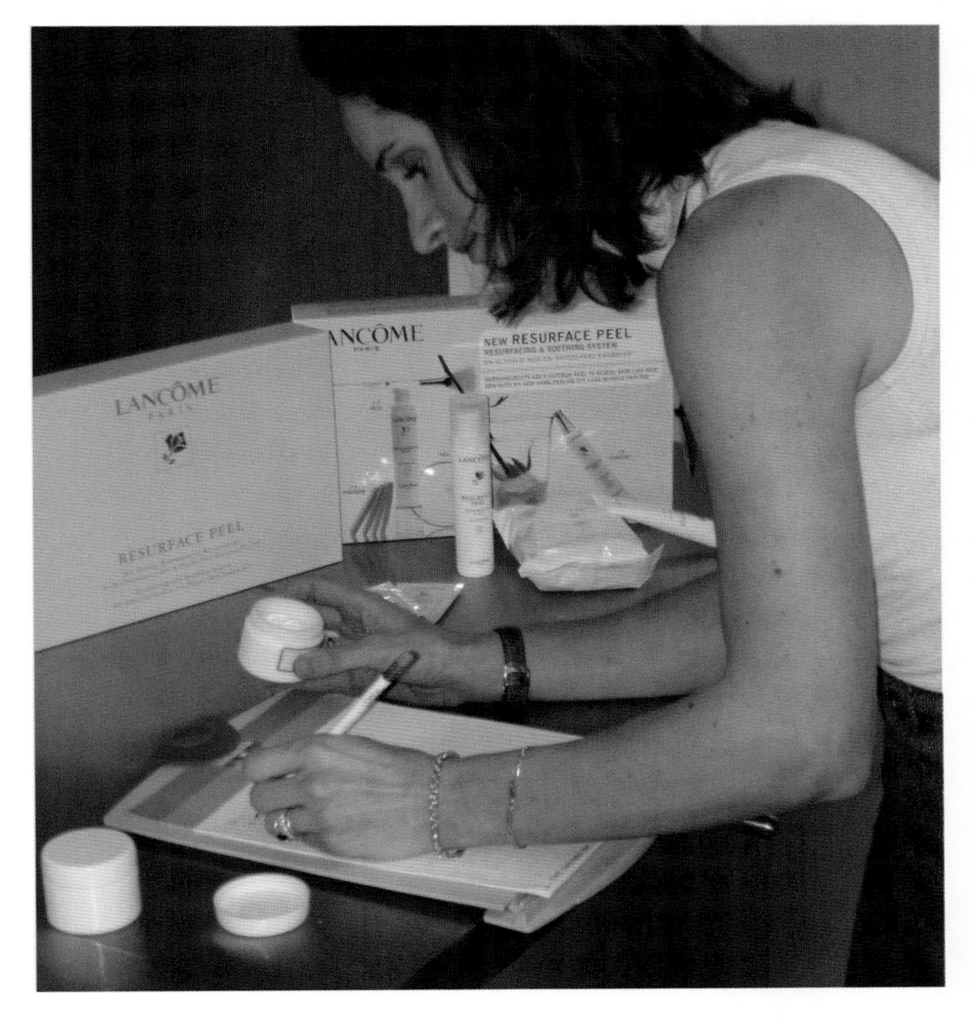

The art of formulation produces products that are attractive, pleasant to use, and effective.

Organic acids can also help people who do not have to worry about wrinkles but have younger, oilier skin prone to spots. Benedicte formulates a range of products containing salicylic acid. 'Salicylic acid is related to aspirin. It is a larger molecule than glycolic acid, so it doesn't penetrate the skin so easily.

'The salicylic acid exfoliates the skin and also acts as an antibacterial. It is added to a whole range of products – cleansers, oil-free moisturizers, and blemish gels.'

The art of formulation

Benedicte originally trained as a pharmacist in Paris, but when a job came up in the labs at L'Oreal, she jumped at the chance. 'Without a knowledge of chemistry, formulating cosmetics would be difficult. Organic chemistry when I was at school could be boring, but now I'm using it every day, and it's great.'

Find out about:

- esters from acids and alcohols
- synthesis of an ester

1D Esters

Fruity-smelling molecules

When you eat a banana, strawberry, or peach, you taste and smell the powerful odour of a mixture of **esters**. A ripe pineapple contains about 120 mg of the ester ethyl ethanoate in every kilogram of its flesh. There are smaller quantities of other esters together with around 60 mg of ethanol.

Esters are very common. Many sweet-smelling compounds in perfumes and food flavourings are esters. Some drugs used in medicines are esters, including aspirin and paracetamol. The plasticizers used to make polymers such as PVC soft and flexible are also esters (see Module C2 *Material choices*, Section G).

Compounds with more than one ester link include fats and vegetable oils such as butter and sunflower oil (see pages 192–195). The synthetic fibres in many clothes are made of a polyester. The long chains in laminated plastics and surface finishes in kitchen equipment are also held together by ester links.

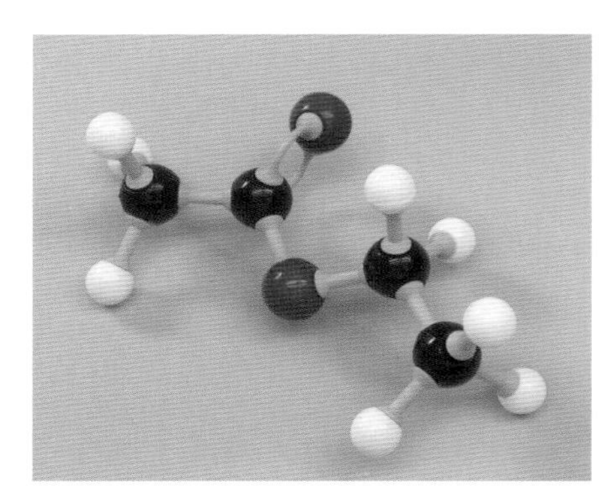

Ethyl ethanoate (ethyl acetate) is a colourless liquid at room temperature. It has many uses. As well as flavouring food, it is a good solvent. Uses as a solvent include decaffeinating tea and coffee and removing coatings such as nail varnish. It is an ingredient of printing inks and perfumes.

The ester with a very strong fruity smell, 3-methylbutyl acetate. It smells strongly of pear drops.

Ester formation

An alcohol can react with a carboxylic acid to make esters. The reaction happens on warming the alcohol and the acid in the presence of a little sulfuric acid to act as catalyst.

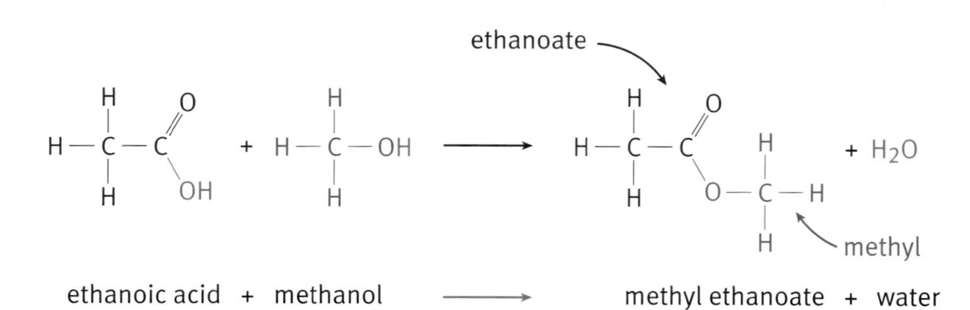

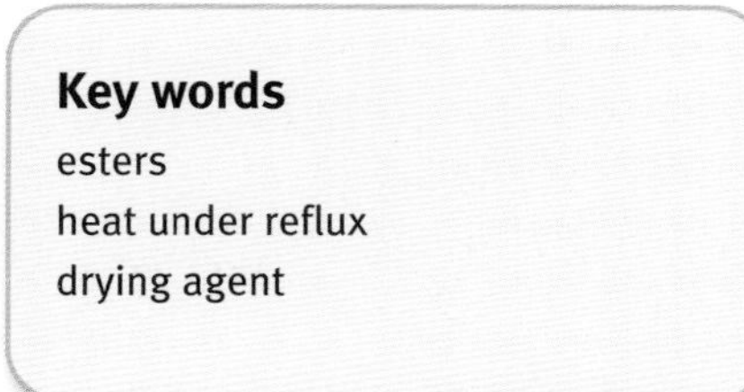
Key words
esters
heat under reflux
drying agent

H Making an ester

The synthesis of ethyl ethanoate on a laboratory scale illustrates techniques used for making a pure liquid product.

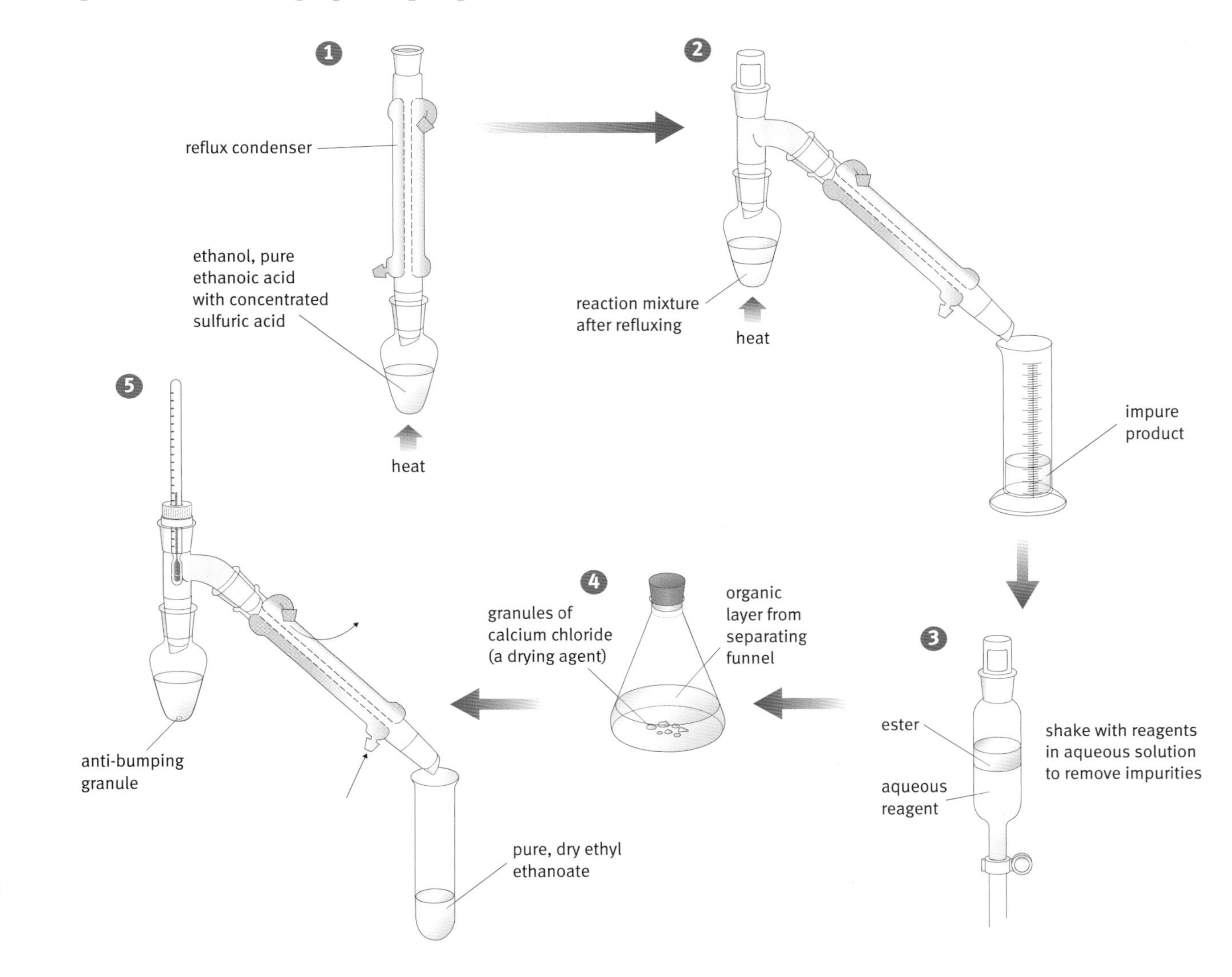

Stages in the laboratory preparation of ethyl ethanoate

Questions

1 a Are the hydrogen atoms bonded to carbon or to oxygen in ethyl ethanoate?

b Would you expect ethyl ethanoate to react with sodium?

c Would you expect ethyl ethanoate to be an acid?

2 Match the steps in the synthesis of ethyl ethanoate with the main stages of a chemical synthesis given in Module C6, Section H.

3 In step A of the synthesis of ethyl ethanoate:

a what is the purpose of the condenser?

b what is the purpose of the sulfuric acid?

4 Calculate the percentage yield if the yield of ethyl ethanoate is 50 g from a preparation starting from 42 g ethanol and 52 g ethanoic acid.

Looking and smelling good

Mimicking nature

Tony Moreton is a chemist working for the Body Shop. Fruit esters crop up all the time in his work: 'We use fruit esters in products that have a fruity smell. They have low molecular masses and low boiling points, which give them the volatility that makes them easy to smell. Their high volatility means they don't linger around for long, and they're referred to as "top notes" in a perfume.

Fruit esters flavour these products.

'As with many organic chemicals, they're flammable, and with their high volatility as well, precautions have to be taken during manufacture.

'The esters used in the industry are "nature identical", which means they are identical to materials found in nature but are made synthetically. Extracting the natural esters would cost about 100 times as much as synthetic ones.'

Esters with a fruity smell

One of the suppliers of ingredients to the Body Shop is a Manchester-based company called Fragrance Oils. One of their products is a blackcurrant perfume concentrate that contains the ester ethyl butanoate. Their perfumery director, Philip Harris, has been familiar with this chemical for a very long time: 'I remember buying strongly flavoured sweets called Pineapple Chunks, which tasted more or less exclusively of ethyl butanoate.

Philip Harris and Farzana Rujidawa working in front of the smelling booths at Fragrance Oils. Philip designed many of the Body Shop's fruit-based fragrances.

'Ethyl butanoate has a strong pineapple aroma, but it's also reminiscent of all sorts of fruity aromas, so we use it in blackcurrant, strawberry, raspberry, apple, mango . . . everything fruity!

'It's quite simply made from ethanol and butanoic acid. The only trouble is that it can hydrolyse back to these starting products, and as butanoic acid is extremely smelly, this isn't so good. We have to avoid using it in alkaline products.

'I've always found it amazing that such an unpleasant smelling material could be used to produce such a delicious fruity smell.'

From fruit esters to soaps

Tony Moreton also uses esters with larger molecules found in various nuts. 'They are formed from glycerol and long-chain fatty acids. They are oily or waxy so are water-resistant, soften the skin, and help retain moisture.

'We can use these fruit esters to produce soap. Synthetic versions of these kinds of esters would be very expensive to make, so in this case we use the natural material.'

Find out about:

- structures of fats and oils
- saturated and unsaturated compounds

1E Fats and oils

It is possible to have molecules with more than one ester link between alcohol and acid. Important examples are fats and oils. These compounds release more energy when oxidized than carbohydrates. This makes them important to plants and animals as an energy store.

The structures of fats and oils

The alcohol in fats and oils is **glycerol**. This is a compound with three —OH groups.

```
    O     O
    |     ||
H — C — O — C — CH2 — CH2 — CH2 — CH2 — CH2 — CH2 — CH2 — CH2 — CH2 — CH2 — CH2 — CH2 — CH3
    |     O
    |     ||
H — C — O — C — CH2 — CH2 — CH2 — CH2 — CH2 — CH2 — CH2 — CH2 — CH2 — CH2 — CH2 — CH2 — CH3
    |     O
    |     ||
H — C — O — C — CH2 — CH2 — CH2 — CH2 — CH2 — CH2 — CH2 — CH2 — CH2 — CH2 — CH2 — CH2 — CH3
    |
    H
```

The general structure of a compound in which glycerol has formed three ester links with fatty acids. In natural fats and oils the fatty acids may all be the same or they may be a mixture.

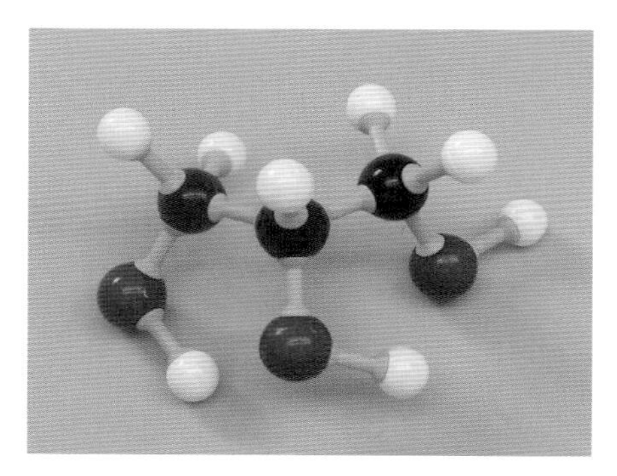

Glycerol, which is also called propan-1,2,3-triol

The carboxylic acids in fats and oils are often called fatty acids. These are compounds with a long hydrocarbon chain attached to a carboxylic acid group.

Saturated and unsaturated fats

Animal **fats** are generally solids at room temperature. Butter and lard are examples. **Vegetable oils** are usually liquid, as illustrated by corn oil, sunflower oil, and olive oil.

Chemically the difference between fats and oils arises from the structure of the carboxylic acids. Stearic acid is typical of the acids combined in animal fats. All the bonds in its molecules are single bonds. Chemists use the term **saturated** to describe molecules like this because the molecule has as much hydrogen as it can take. These saturated molecules are straight.

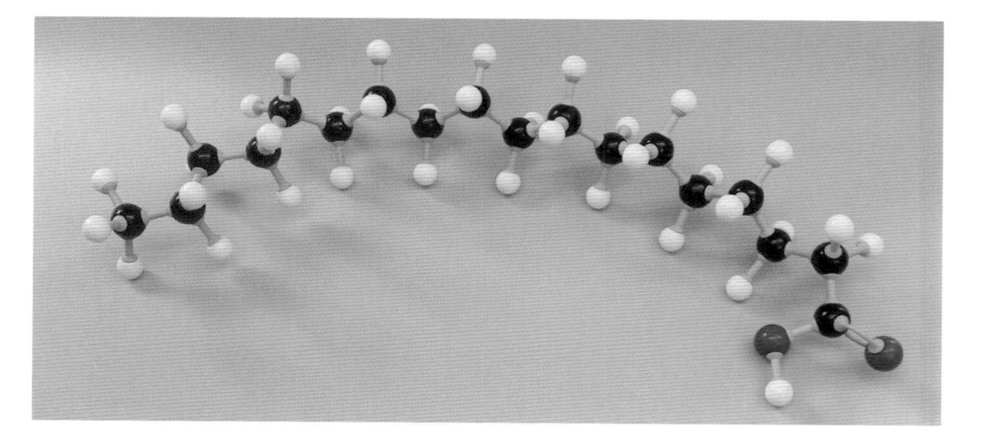

A molecule of stearic acid. It is a saturated compound. The molecule has a very regular shape.

Esters made of glycerol and saturated fats have a regular shape. They pack together easily and are solid at room temperature.

Oleic acid is typical of the acids combined in vegetable oils. There is a double bond in each molecule of this acid. Oleic acid is **unsaturated**. The double bond means that the molecules are not straight. It is more difficult to pack together molecules made of glycerol and unsaturated fats. This means that they are liquid at room temperature.

A molecule of oleic acid. The double bond means that there are carbon atoms that do not form four bonds with other atoms. Because there is not as much hydrogen as there would be with all single bonds, these molecules are 'unsaturated'.

Making soap from fats and oils

An ester splits up into an acid and an alcohol when it reacts with water. In the absence of a catalyst this is a very slow change. Chemists call this type of change hydrolysis. 'Hydro-lysis' is derived from two Greek words meaning 'water-splitting'.

$$\text{ester + water exactly} \longrightarrow \text{acid + alcohol}$$

A strong alkali, such as sodium hydroxide, is a good catalyst for the hydrolysis of esters. Hydrolysis of fats and oils by heating with alkali produces soaps. Soaps are the sodium or potassium salts of fatty acids.

Questions

1 Chemists sometimes describe fats and oils as 'triglycerides'. Why is this an appropriate name for these compounds?

2 The molecular formula of an acid can be written C_xH_yCOOH. What are the values of x and y in:

   a stearic acid?

   b oleic acid?

3 Manufacturers state that some spreads are high in polyunsaturated fats. Suggest what the term 'polyunsaturated' means.

Key words

glycerol
fats
vegetable oils
saturated
unsaturated

Fats, oils, and our health

A range of edible oils from plants

Three types of spread made from vegetable oils

The chemistry of fats and oils has triggered food scares. 'Saturated' fats and 'trans' fats are thought to be bad for people, while 'unsaturated' fats (especially polyunsaturated fats) and some 'omega' fatty acids are good. Many people use these terms with little idea of what they mean and limited understanding of the effects of these fats and fatty acids on health.

Hydrogenated vegetable oil

Margarine was originally a cheap substitute for butter. It can be made from vegetable oils that have been altered to 'harden' them so that they are solid at room temperature. The easiest way to do this is to turn unsaturated fats into saturated fats by adding hydrogen.

The first margarines were made by bubbling hydrogen through an oil with a nickel catalyst. The hydrogen hardened the oil by adding to the double bonds and turning it into a saturated fat. This is a relatively cheap and easy process. However, research in the 1960s started to show that these saturated fats could contribute to heart disease.

Trans and *cis* fats

Some margarine tubs, and bottles of vegetable oil, give information about *cis*- and *trans*-fatty acids. This refers to how the hydrogen atoms are arranged either side of the double bond.

Trans-fatty acids can form during the hydrogenation process used to make some 'hard' margarines. In the 1990s an American scientist, Walter Willet, found evidence that too much *trans*-fat could aggravate heart disease. Not all scientists agreed with him, and the research continues.

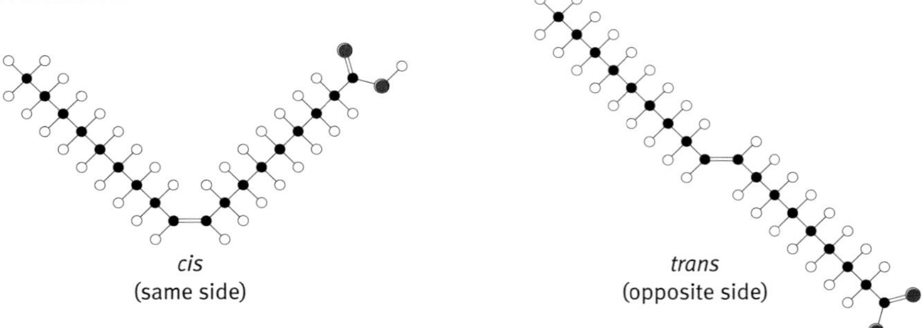

If both parts of the hydrocarbon chain are on the same side of the double bond between carbon atoms, it is *cis*. Nearly all naturally occurring unsaturated fatty acids contain *cis* double bonds. If the two parts of the chain are on the opposite sides of the double bond, it is *trans*. The shape of a *trans*-unsaturated molecule is a bit like a saturated fatty acid. They are not as runny as *cis*-molecules.

Hardening without hydrogen

Food scientists have found ways to turn vegetable oils into solid spreads without adding hydrogen. The fatty acids in vegetable oils are polyunsaturated. The molecules of these fatty acids are not straight because of all the double bonds. So they do not pack together easily.

Imagine what would happen if you could swap the fatty acid chains around so that they all stack together more neatly and make a denser material. This is exactly what modern margarine makers do – using a catalyst, they make the fatty acids rearrange themselves. David Allen works for one of the suppliers to a big supermarket chain: 'It's a bit like musical chairs in that they all change places. Instead of music you have a catalyst, or an enzyme, and the right conditions.'

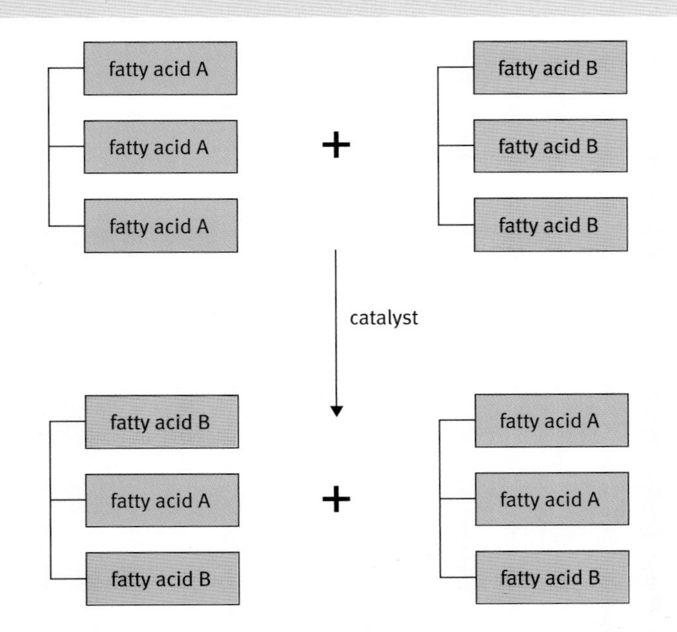

In the presence of a catalyst the fatty acids can swap places between molecules of fats and oils. This results in a mixture. The diagram shows just two of the possible products. This can raise the melting point by several degrees.

Omega-3 and omega-6 oils

Omega is the last letter in the Greek alphabet. The letter is used in a naming system which counts from the last carbon atom in the chain of a fatty acid molecule – the carbon atom furthest from the —COOH group. If you count in this way, omega-3 fatty acids have the first double bond between carbon atoms 3 and 4. Omega-6 fatty acids have the first double bond between carbon atoms 6 and 7.

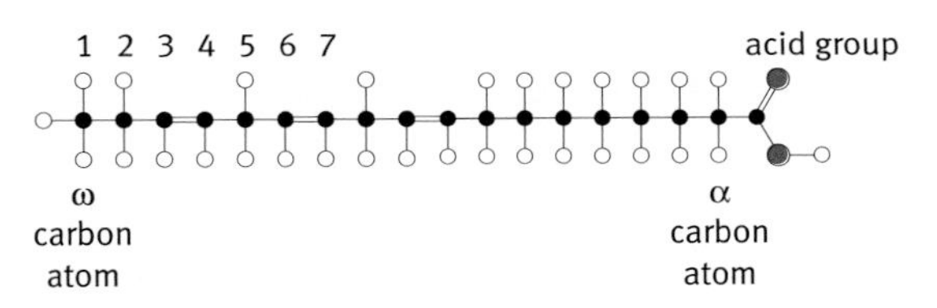

Linolenic acid is a fatty acid with three double bonds. They are all *cis*. This form of linolenic acid is an omega-3 fatty acid. Another form of linolenic acid is an omega-6 fatty acid. If you count from the carbon atom, the omega-6 acid has double bonds between carbon atoms numbered 6 and 7, 9 and 10, and 12 and 13.

Some of these omega compounds are essential fatty acids because our bodies need them but cannot make them. People need to take them in through our diet. According to Professor John Harwood of Cardiff University, it is important to have the right ratio of omega-6 to omega-3: 'You need a ratio of about three times omega-6 to one omega-3. But most people in Western countries take in far too much omega-6 and not enough omega-3 – more like a ratio of 15 to 1.'

Omega-6 fatty acids are found in margarines and most plant oils. Omega-3 acids are found in oily fish and in flax oil. 'Omega-3 can reduce pain in joints,' says Professor Harwood. 'It is also important in the developing brain of the very young and in the brains of the very old, and has implications for cardiovascular disease.'

Topic 2 How much? How fast? How far?

This topic tackles three challenging questions which scientists try to answer when explaining changes to chemicals and materials:

Chemical analysis helps to answer the question 'How much?'.

How much? How fast? How far? These questions matter to the people that control chemical processes. Instruments in the control room help them to monitor the process.

How much?

How much? refers not only to the amounts of reactants and products but also to the quantities of energy given out or taken in during a change.

How fast?

How fast? is important to anyone trying to control changes from cooks in a kitchen to chemical engineers making chemicals on a large scale. Topic 4 on pages 240–251 shows how scientists today are using their understanding of catalysts to make the chemical industry more sustainable.

How far?

How far? is also important to anyone trying to get the maximum yield from a chemical change. The study of carboxylic acids in Topic 1 (pages 184–185) showed that these are weak acids. They do ionize but only to a slight extent. Answering the question 'How far?' helps to explain why some acids, like the organic acids, are weak, while others, such as the mineral acids, are strong.

Molecular theories

Chemists explain their answers to these questions with the help of theories about the behaviour of atoms and molecules. Atoms and molecules are too small to see, so chemists use models to help develop their theories.

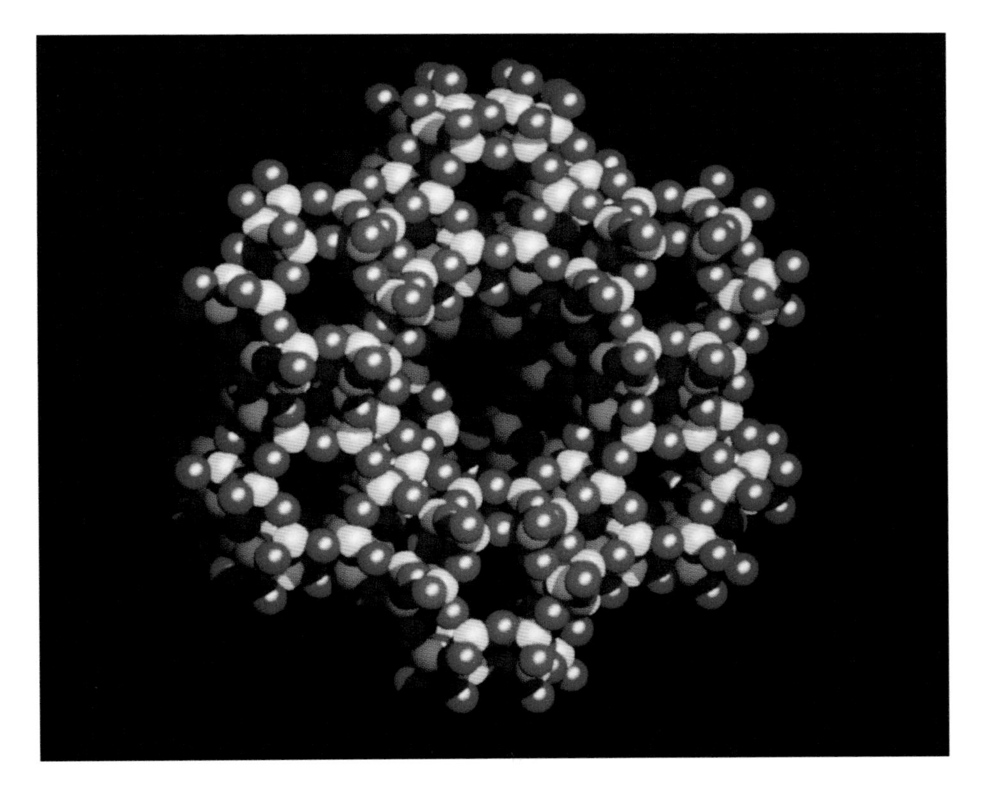

Computer graphic of a model of a zeolite crystal. The yellow atoms are either silicon or aluminium atoms. The red atoms are oxygen. Zeolites are catalysts used to control reactions in the petrochemical industry. Chemists can make synthetic zeolites with crystal structures designed to catalyse particular reactions.

Find out about:

- exothermic and endothermic changes
- energy level diagrams
- bond breaking and bond forming
- calculating energy changes

A forest fire raging in California, USA

2A Energy changes and chemical reactions

All chemical changes give out or take in energy. The study of energy and change is central to the science of explaining the extent and direction of a wide variety of changes. Understanding these energy changes also helps chemists to control reactions.

Changes that give out energy

Many reactions give out energy to their surroundings. This is obvious during burning. Most of the energy we use to keep warm, cook food, and drive machinery comes from fuels reacting with oxygen.

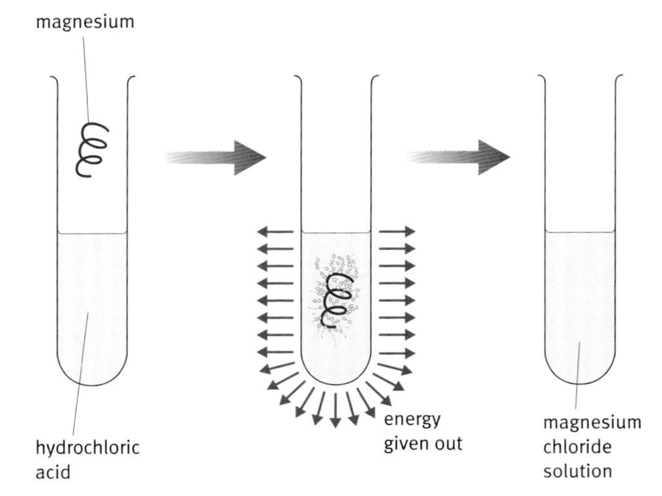

An example of an exothermic reaction

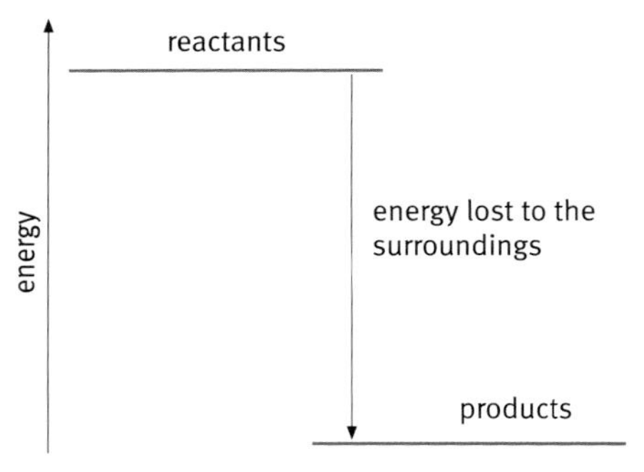

Energy level diagram for an exothermic reaction

Reactions that give out energy are **exothermic**. Respiration is an example of an exothermic change. During respiration, oxygen and glucose change to carbon dioxide and water in ways that provide the energy for growth, movement, and warmth in living things.

The reaction of magnesium with dilute hydrochloric acid is exothermic. It heats up the test tube and surrounding air.

The change can be described with the help of an **energy level diagram**. Energy is released to the surroundings. So the energy of the product, a solution of magnesium chloride, is less than the energy of the reactants (magnesium and hydrochloric acid).

Changes that take in energy

There are some changes that take in energy from their surroundings. These are **endothermic** reactions. Melting and boiling are endothermic changes of state. Photosynthesis is endothermic: plants take in energy from the Sun to convert carbon dioxide and water to glucose. This is the reverse of respiration.

Key words
exothermic
energy level diagram
endothermic

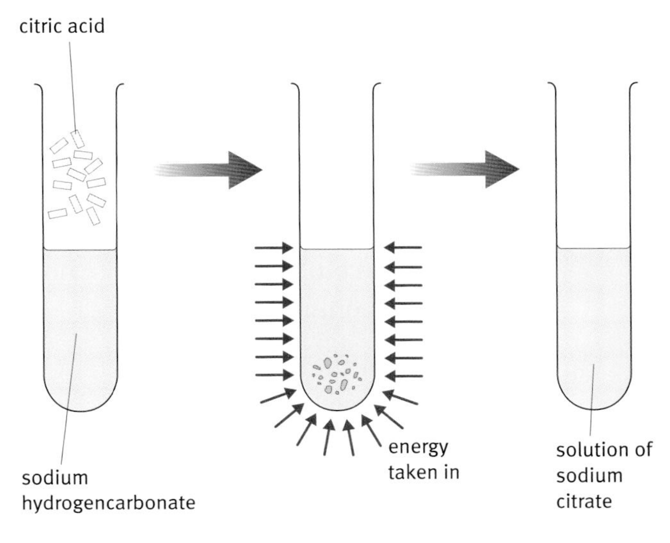

An example of an endothermic reaction

Traditional sherbet sweets were a mixture of sugar, citric acid, and sodium hydrogencarbonate. The sensation of eating sherbet mixture is a combination of the sweetness of the sugar, the acidity of citric acid, and best of all the fizzing as the acid reacts with the carbonate. This sensation is caused by the chilling effect of an endothermic reaction.

The change can also be described with the help of an energy level diagram. Energy is taken in from the surroundings. So the energy of the product, the solution of sodium citrate, is greater than the energy of the reactants (citric acid and sodium hydrogen carbonate.).

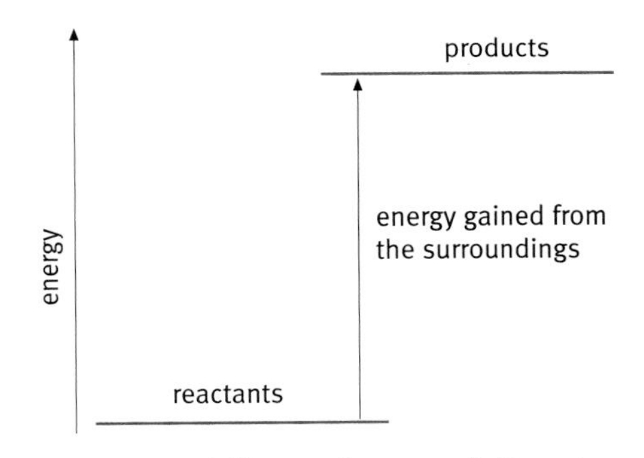

Energy level diagram for an endothermic reaction

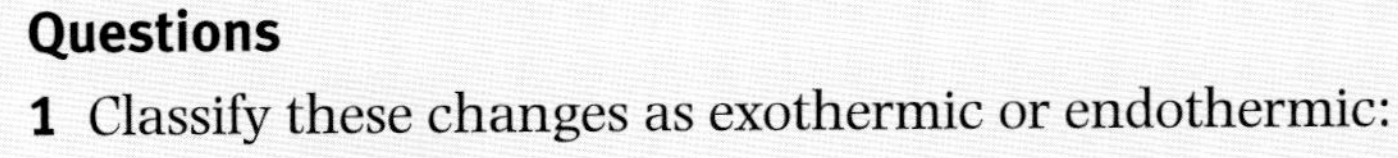

Questions

1 Classify these changes as exothermic or endothermic:

- **a** petrol burning
- **b** water turning to steam
- **c** water freezing
- **d** sodium hydroxide neutralizing hydrochloric acid

2 Turning 18 g ice into water at 0 °C requires 6.0 kJ of energy. Use an energy level diagram to show this change.

How much energy?

There is growing interest in hydrogen as a fuel because water is the only product of its reaction with oxygen. This is a highly exothermic reaction, but even so, a mixture of hydrogen and oxygen has to be coaxed into reacting. At room temperature the two gases do not react. It takes a hot flame or an electric spark to heat up the mixture enough for the reaction to start.

Filling a test car with liquid hydrogen fuel in Germany. In the background is the array of solar cells that generate the electricity to produce the hydrogen at this solar hydrogen filling station.

This is because, in all reactions, regardless of whether they are exothermic or endothermic, some of the chemical bonds in the reactants have to be broken before new chemical bonds in the products can be formed. If the reaction is sufficiently exothermic, then the energy given out keeps the mixture hot enough for the reaction to continue.

The strengths of the chemical bonds that break and form during a reaction determine the size of the overall energy change and whether it is exothermic or endothermic. Think of chemical bonds as tiny springs. In order to get hydrogen to react with oxygen, the tiny springs joining the atoms in the molecules have to be stretched and broken. This takes energy, as you will know if you have ever tried to stretch and break an elastic band.

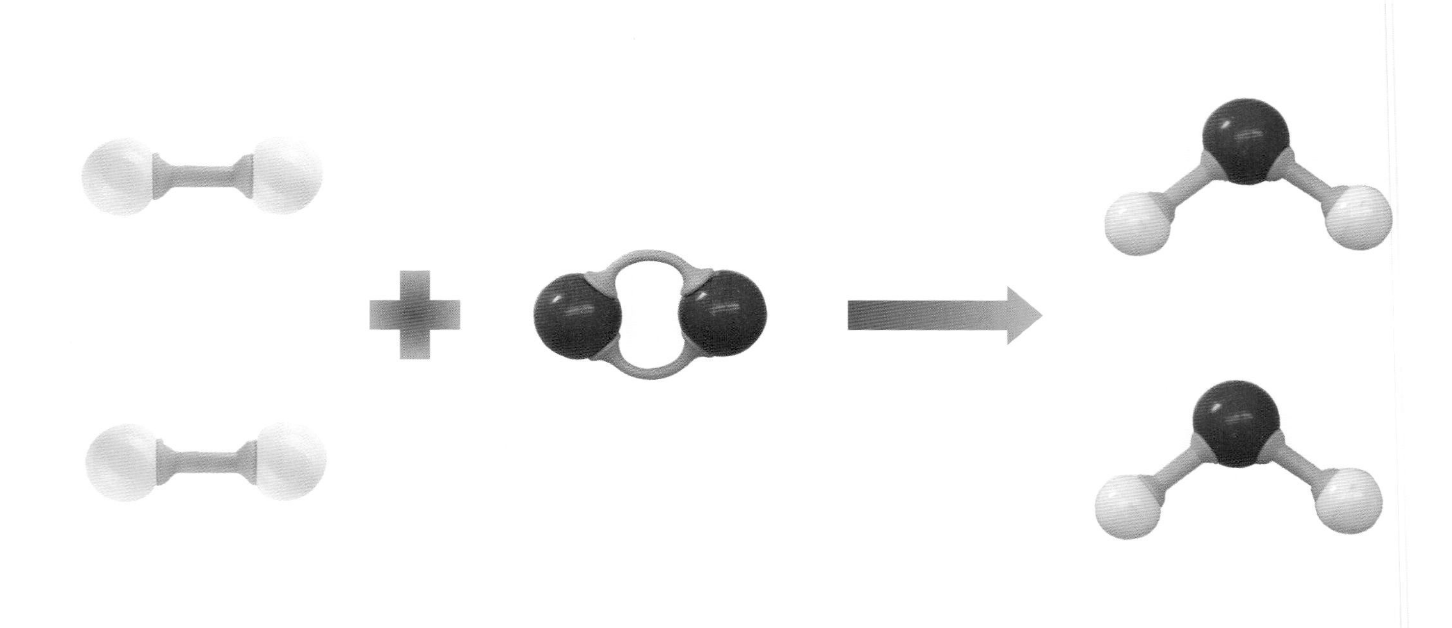

Two H—H bonds and one O=O bond break when hydrogen reacts with oxygen. The atoms recombine to make water as four new O—H bonds form.

The product is water. This is created as new bonds form between oxygen atoms and hydrogen atoms. Bond formation releases energy – just like relaxing a spring. So what decides whether a chemical reaction is exothermic or endothermic is the difference between the energy taken in to break bonds and the energy given out as new bonds form.

H

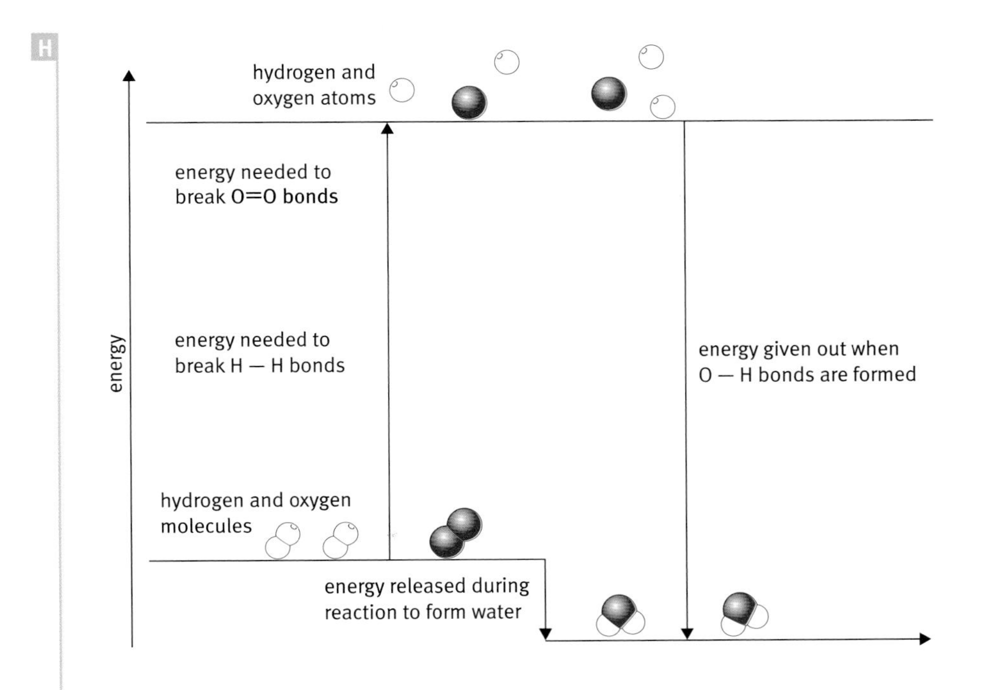

Two H—H bonds and one O=O bond break when hydrogen reacts with oxygen. The atoms recombine to make water as four new O—H bonds form.

Questions

3 Hydrogen burns in chlorine.

$$H_2(g) + Cl_2(g) \longrightarrow 2HCl(g)$$

a Which bonds break during the reaction?

b Which bonds form during the reaction?

c Use the data in the table to calculate the overall energy change for the reacting masses shown in the equation. The energy units are kilojoules (kJ).

Process	Energy change for the formula masses
breaking one H—H bond	434 kJ needed
breaking one Cl—Cl bond	242 kJ needed
forming one H—Cl bond	431 kJ given out

d Is the reaction exothermic or endothermic?

e Draw an energy level diagram for the reaction.

Find out about
- collisions between molecules
- activation energies

2B How fast?

Molecular collisions

In a mixture of hydrogen gas and oxygen gas the molecules are constantly colliding. Millions upon millions of collisions happen every second. If every collision led to a reaction, there would immediately be an explosive reaction.

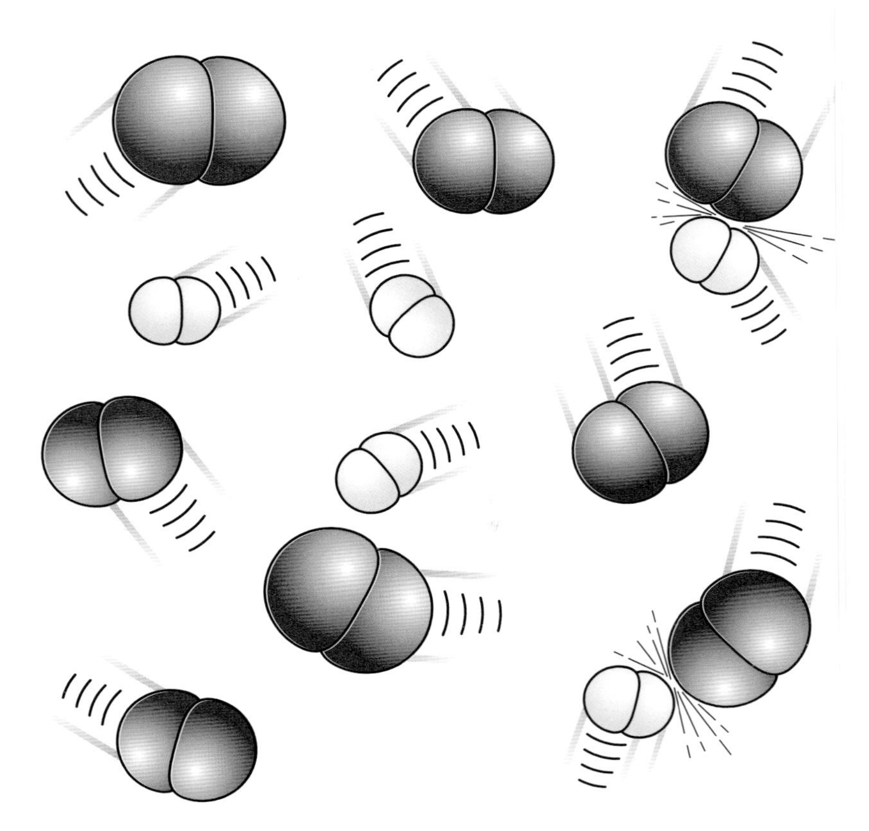

A mixture of hydrogen and oxygen molecules. The molecules that are colliding may react to form new molecules, but only if they have enough energy to start breaking bonds.

Activation energies

It is not enough for the hydrogen and oxygen atoms to collide. Bonds between atoms must break before new molecules can form. This needs energy. For every reaction, there is a certain minimum energy needed before the process can happen. This minimum energy is called the **activation energy**. It is like an energy hill that the reactants have to climb before a reaction will start. The higher the hill, the more difficult it is to get the reaction started.

The collisions between molecules have a range of energies. Head-on collisions between fast-moving particles are the most energetic. If the colliding molecules have enough energy, the collision is 'successful', and a reaction occurs.

Fast and slow reactions

The course of a reaction is like a high-jump competition. The bar is set at a height such that only a few competitors with enough energy can jump it and land safely the other side. The chemical equivalent is shown in the diagram below, where the height of the bar is represented by the activation energy and the landing area by the products of reaction.

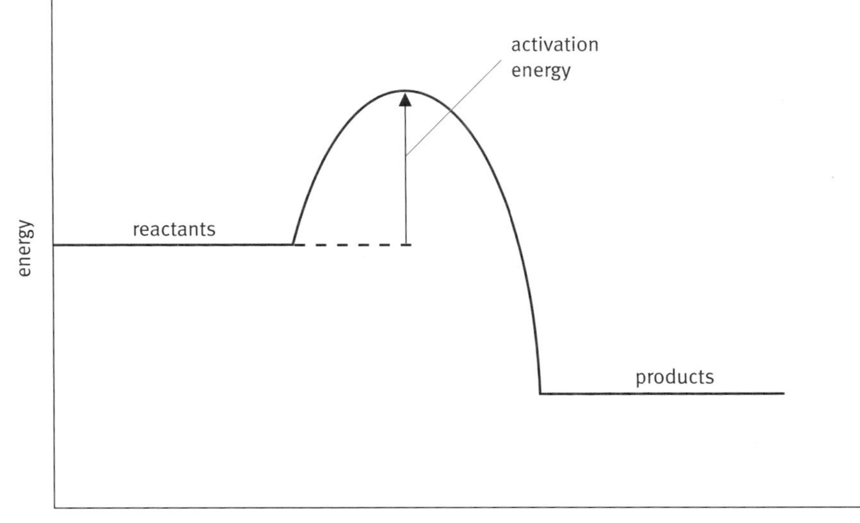

The activation energy for a reaction. The size of the activation energy is usually less than the energy needed to break all the bonds in the reactant molecules because new bonds start forming while old bonds are breaking.

If the high-jump bar is low, many competitors are successful. If it is high, the success rate is much less. In chemical reactions, if the activation energy is low, a high proportion of collisions have enough energy to break bonds, and the reaction is fast even at low temperatures.

Reactions in which the activation energy is high are very slow at room temperature, because only a small fraction of collisions have enough energy to cross the activation energy barrier. Heating the mixture to raise the temperature gives the molecules more energy. In the hot mixture more molecules have enough energy to react when they collide.

Key words

activation energy

Questions

1 a Why is a spark or flame needed to light a Bunsen burner?

b Why does the gas keep burning once it has been lit?

2 Adding a catalyst to a reaction mixture means that the activation energy for the change is lower. Suggest why this speeds up the reaction even if the temperature does not change.

The explosives expert

All chemists secretly love controlled explosions, even if they do not admit it openly. The word 'control' is important. To be useful, explosions have to be controlled. Designing and understanding that control is part of Jackie Akhavan's job.

Jackie Akhavan works for Cranfield University as an explosives chemist. Her work has applications in quarrying and mining, bomb disposal, demolition, and fireworks.

Two types of explosive

'There are two types of high explosive,' explains Jackie. 'There are primary explosives, which have low activation energies, and secondary explosives, which have higher activation energies.

'Therefore it takes less energy to initiate primary explosives. They are more sensitive to an external stimulus such as friction or impact. This makes them more dangerous to handle than secondary explosives. Secondary explosives are more difficult to initiate.'

A fireball from detonating gunpowder

Making explosives

The manufacture of old-fashioned primary explosives such as gunpowder and dynamite is very dangerous. Alfred Nobel invented dynamite. He started off by making nitroglycerine from nitric acid, glycerol, and sulfuric acid, but it was fraught with danger. In 1864 his factory blew up, killing several people, including his own younger brother, Emil. After this he combined nitroglycerine with nitrocellulose and silica in solid sticks, which was marginally safer. When Nobel died, he left the money he had made to set up the Nobel prizes.

Emulsion explosives are much safer. They are made of an emulsion of ammonium nitrate powder in a mixture of a saturated solution of ammonium nitrate, with oil and air. 'None of these materials are classed as explosives on their own, and it is only when they are mixed together that they become an explosive hazard,' explains Jackie. So the materials can be transported separately to a quarry and mixed on site.

New explosives and detonators

Jackie is a polymer chemist by training, and she has worked on making polymer-bonded explosives. These are secondary explosives and are much easier to control. 'A polymer explosive can be manufactured in a variety of forms, including sheets which resemble plasticine. The explosive can be wrapped around an old bomb or a pipe, and a detonator pushed into the sheet. On initiation, the explosive cuts the metal into two pieces.'

'Some detonators, particularly in mining, still use primary explosives because they are easy to initiate and burn to detonation, resulting in a shock wave which provides the energy to set off the main explosive charge.' Jackie is working on replacements that use a less sensitive secondary explosive.

A controlled explosion in a quarry. There is a puff of smoke, a lot of noise, and a shock wave that loosens the block of rock.

Jackie has also been involved in new ways of providing the activation energy to make detonators start explosions. This used to be done using an electrical current, but accidents can happen: 'These types of detonator are vulnerable to initiation by unwanted electromagnetic radiation from overhead pylons or thunderstorms. We've recently developed a new detonator that uses secondary explosives and is initiated with a laser pulsed through a fibre optic cable. It is safe to handle and can't be set off by unwanted electromagnetic radiation.'

Safety

Safety is always important in chemistry and is essential in Jackie's work: 'I am not allowed to detonate an explosion myself as I don't have the specific training – even though I understand the chemistry.' It is in fact illegal as well as dangerous to carry out unauthorised experiments with explosive chemicals. Jackie is well aware that it is serious business: 'I know the power of explosives and the damage they can do. I still have all my fingers, and I intend to keep it that way!'

Find out about:

- reactions that go both ways
- factors affecting the direction of change

2C Reversible changes

Some changes go only in one direction. For example, the reactions that happen to a raw egg in boiling water cannot be reversed by cooling the egg. To produce a soft-boiled egg, a cook has to check that it stays in the water for just the right amount of time. Other processes in the kitchen are easily reversed. A table jelly sets as it cools but becomes liquid again on warming. Chemists, like cooks, have to understand how to control conditions to get reactions to go far enough and in the right direction.

Burning methane in air is an example of an irreversible change. The gas burns to form carbon dioxide and water. It is then pretty well impossible to turn the products back into methane and oxygen.

Reversible changes of state

In contrast, melting and evaporating are familiar **reversible processes**. Heating turns water into steam, but water re-forms as steam condenses on cooling.

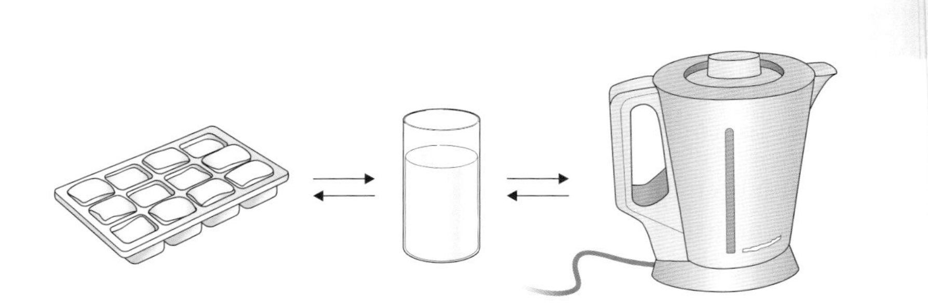

Two familiar reversible processes

Heating turns ice into water:

$$H_2O(s) \longrightarrow H_2O(l)$$

Ice reforms if water cools to 0 °C or below.

$$H_2O(l) \longrightarrow H_2O(s)$$

Combining these two equations gives:

$$H_2O(s) \rightleftarrows H_2O(l)$$

Many chemical reactions are also reversible. A reversible reaction can go forwards or backwards depending on the conditions. The direction of change may vary with the temperature, pressure, or concentration of the chemicals.

Questions

1 a Write a symbol equation to show water turning into steam.

b Write another equation to show steam condensing to water.

c Write a third equation to show the changes in **a** and **b** as a single, reversible change.

2 The pioneering French chemist Lavoisier (1743–1794) heated mercury in air and obtained the red solid mercury oxide. He also heated mercury oxide to form mercury and oxygen.

a How can both these statements be true?

b Why should you not try to repeat the experiment?

3 a Write an equation to show the reversible reaction of carbon monoxide gas with steam to form carbon dioxide and hydrogen.

b In your equation, what happens in the forward reaction?

c In your equation, what happens in the backward reaction?

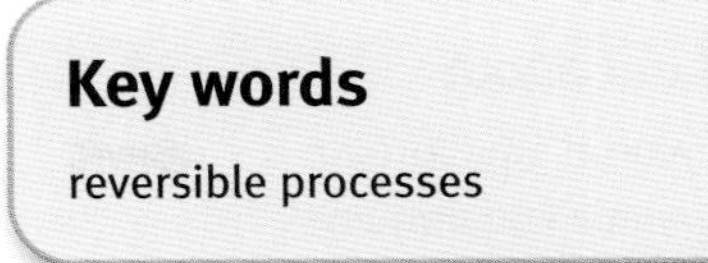

Key words

reversible processes

Temperature and the direction of chemical change

Heating decomposes blue copper sulfate crystals to give water and anhydrous copper sulfate, which is white:

$$CuSO_4.5H_2O(s) \longrightarrow CuSO_4(s) + 5H_2O(l)$$

Add water to the white powder after cooling, and it changes back into the hydrated form of the chemical. As it does so it turns blue again and gets very hot:

$$CuSO_4(s) + 5H_2O(l) \longrightarrow CuSO_4.5H_2O(s)$$

Another example of temperature affecting the direction of change is the formation of ammonium chloride. At room temperature ammonia gas and hydrogen chloride gas react to form a white solid, ammonium chloride:

$$NH_3(g) + HCl(g) \longrightarrow NH_4Cl(s)$$

Gentle heating decomposes ammonium chloride back into ammonia and hydrogen chloride:

$$NH_4Cl(s) \longrightarrow NH_3(g) + HCl(g)$$

Ammonium chloride forming where ammonia and hydrogen chloride gases meet above concentrated solutions of the two compounds on glass stoppers.

Concentration and the direction of chemical change

This equation describes the reaction between iron and steam:

$$3Fe(s) + 4H_2O(g) \longrightarrow Fe_3O_4(s) + 4H_2(g)$$

The change from left to right (from reactants to products) is the forward reaction. The change from right to left (from products to reactants) is the backward reaction.

The forward reaction is favoured if the concentration of steam is high and the concentration of hydrogen is low.

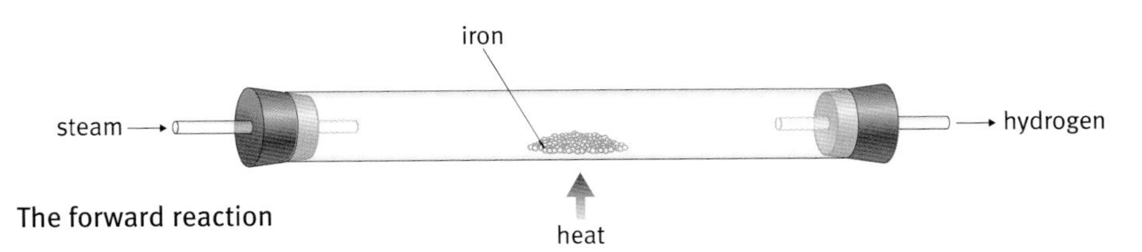

The forward reaction

The backward reaction is favoured if the concentration of hydrogen is high and the concentration of steam is low:

$$Fe_3O_4(s) + 4H_2(g) \longrightarrow 3Fe(s) + 4H_2O(g)$$

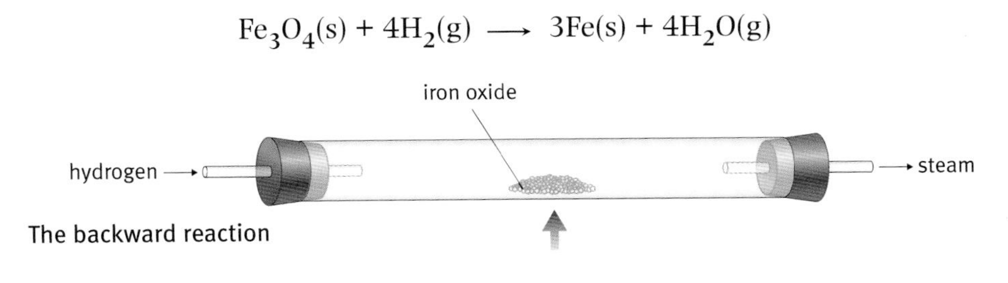

The backward reaction

Find out about

- chemical equilibrium
- dynamic equilibrium
- strong and weak acids

2D How far?

Equilibrium

Reversible changes often reach a state of balance, or **equilibrium**. A solution of litmus in water at pH 7 is purple because it contains a mixture of the red and blue forms of the indicator. Similarly melting ice and water are at equilibrium at 0 °C. At this temperature, the two states of water coexist with no tendency for all the ice to melt or all the water to freeze.

When reversible reactions are at equilibrium, neither the forward nor the backward reaction is complete. Reactants and products are present together and the reaction appears to have stopped. Reactions like this are at equilibrium. Chemists use a special symbol in equations for reactions at equilibrium: $\rightleftharpoons$

So at 0 °C,

$$H_2O(s) \rightleftharpoons H_2O(l)$$

The question 'How far?' asks where the equilibrium point is in a reaction. At equilibrium the reaction may be well to the right (mainly products), well to the left (mainly reactants), or at any point between these extremes.

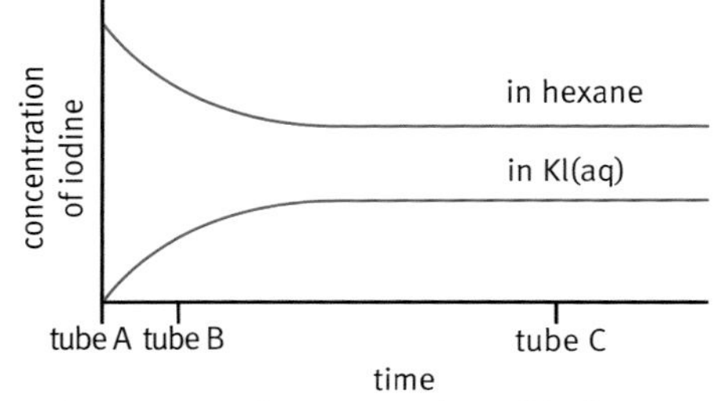

The change of concentration of iodine with time in the mixture shown in the diagram below.

Reaching an equilibrium state

A mixture of two solutions of iodine helps to explain what happens when a reversible process reaches a state of equilibrium. Iodine is slightly soluble in water but much more soluble in a potassium iodide solution in water. The solution with aqueous potassium iodide is yellow-brown. Iodine is also soluble in organic solvents (such as a liquid alkane), in which it forms a violet solution.

Aqueous potassium iodide and the organic solvent do not mix.

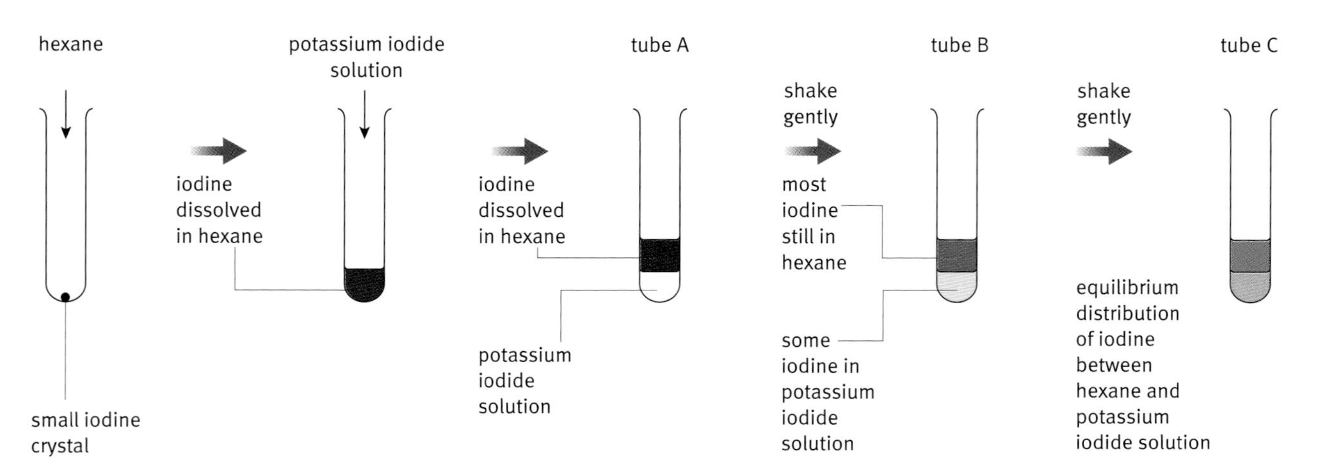

Approaching the equilibrium state starting with all the iodine in the liquid alkane

The graphs show how the iodine concentrations in the two layers change with shaking. In tube C, the iodine is distributed between the organic and aqueous layers and there is no more change. In this tube there is an equilibrium:

$$I_2(\text{organic}) \rightleftharpoons I_2(\text{aq})$$

Key words
equilibrium

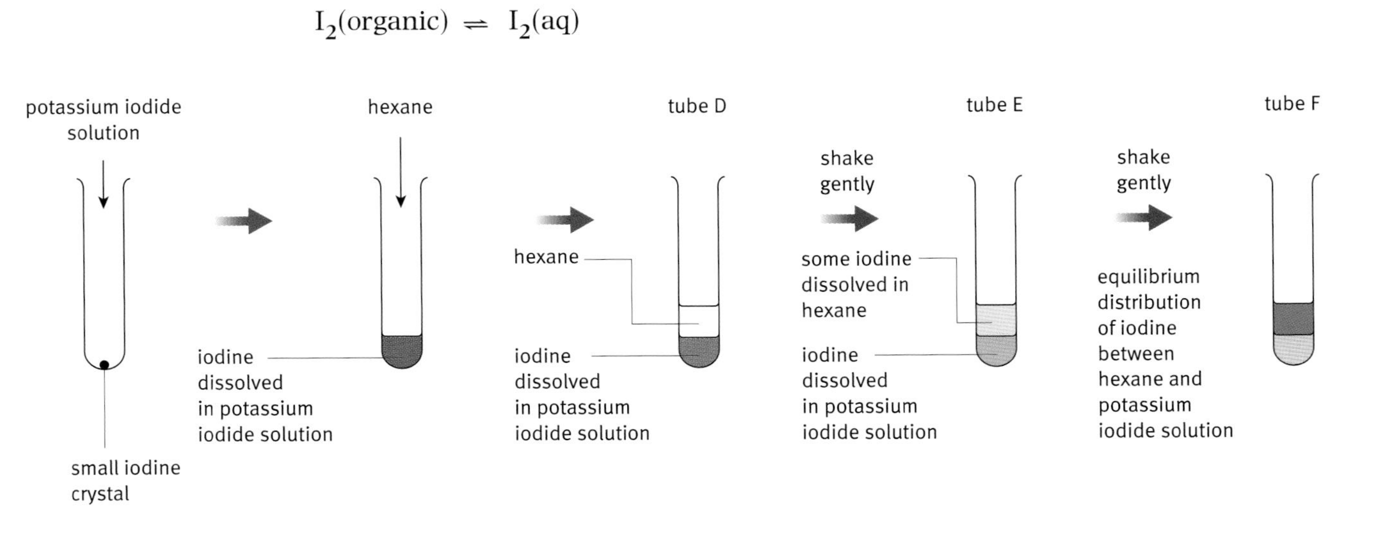

Approaching the equilibrium state starting with all the iodine in the aqueous layer

Tube F looks just like tube C. Tube F is also at equilibrium: equilibrium mixtures in the two tubes are the same. This illustrates two important features of equilibrium processes:

- At equilibrium, the concentrations of reactants and products do not change.
- An equilibrium state can be approached from either the 'reactant side' or the 'product side' of a reaction.

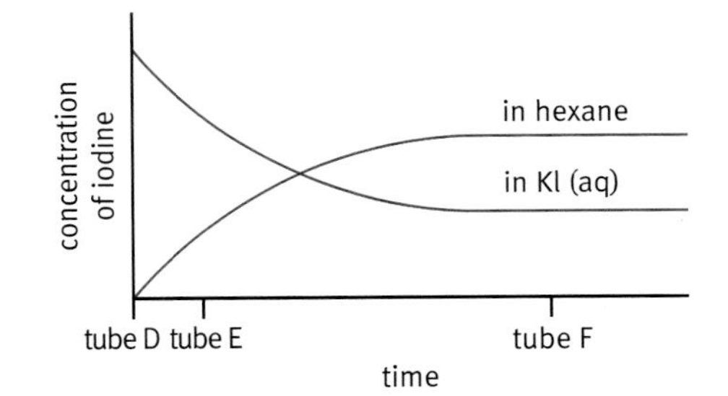

The change of concentration of iodine with time in the mixture shown in the diagram above.

Dynamic equilibrium

The diagram below gives a picture of what happens to the iodine molecules if you shake a solution of iodine in an organic solvent with aqueous potassium iodide (see tube A on page 208).

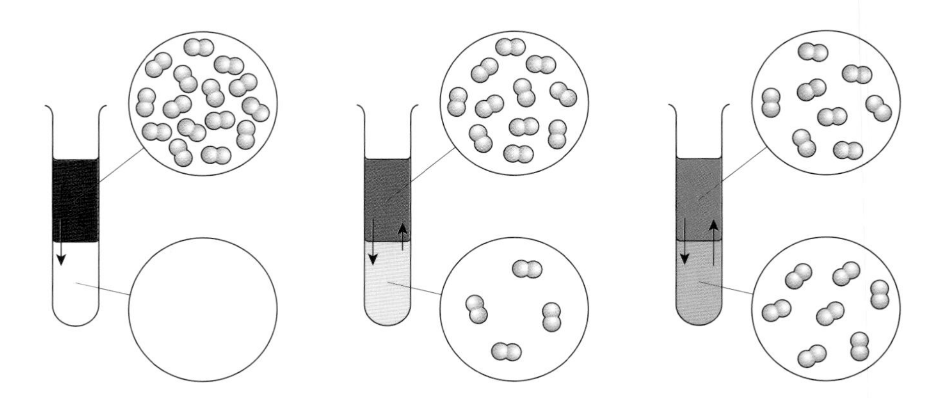

Iodine molecules reaching dynamic equilibrium between two solvents. The solvent molecules are far more numerous. They are not shown.

All the iodine starts in the upper, organic layer. At first, when the solution is shaken, movement is in one direction (the forward reaction) as some molecules move into the aqueous layer. There is nothing to stop some of these molecules moving back into the organic layer. This backward reaction starts slowly because the concentration in the aqueous layer is low. So to begin with, the overall effect is that iodine moves from the organic to the aqueous layer. This is because the forward reaction is faster than the backward reaction.

As the concentration in the organic layer falls, the rate of the forward reaction goes down. As the iodine concentration in the aqueous layer rises, the rate of the backward reaction goes up. There comes a point at which the two rates are equal. At this point both forward and backward reactions continue, but there is no overall change because each layer is gaining and losing iodine at the same rate. This is **dynamic equilibrium**.

Questions

1 Under what conditions are these in equilibrium:

a water and ice?

b water and steam?

c salt crystals and a solution of salt in water?

2 a Why do iron and steam not reach an equilibrium state when they react as shown in the upper diagram on page 207?

b Suggest conditions in which a mixture of iron and steam would react to reach an equilibrium state.

c What chemicals would be present in an equilibrium mixture formed from iron and steam?

Theories of acidity

Understanding chemical equilibrium allows chemists to control the pH in medicines and to formulate 'pH-balanced' shampoos.

Theories of acidity have come a long way since Robert Boyle (1661) gave the name 'acid' to chemicals with a sharp taste. He explained their properties by imagining spikes on the atoms.

In 1816 Humphry Davy suggested that acids all behave the same way because they contain hydrogen. Later in the same century, Svante Arrhenius took this a step further by thinking of acids as compounds which form hydrogen ions when they dissolve in water.

The Arrhenius explanation still provides a useful working definition of an acid, even though ideas about acidity moved on during the twentieth century, as you will learn if you go on to study chemistry at a more advanced level.

Key words
strong acid
weak acids
dynamic equilibrium

As well as his ideas about acidity, Humphry Davy (1778–1829) used electrolysis to discover several new elements, including sodium, potassium, magnesium, and calcium. He showed that chlorine and iodine are elements. He also invented the safety lamp for miners.

Strong acids

Hydrogen chloride gas dissolves in water to give a solution of hydrochloric acid.

$$HCl(g) + water \longrightarrow H^+(aq) + Cl^-(aq)$$

All the molecules of hydrogen chloride ionize when the gas dissolves in water (see Module C6 *Chemical synthesis*, Section C). Hydrogen chloride is a **strong acid**.

Other examples of strong acids are sulfuric and nitric acids.

Weak acids

Carboxylic acids are **weak acids**. In a dilute solution of ethanoic acid only about one molecule in a hundred ionizes. In solution there is a dynamic equilibrium:

$$CH_3COOH(aq) + H_2O(l) \rightleftharpoons CH_3COO^-(aq) + H^+(aq)$$

Questions

3 a How are the reactions of magnesium with hydrochloric acid and nitric acid:
i similar? **ii** different?

b How can the Arrhenius theory explain why solutions of hydrochloric acid and nitric acid react in a similar way with magnesium?

4 Explain what it means to say that there is a dynamic equilibrium in a solution of ethanoic acid in water.

Equilibrium at work

pH-balanced

Shampoos and other toiletries are often labelled as being 'pH-balanced', and you might have wondered what that means. Well, it is not just marketing hype; it is about the science of equilibrium.

Controlling the pH in shampoos, bath bubbles, and shower gels helps to protect the skin and eyes.

The pH of something is a measure of how acid or alkaline it is. The nearer 0 it is, the more acid, and the nearer 14, the more alkaline (see Module C6 *Chemical synthesis*, Section B). A pH of 7 means it is neutral. Some products need to be a particular pH to work effectively, and chemists have to make sure that pH does not change. This might happen, for example, if a product sits on a shop shelf for months.

pH control

Stewart Long works for Boots the Chemists as a formulations manager: 'Particular products do have to be made at specific pH values. Skin care products need to be slightly acid at around 5.5. Shampoos tend to be more alkaline.

'Hair colouring products need to be alkaline too. This makes the scales on the hair shaft open up so the colour can penetrate more easily. Conditioners are acidic and this makes the scales flatten down again making the hair shaft smooth and shiny.'

But how does Stewart formulate a particular pH and make sure the product stays that way? 'Well, we have to control what is essentially a reversible reaction, making sure there's the right concentration of H^+ ions in the solution.'

Buffer solutions

Stewart's formulations include buffer solutions. These are mixtures of molecules and ions in solution that help to keep the pH more or less constant. A buffer solution cannot prevent pH changes, but it can even out the large swings in pH that can happen without a buffer. A typical buffer solution consists of a weak acid mixed with one of its salts.

Topic 3 Chemical analysis

The business of analysis

Analytical measurement is important. It is essential to ensure that the things we use in our everyday lives do us good and not harm. Over £7 billion is spent each year on chemical analysis in the UK.

The LGC in Teddington is the UK's largest independent analytical laboratory. Its work covers food and agriculture, oil and chemicals, the environment, health care, life sciences, and law enforcement. It organizes proficiency tests to check the performance of analytical laboratories.

Health and safety

The responsibility for health and safety in a laboratory is shared by everybody. Observing these regulations helps to keep accidents down to a minimum and the risk of injury low. Laboratories have their own health and safety regulations and codes of practice. Many have health and safety officers.

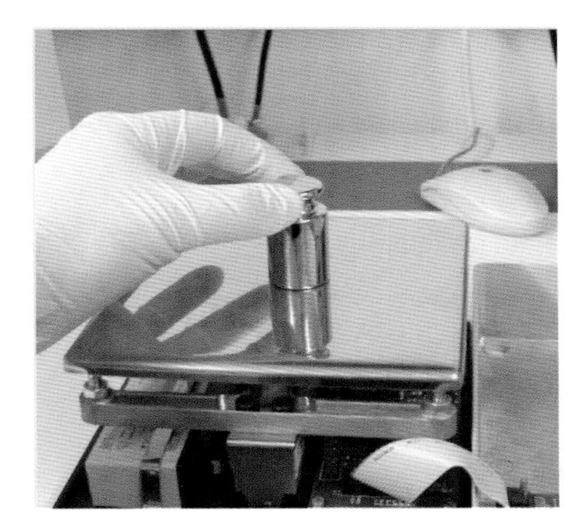

This electronic balance, like all equipment, must be maintained correctly and tested.

Looking after equipment

Equipment must be kept in good working condition. It has to be serviced regularly. Measuring instruments are checked at regular intervals.

Equipment should be cleaned properly after use and stored correctly. This is particularly important for fragile pieces of equipment such as glassware.

Accreditation

Analytical laboratories must show that they can do the job. Like the things tested in them, all laboratories must meet standards. Their standards are checked by the United Kingdom Accreditation Service (UKAS).

Analysts use proficiency tests to assess their work. Each laboratory receives identical samples to analyse. They send their results back to the organizers, who evaluate them. The laboratories are not named in the report, but results are coded so that a laboratory can recognize its results and see how well it has done.

International standards

There are international standards too. The International Olympic Committee (IOC), for example, accredits 27 laboratories to test blood and urine samples from athletes. The laboratories are all over the world and analyse 100 000 samples each year.

An analyst testing for stimulants in one of the laboratories accredited by the IOC

Find out about:

- qualitative and quantitative methods
- steps in analysis
- sampling

3A Stages in an analysis

Choosing an analytical method

The first step is to pick a suitable method of analysis. A **qualitative** method can be used if the aim is simply to find out the chemical composition of the specimen. However, if the aim is to find out how much of each component is present, then a **quantitative** approach is essential.

Analysts study samples of blood and urine to look for the presence of banned substances in athletes and other sports people. This can include qualitative analysis to identify drugs and quantitative analysis to find out how much of a banned substance there is in a sample.

Taking a sample

The analysis must be carried out on a **sample** that represents the bulk of the material under test. This can be hard to achieve with an uneven mixture of solids, such as a soil, but much easier when the chemicals are evenly mixed in solution, as in urine.

Measuring out laboratory specimens for analysis

Analysts take a sample and measure out accurately known masses or volumes of the material for analysis. It is common to carry through the analysis with two or more samples in parallel to check on the reliability of the final result. These are **replicate samples**.

Dissolving the samples

Many analytical methods are based on a solution of the specimens. With the analysis of acids or alkalis that are soluble in water, this is not a problem. It can be much more difficult to prepare a solution when analysts are working with minerals, biological specimens, or polymers.

Measuring a property of the samples in solution

When determining quantities, analysts look for a property to measure that is proportional to the amount of chemical in the sample. With an acid, for example, the approach is to find the volume of alkali needed to neutralize it. The more acid present, the greater the volume of alkali needed to neutralize it.

Calculating a value from the measurements

An understanding of chemical theory allows analysts to convert their measurements to chemical quantities. Given the equation for the reaction, and the concentration of the acid, it is possible to calculate the concentration of the acid from the volume of alkali needed to neutralize it.

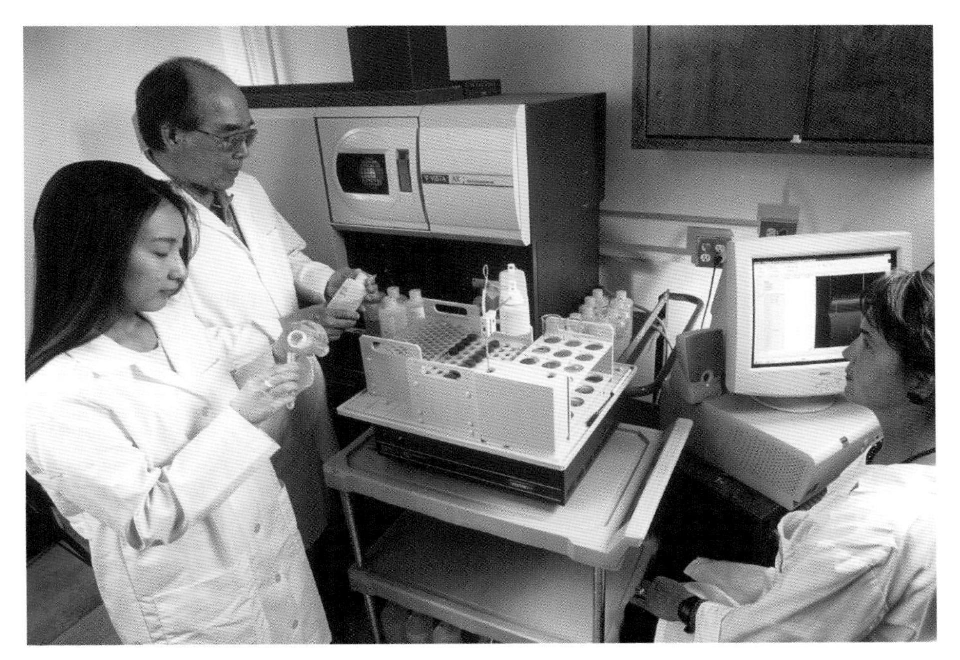

Analysts preparing blood and urine samples for analysis. They are using a type of spectrometer to measure the concentrations of iron and zinc in the samples.

Estimating the reliability of the results

Analysts have to state how much confidence they have in the accuracy of their results. Comparing the values obtained from two or three replicate samples helps.

Key words

qualitative
quantitative
sample
replicate sample

Questions

1 Who might be interested in the results of the following analyses and why?
  a The potassium ion concentration in soil.
  b The concentration of sulfur dioxide in the air.
  c The concentration of alcohol in blood.
  d The percentage by mass of haematite in a rock sample.
  e The concentration of a steroid hormone in urine.
  f The concentration of nitrates in drinking water.

3B Sampling

Analysts work with samples of materials. Rarely do they analyse the whole thing. How big the samples need to be depends on the analyses to be carried out.

Representative samples

The samples the analyst chooses must be **representative**. In other words, the samples should give an accurate picture of the material as a whole.

The composition of a homogeneous material is uniform throughout, like a dairy milk chocolate bar.

The composition of a heterogeneous material varies throughout it, like a chocolate bar made in layers.

Scientists have to decide:

- how many samples, and how much of each, must be collected to ensure they are representative of the material
- how many times an analysis should be repeated on a sample to ensure results are reliable
- where, when, and how to collect the samples of the material
- how to store samples and transport them to the laboratory to prevent the samples from 'going off', becoming contaminated, or being tampered with

Analysing water

Think about analysing water from two different sources. One is bottled water bought at a local supermarket. The other is from a local stream.

The bottled water is clear. There are no solids in suspension. It is likely to be tested for dissolved metal salts. The water is homogeneous, so only a single sample is needed. However, to check a batch of bottles, the analyst would take samples from a number of bottles. How much is needed depends on the test. There are no storage or transport problems. The bottle can be opened in the laboratory. This is a straightforward sampling problem.

Water from the stream may be cloudy. It may contain small creatures. It may be tested for a range of things, from the concentrations of dissolved chemicals to the number and variety of living organisms. Samples may vary from one part of the stream to another. They are likely to be heterogeneous. The time of year when samples are collected will have an effect on the water's composition. Also, samples need to be stored and taken back to the laboratory for analysis. This is a complex sampling problem that needs careful planning.

Key words

representative sample

Questions

1. Suggest how an analyst should go about sampling when faced with the following problems? In each case, identify the difficulties of taking representative samples. Suggest ways of overcoming the difficulties.
   - **a** Measuring the concentration of chlorine disinfectant in a swimming pool.
   - **b** Checking the purity of citric acid supplied to a food processor.
   - **c** Detecting banned drugs in the urine of athletes.
   - **d** Monitoring the quality of aspirin tablets made by a pharmaceutical company.
   - **e** Determining the level of nitrates in a farmer's field.
2. Why is the way that samples are stored important?

Collecting samples

A life of grime

All local authorities have Environmental Health Officers. Most of their work involves checking restaurants, cafes, and food shops, but they do many other things as well, including checking pollution levels. The photograph shows Ralph Haynes of Camden Council in London taking soil samples.

Ralph Haynes taking a soil sample in Camden

Ralph says: 'There is concern that chromium salts from an old metal plating factory nearby may have contaminated the soil. I am taking 1 kg samples and putting them in plastic containers. I am taking care to label it properly. There is a British Standard on labelling, you know: it's called BS5969.

'I'm taking care to take a sample from where I can actually see the change in the soil. I'm also going to take a sample from where I can't see it. Then I'm going to take samples from anywhere where I think people could be at risk – such as gardens where children play.

'Soil samples will keep for a while, but I'm also going to take water samples, and these must get to the lab within a couple of days.'

Sporting samples

Sports men and women are often asked to provide urine samples to check if they have been taking drugs. Sometimes these cases hit the headlines, and allegations have been made in the past that urine samples have been tampered with.

The scientists who carry out the analysis at Kings College, London, have to be sure that they have the right sample and that it has been correctly stored and labelled.

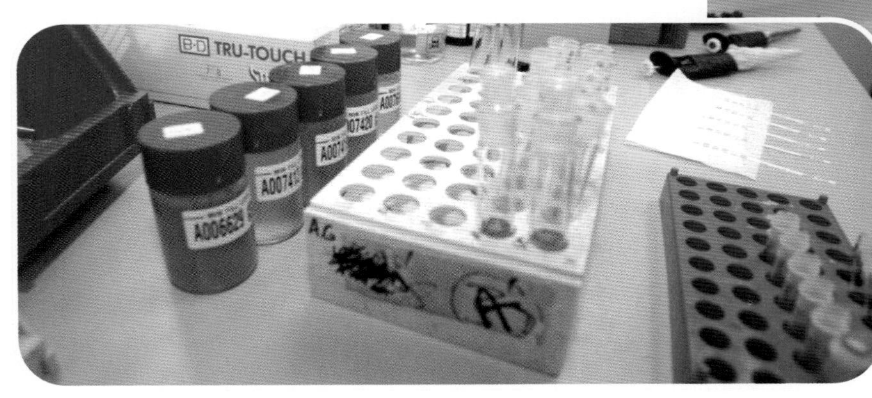

Labelled urine samples from athletes ready for testing

- First, the athlete has to produce the sample in front of a testing officer, who has to actually see the urine leaving the athlete's body and ensure there has been no cheating.
- With the testing officer watching, the athlete is allowed to pour the sample into two bottles. They seal the bottles themselves so that they feel assured no one else has tampered with them.
- The bottles are labelled with a unique code rather than the athlete's name, so the lab does not know the identity of the athlete. The bottles are sent to the lab by courier in secure polystyrene packaging.
- At the lab, one bottle is analysed immediately and the other stored in the freezer in case there is a query at a later date.

'We send the results to the Sports Council,' says Richard Caldwell, one of the analysts. 'It is someone at the Council who tallies up which bottle was collected from which person. It's quite interesting when we have a positive result and we find out from the press a few months later who it was!'

Find out about:

- principles of chromatography
- paper chromatography
- thin-layer chromatography

3C Chromatography

There are several types of **chromatography.** At the cheap-and-simple end is paper chromatography, which can be done with some blotting paper and a solvent. At the expensive end is gas chromatography, which involves high-precision instruments. All types of chromatography, however, work on similar principles.

Chromatography can be used to:

- separate and identify the chemicals in a mixture
- check the purity of a chemical
- purify small samples of a chemical

Principles of chromatography

In chromatography a **mobile phase** moves through a medium called the **stationary phase**. The analyst adds a small sample of the mixture to the stationary phase. As the mobile phase flows along, the chemicals in the sample move through the stationary phase. Different chemicals move at different speeds, so they separate.

- The chemical moves quickly if the position of equilibrium favours the mobile phase.
- The chemical moves slowly if the position of equilibrium favours the stationary phase.

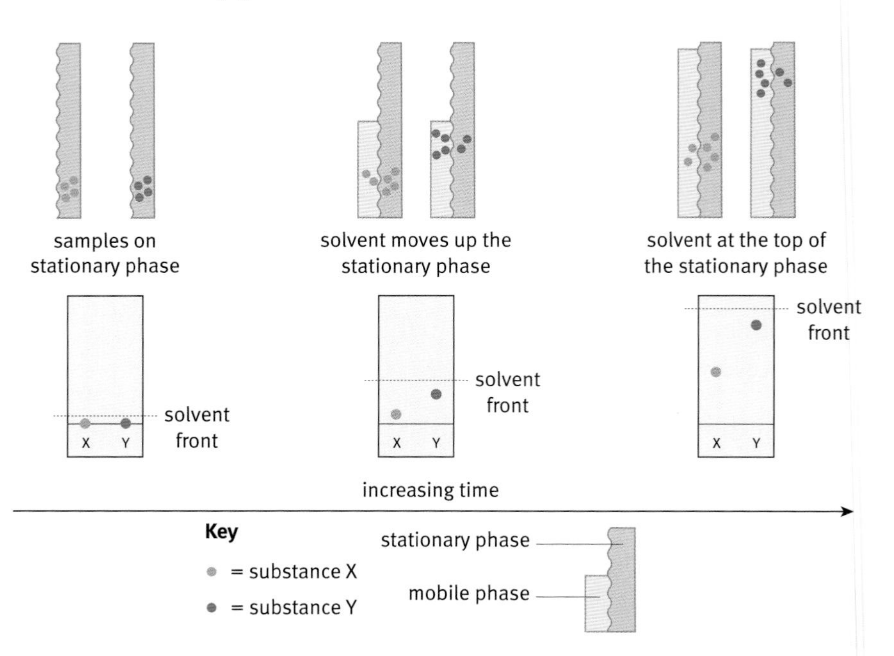

For each chemical there is a dynamic equilibrium for the molecules as they distribute themselves between the two phases. How quickly substances move through the stationary phase depends on the position of equilibrium.

Paper and thin-layer chromatography

Paper chromatography and thin-layer chromatography (TLC) are used to separate and identify substances in mixtures. The two techniques are very similar.

These techniques do not require expensive instrumentation but are limited in their use. Paper chromatography is very rarely used. Thin-layer chromatography (TLC) is 'low technology', but it can be useful before moving on to more complex techniques. TLC is quick, cheap, and only requires small volumes of solution. A large number of samples can be run at once.

TLC is simple and quick, so it is often used to monitor the progress of organic reactions and to check the purity of products.

Forensic laboratories may use TLC to analyse dyes extracted from fibres and when testing for controlled drugs, cannabis in particular.

Stationary and mobile phases

In paper chromatography the stationary phase is the paper, which does not move. In TLC the stationary phase is an absorbent solid supported on a glass plate or stiff plastic sheet.

In both paper chromatography and TLC the mobile phase is a solvent, which may be one liquid or a mixture of liquids. Substances are more soluble in some solvents than others. For example, some substances dissolve well in water, while others are more soluble in petrol-like, hydrocarbon solvents. With the right choice of solvent it becomes possible to separate complex mixtures.

Chemists call solutions in water **aqueous** solutions. The term **non-aqueous** describes solvents with no water in them.

Key words

chromatography
mobile phase
stationary phase
aqueous
non-aqueous

Questions

1 Name two substances that dissolve better in water than in hydrocarbon (or other non-aqueous) solvents.

2 Name two substances that dissolve better in non-aqueous solvents than in water.

Chromatography plates must be spotted carefully. Small, concentrated spots are needed. Their starting position should be marked.

Preparing the paper or plate

The sample is dissolved in a solvent. This solvent is not usually the same as the mobile phase.

A small drop of the solution is put on the paper, or TLC plate, and allowed to dry, leaving a small 'spot' of the mixture.

If the solution is dilute, further drops are put in the same place. Each is left to dry before the next is added. This produces a small spot with enough material to analyse. The separation is likely to be poor if the spot spreads too much.

One way of identifying the chemicals in the sample is to add separate spots of solutions of substances suspected of being present in the unknown mixture. These are called **reference materials**.

Running the chromatogram

The analyst adds the chosen solvent (the mobile phase) to a chromatography tank and covers it with a lid. After the tank has stood for a while, the atmosphere inside becomes saturated with solvent vapour.

The next step is to place the prepared paper or TLC plate in the tank, checking that the spots are above the level of the solvent.

The solvent immediately starts to rise up the paper or plate. As the solvent rises, it carries the dissolved substances through the stationary phase. Covering the tank ensures that the solvent does not evaporate.

The paper, or TLC plate, is taken from the tank when the solvent gets near the top. The analyst then marks the position of the **solvent front**.

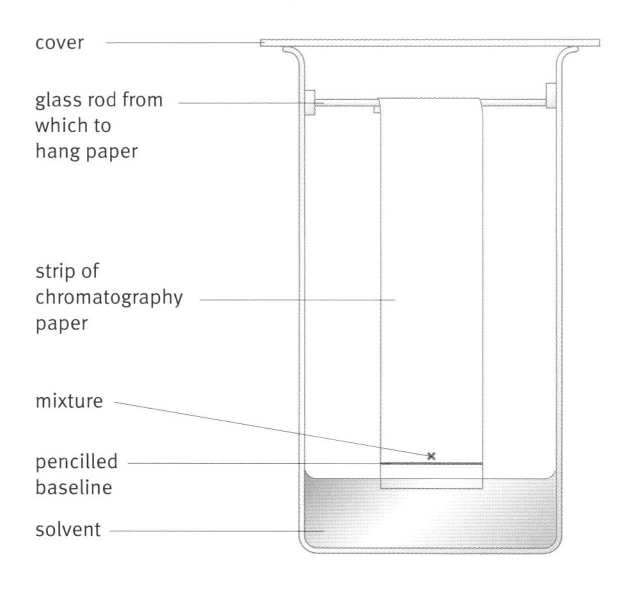

A chromatography tank. The sample spots on the paper or plate must be above the level of the solvent.

Locating substances

There is no difficulty marking the positions of coloured substances. All the analyst has to do is outline the spots in pencil and mark their centres before the colour fades.

There are two ways to locate colourless substances:

- Develop the chromatogram by spraying it with **a locating agent** that reacts with the substances to form coloured compounds.
- Use an ultraviolet lamp with TLC plates that contain fluorescers, so that the spots appear violet in UV light.

Key terms

reference materials
solvent front
locating agent
retardation factor

Interpreting chromatograms

Chemicals may be identified by comparing spots with those from standard reference materials.

A chemical may also be identified by its **retardation factor** (R_f). This does not change, provided the same conditions are used. It is calculated using the following formula by measuring the distance travelled by the substance:

$$R_f = \frac{\text{distance moved by chemical}}{\text{distance moved by solvent}} = \frac{y}{x}$$

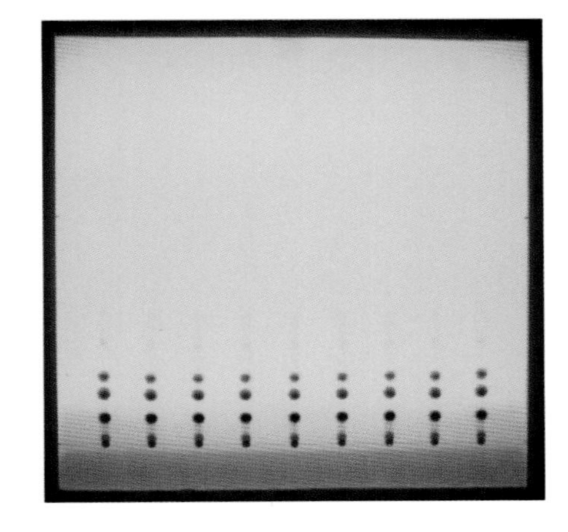

'Invisible' spots can often be seen under a UV lamp.

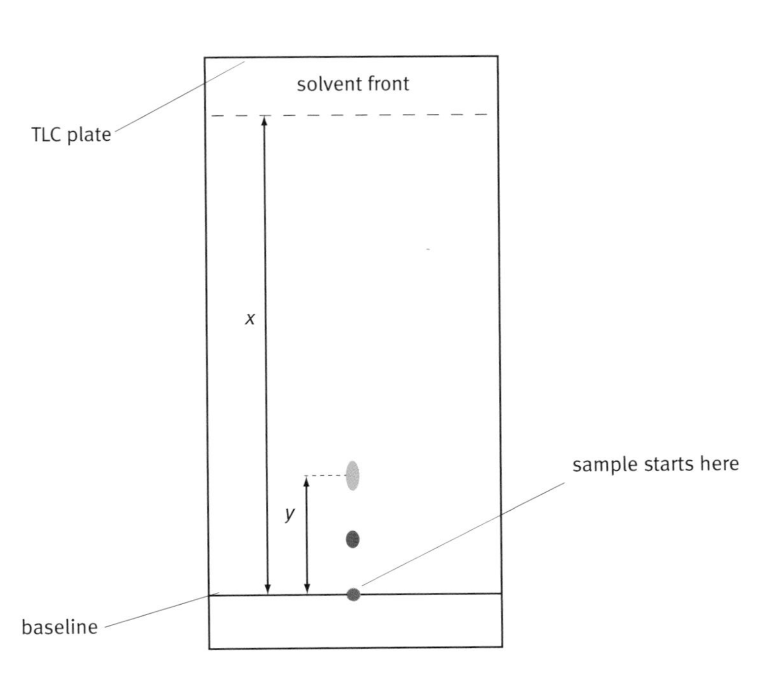

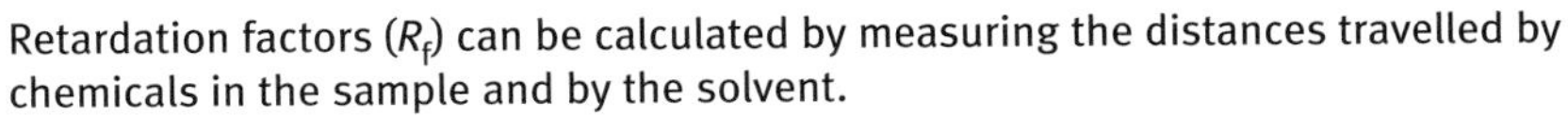

Retardation factors (R_f) can be calculated by measuring the distances travelled by chemicals in the sample and by the solvent.

Questions

3 Why is it sometimes necessary to 'develop' a chromatogram? How can this be done?

4 Why is it sometimes useful to use thin-layer chromatography plates that have been impregnated with fluorescers?

Improving performance

Thin-layer chromatography usually gives better separation than paper chromatography. Improved separation might be achieved by:

- using a different solvent for the mobile phase
- leaving the chromatogram to develop for longer (though this is limited by how far the solvent front can travel)

An estimate of how much of a substance is present may be made from the intensity of its coloured spot. Sometimes the spot is removed from the chromatogram with a solvent and the intensity of the coloured solution formed matched against reference solutions.

There are instruments that turn TLC into a quantitative technique. The TLC plate is inserted into a spectrophotometer, which measures the intensities of spots. The quantities of each chemical in the mixture can then be calculated.

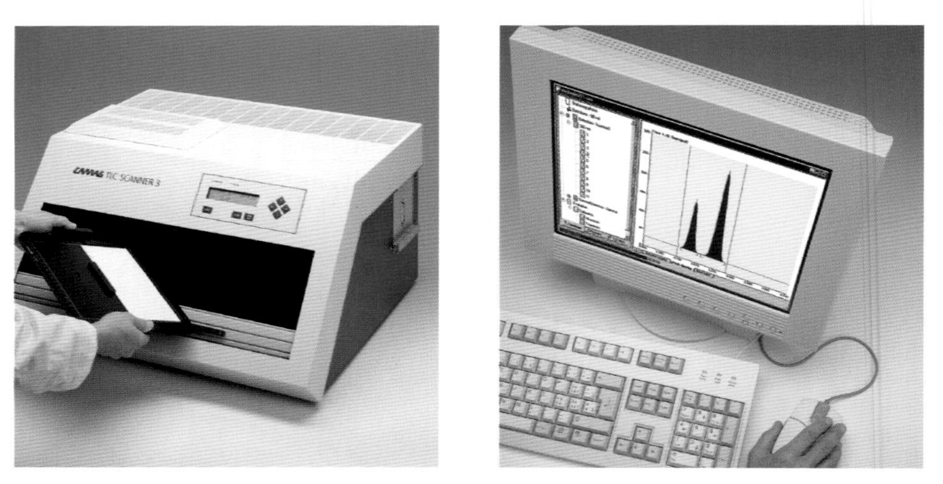

A spectrophotometer can detect and measure the intensities of coloured or colourless spots on a TLC plate.

Questions

5 Paper chromatography is used to separate a mixture of a red and a blue chemical. The blue compound is more soluble in water while the red chemical is more soluble in the non-aqueous chromatography solvent. Sketch a diagram to show the chromatogram you would expect to form.

6 How can quantitative data be obtained from thin-layer chromatograms?

7 Suppose you were testing a white powder that was a mixture of several compounds. Describe how you would set about identifying the compounds using either paper chromatography or thin-layer chromatography.

Faster cheese making

In cheese making, the enzyme rennet coagulates the milk to make solid casein. As the cheese matures, enzymes break up some of the proteins into their amino acids. This gives the cheese its characteristic flavour and texture. The trouble is that the ripening process takes several weeks, and time is money to manufacturers. Asana Nemat Tollahi, a PhD student at London Metropolitan University, is studying the ripening process of Iranian Brined Cheese, which is similar to Greek Feta.

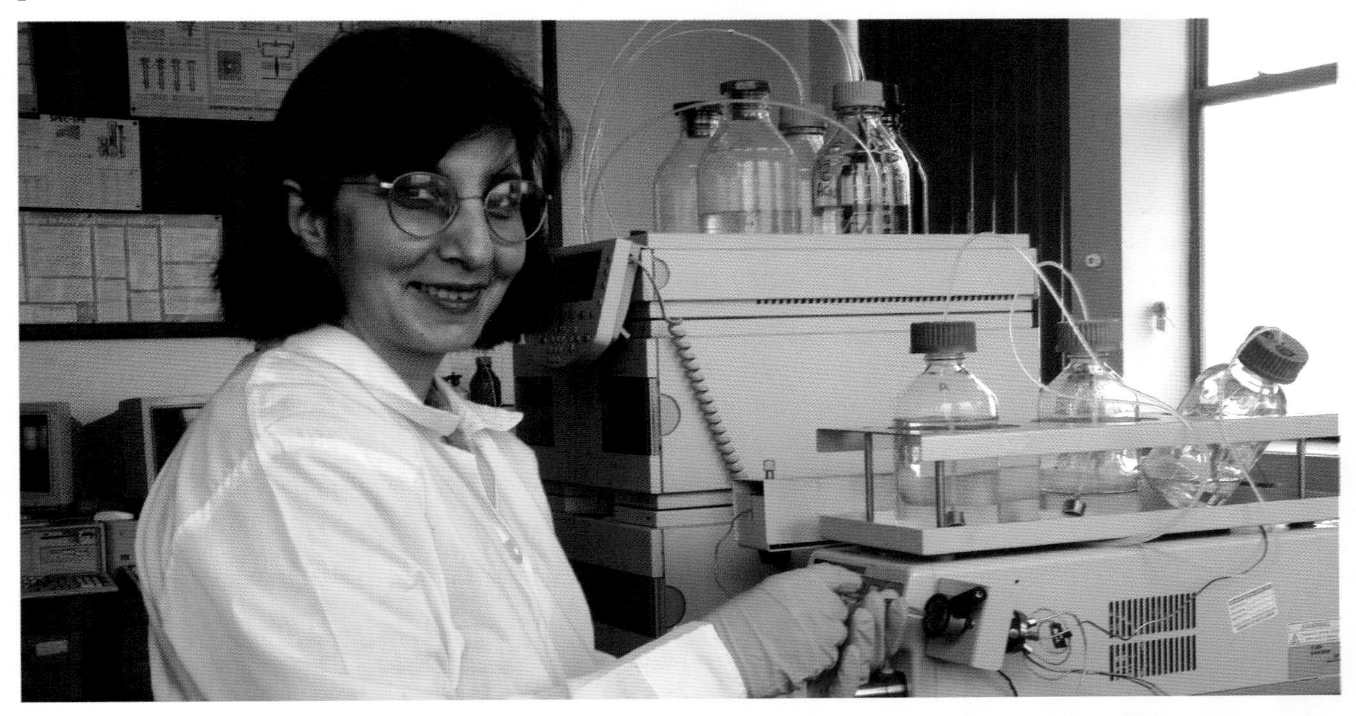

Asana injecting a sample for analysis into her liquid chromatography instrument

Asana takes samples of the cheese at different stages of its ripening. First she has to get the amino acids into solution by pulverizing the cheese, and mixing it with water. Then she injects a sample of this solution into a high-performance liquid chromatography machine. This is a chromatography technique for separating chemicals in solution. It is widely used in the food industry and in food research. The stationary phase is a solid, as in TLC, but instead of being spread on a plate, it is enclosed in a long, thin metal tube called a column. As in TLC, the mobile phase is a liquid. The liquid is pumped at high pressure through the column to separate the chemicals in a mixture.

Asana separating the curds while making cheese

Asana uses the results to determine which amino acids are present and in what quantities. She is hoping that this will help to find new ways to make the enzymes speed up the reaction.

Find out about:
- gas chromatrography
- retention times

3D Gas chromatography

Gas chromatography (GC) is used to separate complex mixtures. The technique separates mixtures much better than paper or thin-layer chromatography.

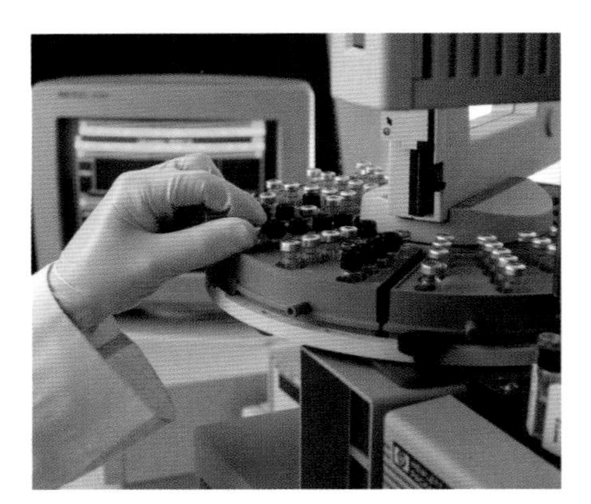
This scientist uses GC to check the quality of water from a river.

This technique is also more sensitive than paper or thin-layer chromatography, which means it can detect small quantities of compounds. That is why it is usually preferred to paper or thin-layer chromatography. The technique not only identifies the chemicals in a mixture but can also measure how much of each is present.

Understanding the limits of detection for a technique can be very important, otherwise an analyst can report that a contaminant is absent when it is in fact present but at too low a concentration to be detected. Careful research has been necessary to find out the detection limits for such chemicals as pesticide residues in food.

Stationary and mobile phases

The principles of GC are the same as for paper and thin-layer chromatography. A mobile phase carries a mixture of compounds through a stationary phase. Some compounds are carried through more slowly than others. This is because they have different boiling points or a greater attraction for the stationary phase. Because they travel at different speeds, the compounds can be separated and identified.

The mobile phase is a gas such as helium. This is the **carrier gas**.

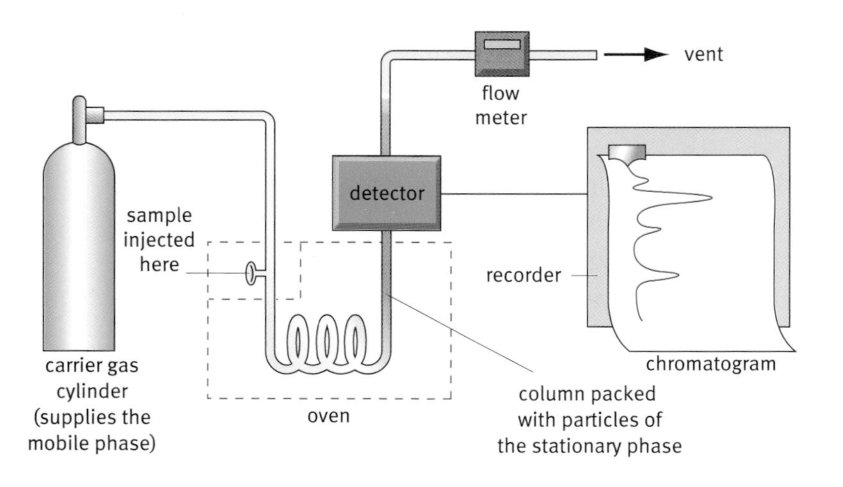

A carrier gas takes the mixture through the column containing the stationary phase. As compounds in the mixture come out the other end, they are detected and recorded.

The stationary phase is a thin film of a liquid on the surface of a powdered solid. The stationary phase is packed into a sealed tube, which is the column. The column is long and thin. Some columns are 25 m long but only 0.25 mm in diameter.

Only very small samples are needed. The analyst uses a syringe to inject a tiny quantity of the sample into the column. Samples are generally gases or liquids.

The column is coiled inside an oven, which controls the temperature of the column. This means that it is possible to analyse solids if they can be injected in solution and then turn to a vapour at the temperature of the column.

Separation and detection

Once the column is at the right temperature, the carrier gas is turned on. Its pressure is adjusted to get the correct flow rate through the column. The analyst injects the sample at the start of the column where it enters the oven. The chemicals in the sample turn to gases and mix with the carrier gas. The gases pass through the column.

In time, the chemicals from the sample emerge from the column and pass into a detector. The chemicals can be identified using mixtures of known composition.

Interpreting chromatograms

The detector sends a signal to a recorder when a compound appears. A series of peaks, one for each compound in the mixture, make up the chromatogram. The position of a peak is a record of how long the compound took to pass through the column. This is its retention time. The height of the peak indicates how much of the compound is present.

Key terms

carrier gas

retention time

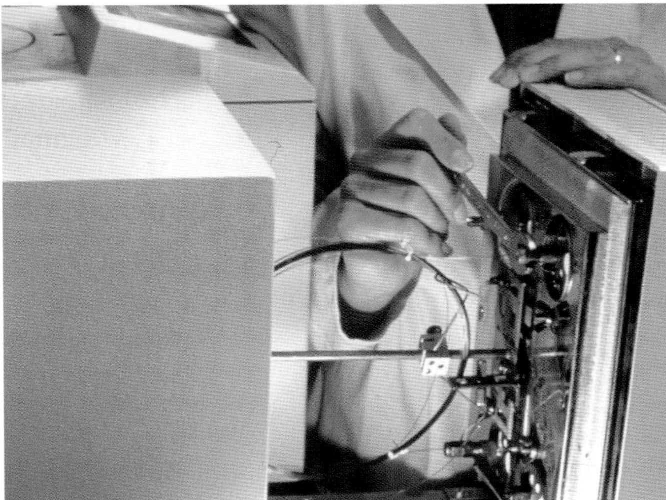

The coiled column inside the oven of a GC instrument. The detector is connected to a computer and the chromatogram appears on the screen.

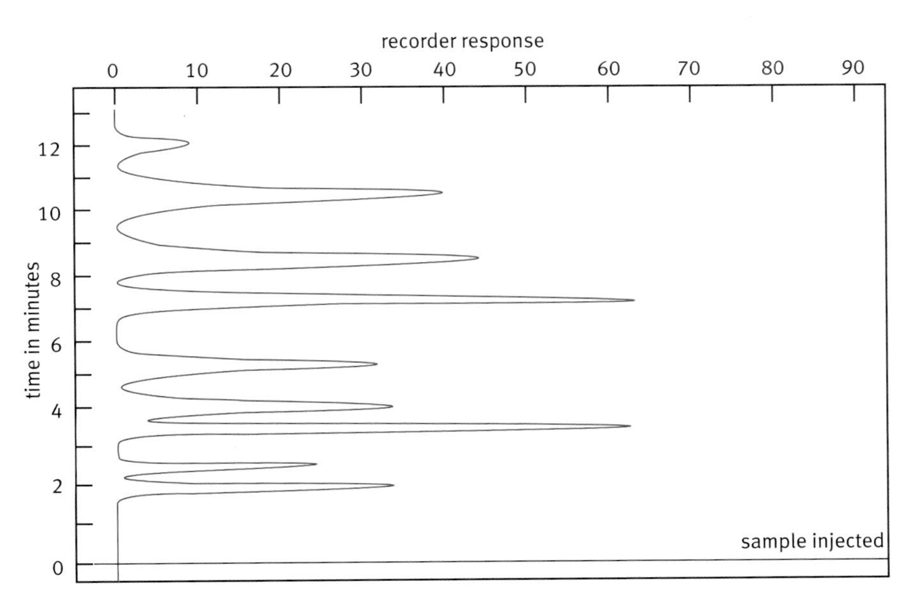

Each compound in the mixture appears as a peak in the chromatogram. The time it takes to get through helps the scientist to identify it. The height of the peak enables the scientists to say how much there is.

Questions

1 Look at the chromatogram on this page.

- **a** How many components have been separated?
- **b** Estimate the retention time of each component.
- **c** Which component was present in the largest quantity and which one was present in the smallest quantity?

Chemistry in action

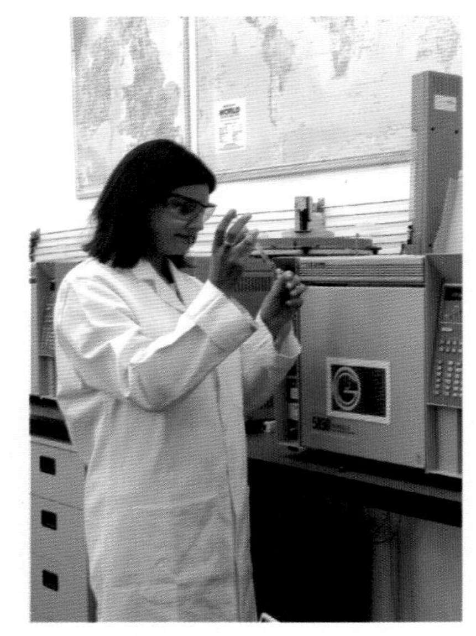

Analytical chemist Anna Mukherjee carrying out a GC analysis at the University of Bristol

Chemical archaeology

Richard Evershed is a professor at Bristol University. He has used gas chromatography all his career: 'During my PhD, I used gas chromatography to study the chemical messages insects use to communicate with. After my PhD, I moved to the University of Bristol, where I used gas chromatography and high-performance liquid chromatography to look at the organic chemicals preserved in ancient rocks originating from organisms that lived many millions of years ago.'

Ancient traces

Richard recognized that gas chromatography could also be used to identify the remains of fats, waxes, and resins preserved at archaeological sites. 'We used gas chromatography linked to mass spectrometry for the first time in archaeology to determine the origin of samples of wood tar from King Henry VIII's flagship, the *Mary Rose*.'

Richard has been studying very old cooking pots to find out what people ate in the past. He analyses organic residues trapped in the walls of unglazed vessels. The residues are often mixtures of fats and waxes. He uses gas chromatography to separate the chemicals in the mixtures. By gradually increasing the temperature of the gas chromatography column, he is able to separate compounds with different boiling points. When the temperature reaches the boiling point of a chemical, it turns into gas and is carried by the carrier gas to a flame, where it burns to produce an electrical signal. The separated compounds appear as a series of peaks on the chromatogram.

'We've identified degraded residues of animal fats in pots from Turkey going back 8000 years,' says Richard. 'We've also found plant oils, animal fats, and beeswax in ancient lamps. We made a real breakthrough when we found that we could identify traces of butter in 6000-year-old pottery from prehistoric Britain. This showed us that milking animals is a very ancient practice.'

Analysis of the chemicals absorbed into old pots gives clues to the food that our ancestors cooked.

A whiff of cabbage

Traces of cooked cabbage can survive for a long time. Richard found this to be true when investigating a set of pots dating from late Saxon times. The pots are over a thousand years old. 'I've found traces of cabbage preserved in the pot wall. It's the natural wax you can see on the surface of the cabbage, which is released during boiling. We can extract the same waxes from modern supermarket cabbages, and the gas chromatography traces look pretty much identical.'

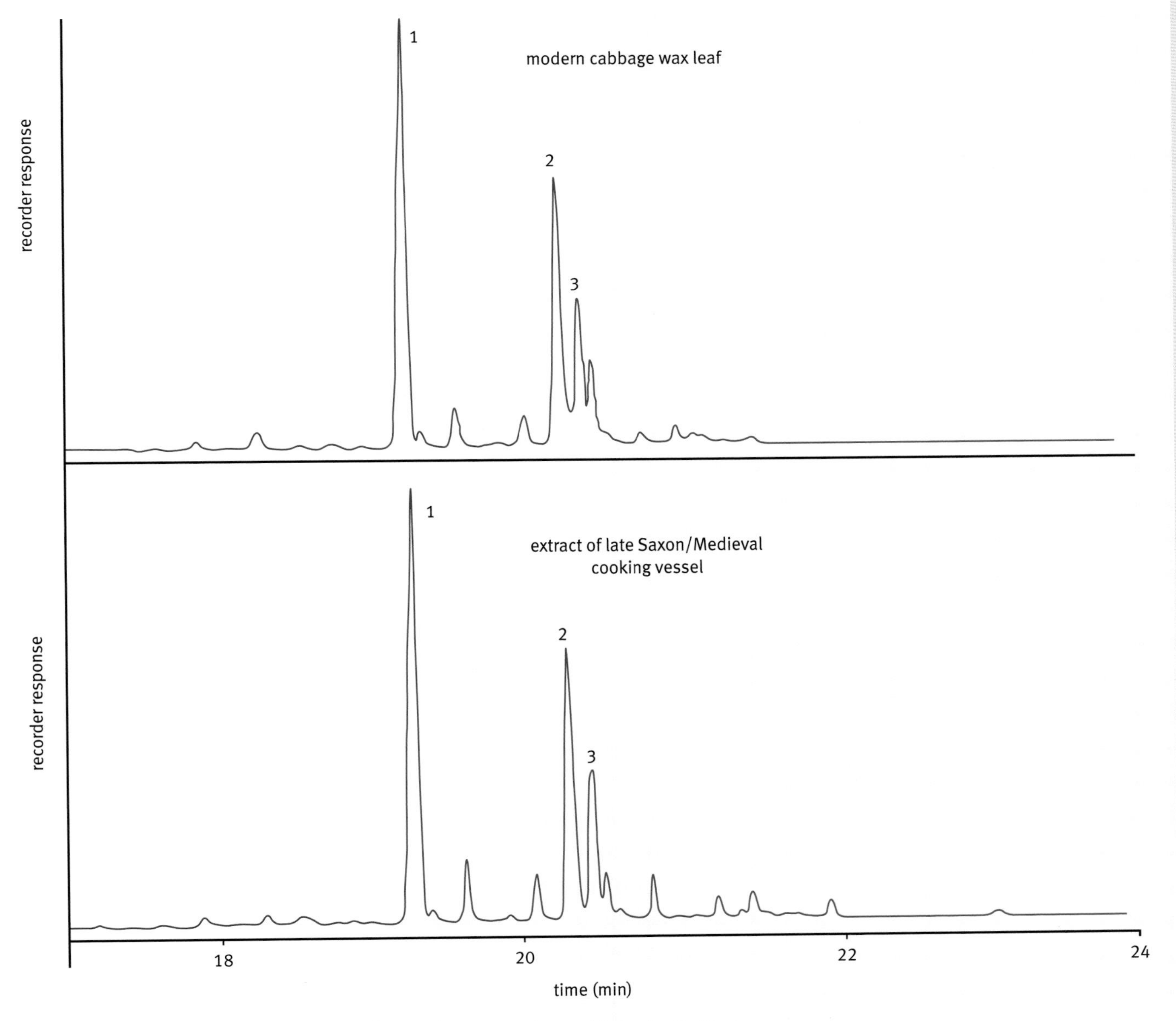

Gas chromatography traces comparing the wax from a modern cabbage leaf with the lipids extracted from a late Saxon cooking pot

Find out about:
- acid–alkali titrations
- standard solutions

3E Titrations

A **titration** is a quantitative technique based on measuring the volumes of solutions that react with each other. Chemists use titrations to measure concentrations and to investigate the quantities of chemicals involved in reactions. Titrations are widely used because they are quick, convenient, accurate, and easy to automate.

Titration procedure

In a typical titration, an analyst uses a **pipette** (or a burette) to transfer a fixed volume of liquid to a flask. In an acid–base titration to find the concentration of an acid, this might be 20 cm^3 of the solution of acid.

Next, the analyst adds one or two drops of a coloured indicator. The indicator is chosen to change colour sharply when exactly the right amount of alkali has been added to react with all the acid. The indicator works because there is a very sharp change of pH at this point, which is called the **end point.**

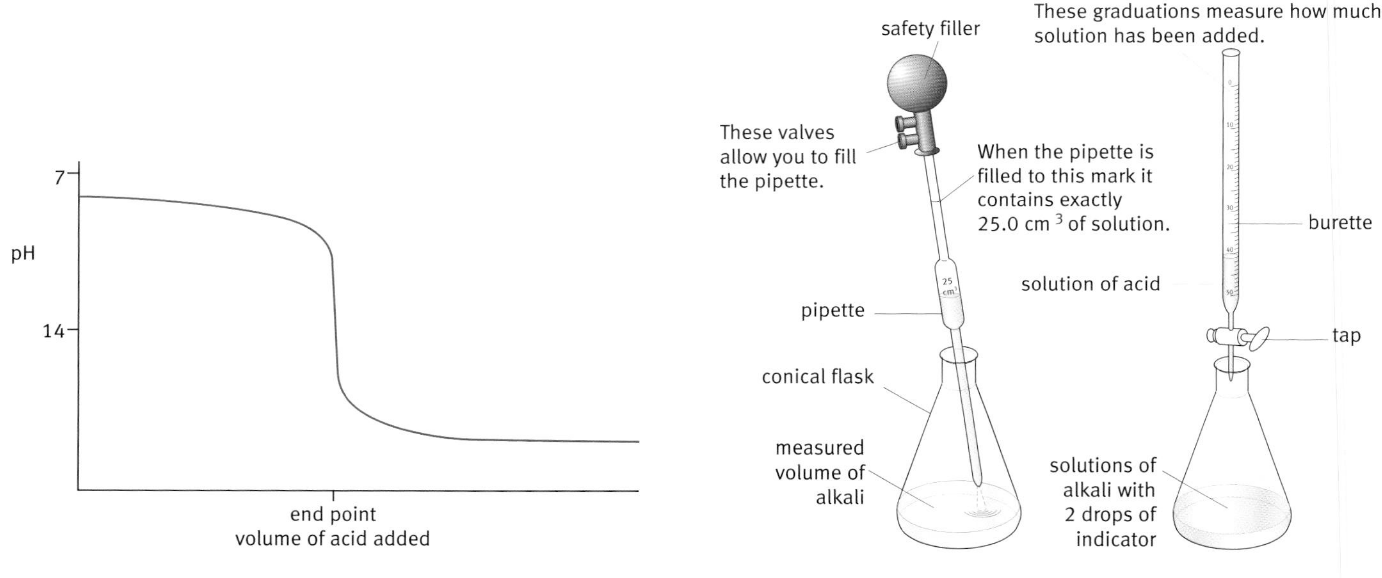

pH change when titrating an alkali with an acid

Apparatus for an acid–base titration

The analyst has a **burette** ready containing a solution of acid with a concentration that is known accurately. Then the analyst runs the acid from a burette into the alkali a little at a time until the indicator changes colour. Reading the burette scale before and after the titration shows the volume of alkali added.

It is common to do a rough titration first to get an idea where the end point lies and then to repeat the titration more carefully, adding the acid drop by drop near the end point. The analyst repeats the titration two or three times as necessary to achieve consistent values for the value of alkali added.

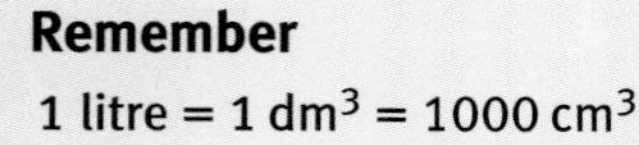

Remember
1 litre = 1 dm^3 = 1000 cm^3

Preparing accurate solutions

The accuracy of a titration can be no better than the accuracy of the solutions used to make the measurements. If chemists know the concentration of a solution accurately, they call it a **standard solution**, because it can be used in analysis to measure the concentrations of other solutions.

Key terms
- titration
- pipette
- end point
- burette
- standard solution

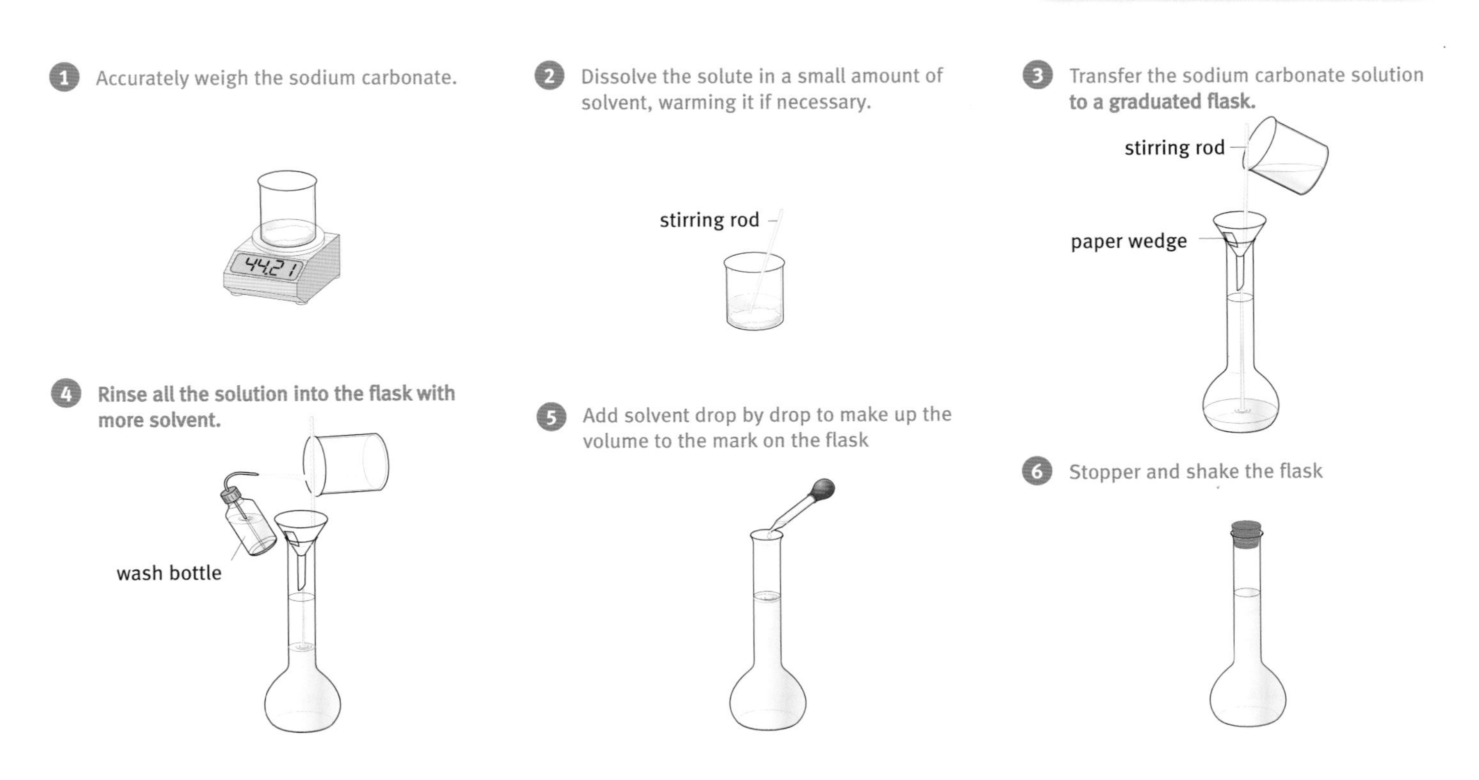

The procedure for making up a standard solution with an accurately known concentration

H

Questions

1 What is the concentration of these solutions in grams per litre (g/dm^3)?

a A solution of sodium carbonate made by dissolving 4.0 g of the solid in water and making the volume up to 500 cm^3 in a graduated flask.

b A solution of nitric acid made by diluting 6.4 g of the acid and making the volume up to 500 cm^3 in a graduated flask.

c A solution of citric acid made by dissolving 2.25 g of the solid in water and making the volume up to 250 cm^3 in a graduated flask.

2 What is the mass of solute in these samples of solutions?

a A 10 cm^3 sample of a solution of silver nitrate with a concentration of 2.55 g/dm^3.

b A 25 cm^3 sample of a solution of sodium hydroxide with a concentration of 4.40 g/dm^3.

3 Suggest reasons why it is possible to prepare a solution with an accurately known concentration by weighing out anhydrous sodium carbonate and dissolving it in water but not possible to do the same thing with pellets of sodium hydroxide.

4 What volume of a solution of hydrochloric acid containing 18.2 g/dm^3 is needed to remove 5.0 g of limescale ($CaCO_3$)?

Titrating acids in food and drink

Rachel Woods is the quality control manager for Danisco, a company that manufactures ingredients for food and soft drinks. Acids are very important in her work. Some acids occur naturally in the fresh ingredients; others are added to improve flavour and keeping quality.

The quantity of acid in any food or drink is important. Think of a soft drink – too much acid and it tastes sour; too little and it might be insipid. With just the right amount it tastes refreshing and fruity, and just the right amount of acid makes a drink seem more thirst-quenching and satisfying.

Rachel measures the mass of a sample of blueberry juice.

Automated titrations

One of the most important acids in Rachel's work is citric acid. She regularly has to test ingredients and finished products to check their citric acid content. She uses a titration machine. Here she is testing a sample of blueberry juice. First, using a dropping pipette, she takes a sample of the juice from the container into a beaker. It is the same beaker in which she will carry out the titration, which makes things so much quicker and simpler. She has weighed exactly 4.30 grams.

She adds boiled water to the juice to bring it up to the 300 cm^3 mark. She does not use water straight from the tap as it has dissolved calcium hydrogencarbonate in it. Boiling the water removes this. For many titrations distilled water is necessary, but for these food samples ordinary boiled water is fine.

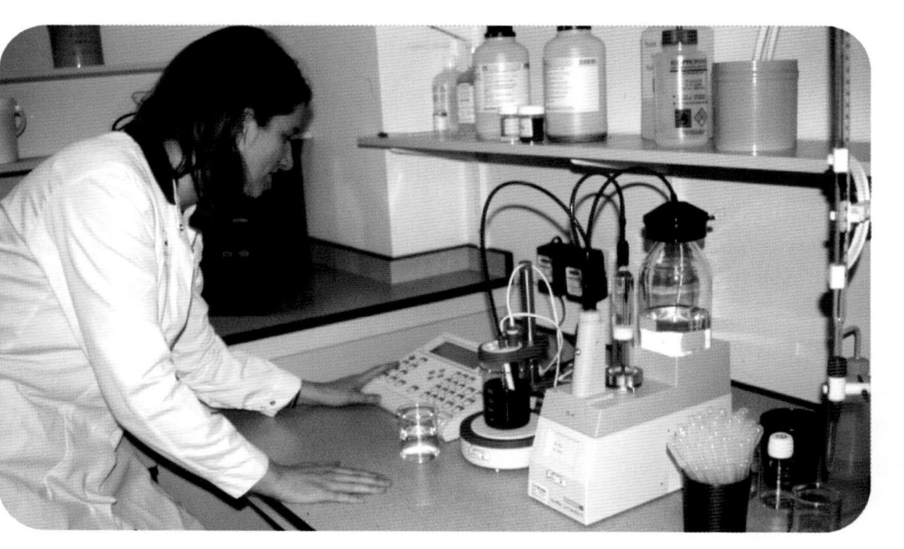

Rachel sets up her automatic titration machine. The burette tube in the middle fills from the reservoir of alkali on the right.

All Rachel has to do is put the pH probe and tube from the burette into the beaker and put in on the stand. It has a magnetic stirrer, so she does not even have to shake the flask. She presses a button, and the burette tube starts to fill from the reservoir bottle. The concentration of her sodium hydroxide, NaOH, solution is 8.0 g/dm^3.

When the burette is full, it slowly pumps the NaOH solution into the beaker. Because blueberry juice contains a natural indicator, you can see a colour change from purply-red to dark blue-grey as it reaches the end point.

The exact point is not easy to see, but that does not matter, as the machine works by measuring the pH of the solution. It measures the quantity of NaOH required to bring the blueberry juice up to a pH of 8.3. Not that Rachel has to measure that volume – far from it: all she has to do is look at the readout on the screen. It shows a little graph and calculates the percentage of citric acid in the juice. No calculation is needed. This sample is 6.08% citric acid, which is exactly what it should be, and using the machine, Rachel can test as many samples as she needs very quickly and simply.

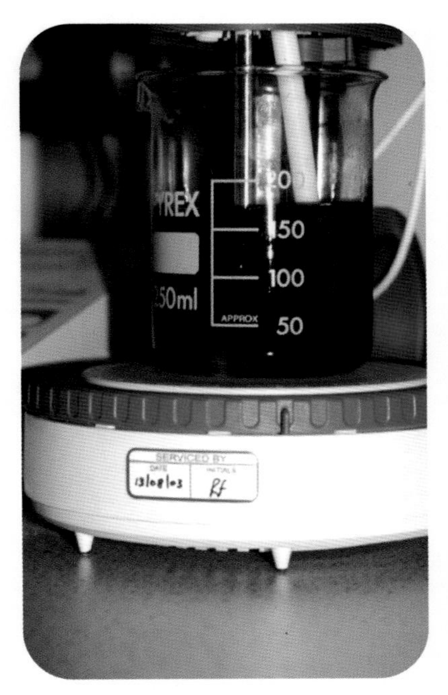

The beaker with the blueberry juice in the titration machine. A magnetic stirrer mixes the juice with the alkali added by the burette. The probe dipping into the solution measures the pH.

The hands-on method

Not all Rachel's titrations can be done on the machine. Here she is using the traditional method to test the concentration of a sample of butyric acid.

Butyric acid (C_3H_7COOH) is an important part of butter and cheese flavours, so it used in things like cheese-and-onion crisps, and in those 'I-think-it-tastes-like-butter' margarines you see advertised on the TV. Used in this way, butyric acid is great, but unfortunately in large quantities it smells terrible.

It is not Rachel's favourite titration, but she is so used to it she manages to keep smiling. Butyric acid is associated with fats, so it is not soluble in water. Instead it is dissolved in ethanol.

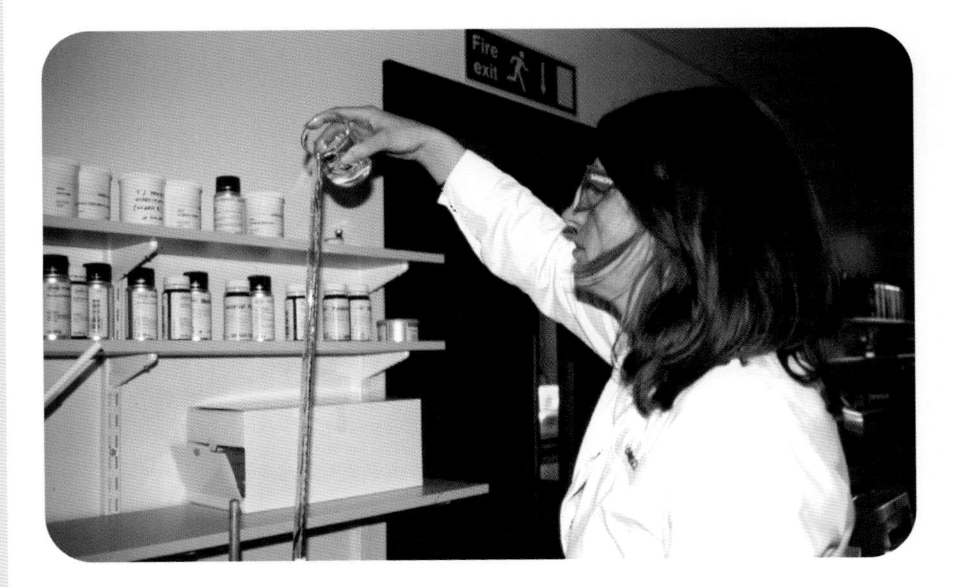

Rachel filling the burette with the acid solution. She keeps her eye level with the zero mark on the burette. She wants the meniscus to sit exactly on the line.

Using a pipette to run a measured volume of the standard potassium hydroxide into a titration flask

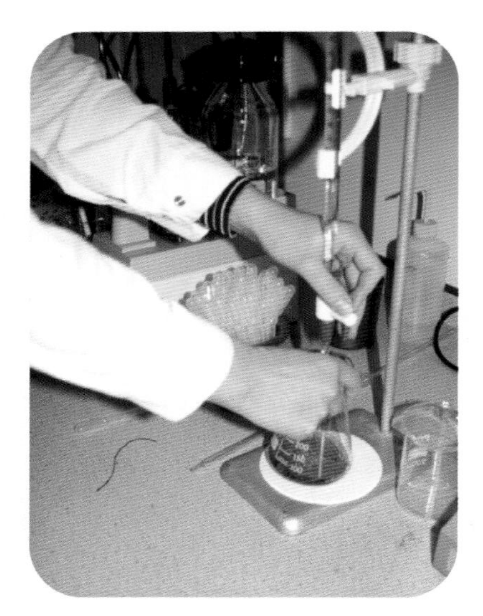

Rachel runs the butyric acid from the burette into the flask during the titration.

The next job is to pipette exactly 20.0 cm^3 of warm potassium hydroxide, KOH, solution into a clean flask. The concentration of the alkali is accurately known. Again Rachel keeps her eye level with the mark on the pipette as she uses the valve on the filler to adjust the level. A tiny drop of solution is always left in the end of the pipette, but Rachel resists the temptation to blow it out. As she adds a few drops of the indicator, phenolphthalein, the alkaline solution turns a stunning shade of shocking pink.

Now the titration begins. With her right hand she swirls the flask, and with her left hand she gently releases the tap on the burette to let 1 cm^3 of the acid into the flask at a time. She keeps her eye on the flask all the time. Because she does this titration so regularly, she knows when the end point is coming up. It happens suddenly. First the solution in the centre of the flask goes clear, then the pink colour disappears altogether. At that precise point Rachel closes the tap on the burette and takes a note of the reading: 40.2 cm^3. She repeats the titration at least once more.

H Interpreting the results

Relative formula masses: C_3H_7COOH = 88 and KOH = 56

The equation for the reaction in the titration flask is

$$\underset{88\,\text{g}}{C_3H_7COOH} + \underset{56\,\text{g}}{KOH} \longrightarrow C_3H_7COOK + H_2O$$

Concentration of the potassium hydroxide solution = 11.2 g/dm^3

In 20.0 cm^3 of the KOH solution there is

$$\frac{20}{1000} \times 11.2\,\text{g} = 0.224\,\text{g}$$

If 56 g KOH reacts with 88 g of the acid, then 0.224 g reacts with

$$\frac{0.224}{56} \times 88\,\text{g} = 0.352\,\text{g butyric acid}$$

This amount of acid was present in 40.0 cm^3 = 0.040 dm^3 of butyric acid solution. So the concentration of the butyric acid solution is

$$\frac{0.352\,\text{g}}{0.040\,\text{dm}^3} = 8.80\,\text{g/dm}^3$$

3F Evaluating results

Scientists need to be able to make sense of analyses and tests. This means that they have to be able to interpret their significance and say what they show. Scientists must also judge how confident they are about the accuracy of results.

Find out about:

- systematic and random sources of uncertainty
- accuracy and precision

Measurement uncertainty

All measurements have an uncertainty. This means that scientists usually give results within a range. For example, they may analyse the purity of a drug and give the answer as 99.1 ± 0.2%. This means that the average value obtained from analyses of several samples was 99.1%. The precise value is uncertain. The scientists are confident that the actual value lies between 98.9% and 99.3%. To show this, they quote the results as 99.1 (the mean) ± 0.2%.

Errors of measurement are not mistakes. Mistakes are failures by the operator and include such things as forgetting to fill a burette tip with the solution, or taking readings from a sensitive balance in a draught. Mistakes of this kind should be avoided by people doing the practical work.

Types of uncertainty

There are two general sources of **measurement uncertainty**: systematic errors and random errors.

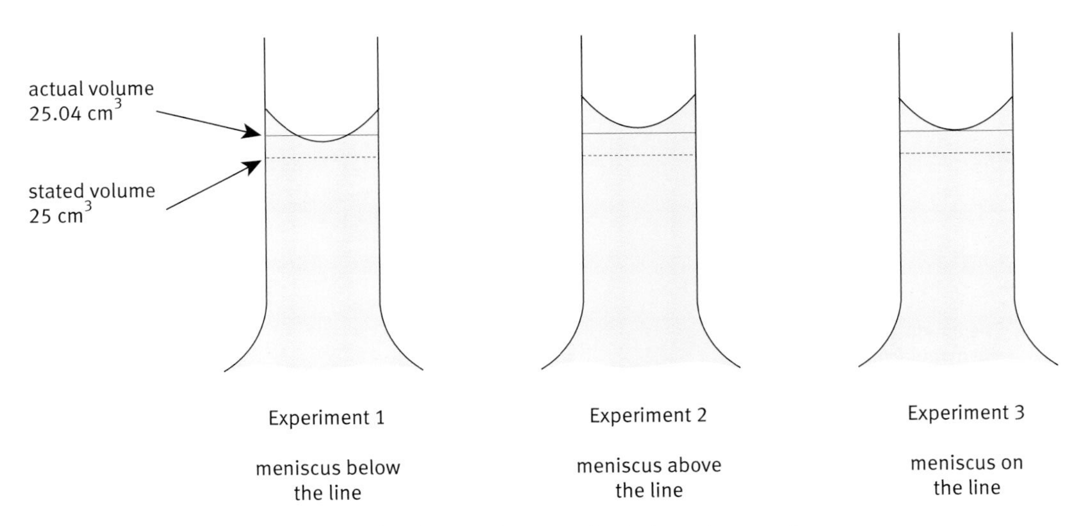

Systematic and random errors in the use of a pipette. The manufacturing tolerance for a 25 cm^3 grade B pipette is ±0.06 cm^3. This can give rise to a systematic error. Every time an analyst uses the pipette, the meniscus is aligned slightly differently with the graduation mark. This gives rise to random error.

Random error means that the same measurement repeated several times gives different values. This can happen, for example, from making the judgements about the colour change at an end point or from estimating the reading from a burette scale.

Systematic error means that the same measurement repeated several times gives values that are consistently higher than the true value, or consistently lower. This can result from incorrectly calibrated measuring instruments or from making measurements at a consistent, but wrong, temperature.

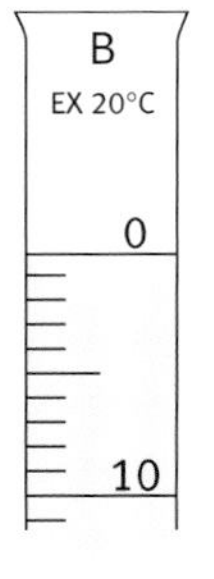

It is difficult to determine accurately the volume of liquid in a burette if the meniscus lies between two graduation marks.

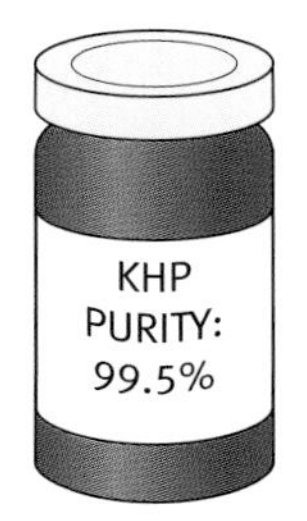

The material used to prepare a standard solution may not be 100% pure.

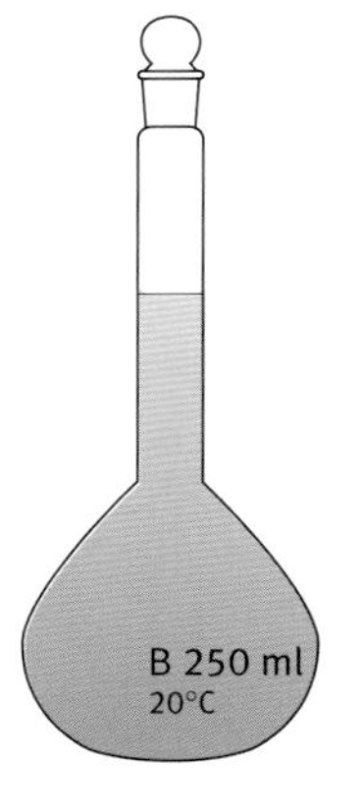

A 250 cm^3 volumetric flask may actually contain 250.3 cm^3 when filled to the calibration mark owing to permitted variation in the manufacture of the flask.

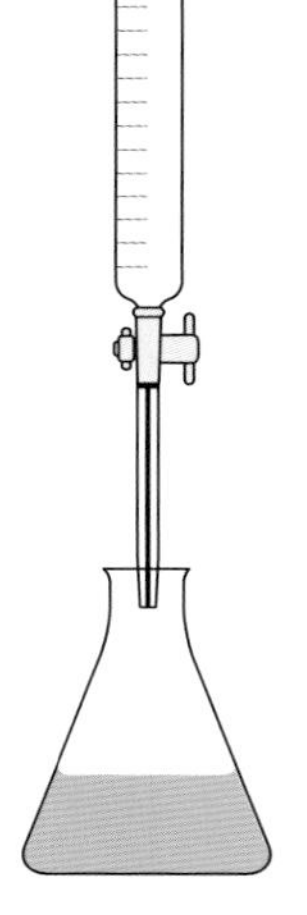

It is difficult to make an exact judgement of the end point of a titration (the exact point at which the colour of the indicator changes).

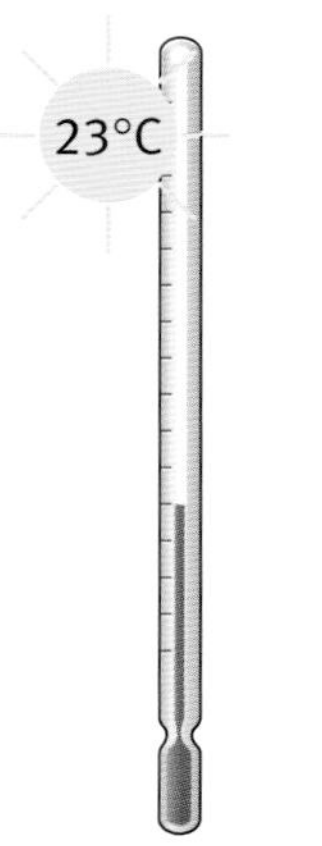

The burette is calibrated by the manufacturer for use at 20 °C. When it is used in the laboratory the temperature is 23 °C. This difference in temperature will cause a small difference in the actual volume of liquid in the burette when it is filled to a calibration mark.

The display on a laboratory balance will only show the mass to a certain number of decimal places.

Sources of uncertainty in analysis by titration

An analysis or test is often repeated to give a number of measured values, which are then averaged to produce the result.

- **Accuracy** describes how close this result is to the true or 'actual' value.
- **Precision** is a measure of the spread of measured values. A big spread indicates a greater uncertainty than does a small spread.

Key terms

measurement uncertainty
accuracy
precision

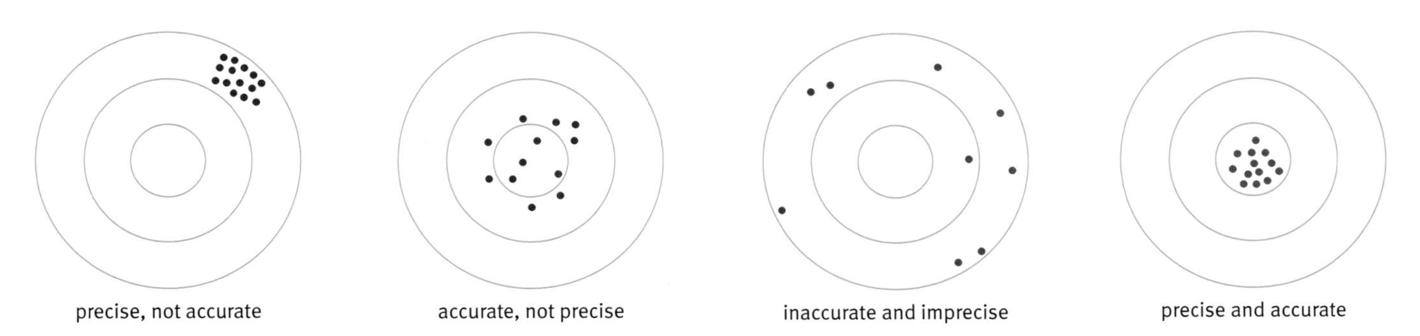

Accuracy and precision are not the same thing.

Conclusions

The conclusions scientists draw from their work must be valid and justifiable.

- Valid means that the techniques and procedures used were suitable for what was being analysed or tested.
- Justifiable means that conclusions reached are backed by sound, reliable evidence.

Questions

1 An analyst determined the percentage of potassium in three brands of plant fertilizer for house plants by making five measurements for each brand.

These are the results of measuring the percentage by mass of potassium in three brands:

A 4.93, 4.89, 4.71, 4.81, 4.74

B 6.76, 7.91, 6.94, 6.71, 6.86

C 4.72, 4.76, 4.68, 4.70, 4.69

a Determine the mean and range for each brand.

b What conclusions can you draw about the three brands?

2 Why would the results be inaccurate if an analyst used hot solutions in graduated glassware?

Topic 4 Green chemistry

The chemical industry

The chemical industry converts raw materials into useful products. The products include chemicals for use as drugs, fertilizers, detergents, paints, and dyes.

The chemical industry takes crude oil, air, sea water, and other raw materials and converts them to pure chemicals such as acids, salts, solvents, compressed gases, and organic compounds.

The industry makes **bulk chemicals** on a scale of thousands or even millions of tonne per year. Examples are ammonia, sulfuric acid, sodium hydroxide, chlorine, and ethene.

On a much smaller scale the industry makes **fine chemicals** such as drugs and pesticides. It also makes small quantities of speciality chemicals needed by other manufacturers for particular purposes. These include such things as flame retardants, food additives, and the liquid crystals for flat-screen televisions and computer displays.

Key words
bulk chemicals
fine chemicals

Greener industry

The chemical industry is reinventing many of the processes it uses. The industry seeks to become 'greener' by:

- turning to renewable resources
- devising new processes that convert a high proportion of the atoms in the reactants into the product molecules.
- cutting down on the use of hazardous chemicals
- making efficient use of energy
- reducing waste
- preventing pollution of the environment

Harvesting a natural resource. Lavender is distilled to extract chemicals for the perfume industry.

Find out about:

- feedstocks for the chemical industry
- products from the chemical industry

4A The work of the chemical industry

Raw materials

The basic raw materials of the chemical industry are

- crude oil
- air
- water
- vegetable materials
- minerals such as metal ores, salt, limestone, and gypsum

The first step in any process is to take the raw materials and convert them into a chemical, or mixture of chemicals, that can be fed into a process. Crude oil, for example, is a complex mixture of chemicals. An oil refinery distils the oil and then processes chemicals from the distillation column to produce purified **feedstocks** for chemical synthesis.

Chemical plants

At the centre of the plant is the reactor, where reactants are converted into products. The feedstock may have to be heated or compressed before it is fed to the reactor. The reactor often contains a catalyst.

Generally, a mixture of chemicals leaves the reactor. The mixture includes the desired product, but there may also be **by-products** and unchanged starting materials. So the chemical plant has to include equipment to separate the main product and by-products and to recycle unchanged reactants.

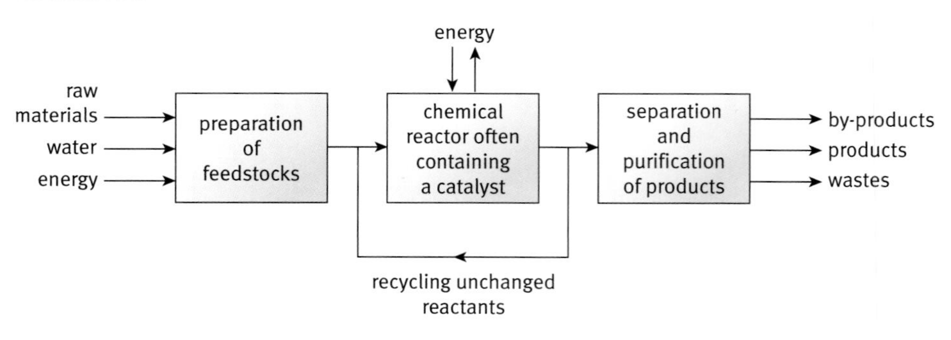

A schematic diagram to summarize a chemical process

Chemical plants need energy. Some of the chemical reactions occur at a high temperature, so energy is often needed for heating. Also, a lot of electric power is needed for pumps to move reactants and products from one part of the plant to another.

Sensors monitor the conditions at all the key points in the plant. The data is fed to computers in the control centre where the technical team controls the plant.

Products from the chemical industry

The chemical industry produces five main types of product. Some of these are made in huge quantities. Other chemicals have high value but are made in much smaller amounts. (See the pie chart in Module C6 *Chemicals of the natural environment*, page 150, which shows the range of products made by the chemical industries in Britain and their relative value.)

Basic inorganics including fertilizers

The industry makes large amounts of these chemicals. Chlorine, sodium hydroxide, sulfuric acid, and fertilizers are all bulk chemicals.

Petrochemicals and polymers

Petrochemical plants use hydrocarbons from crude oil to make a great variety of products, including polymers (see Module C2 *Material choices*, Sections E and G). Among many other chemicals, the industry makes ethene. Ethene, C_2H_4, is then used to make many different products, including polymers such as polythene and PVC, and solvents such as ethanol. The worldwide production of ethene is over 80 million tonnes each year.

Dyes, paints, and pigments

The discovery of a new mauve dye by the young William Perkin in 1856 was the starting point for the development of many new synthetic colours. At that time coal tar was the main source of carbon chemicals for dyes. Modern dyes are made from petrochemicals.

Pharmaceuticals

The pharmaceutical industry grew from the dyestuffs industry. The industry produces drugs and medicines. This is a part of the chemical industry that is changing rapidly to explore the possibilities arising from the growing understanding of the human genome.

Speciality chemicals

Speciality chemicals are used to make other products. They include food flavourings and the liquid crystal chemicals in flat-screen displays.

A maintenance worker inside a large chemical reactor

Key words

feedstocks
by-products

Transport workers bring materials in and out of the chemical plant.

Questions

1 When the tank in the picture on the left is in use, it is filled with a reaction mixture. Suggest the purpose of:

a the rotating paddle in the centre of the tank

b the network of pipes round the edge of the tank

Find out about:
- new sources of chemicals
- measures of efficiency
- safer ways to make chemicals
- energy efficiency
- reducing waste and recycling

4B Innovations in green chemistry

In the second half of the twentieth century the chemical industry found that its reputation with the public was falling. Many people were worried about synthetic chemicals and their impact on health and the environment. Politicians responded by passing laws to regulate the industry. At first, these laws were aimed at dealing with the industry as it was then by treating pollution and minimizing its effects.

Controlling the flow of oil at a refinery. Crude oil is the main source of organic chemicals today.

Crops can be grown to supply feedstocks for the chemical industry rather than for their food value. In Europe this includes growing wheat, maize, sugar beet, and potatoes as sources of sugars or starch.

More recently, legislation has set out to prevent pollution by changing the industry. New laws encourage companies to reduce the formation of pollutants through changes in production and the use of raw materials that are cost effective and renewable.

In the early stages the aim was to cut risks by controlling people's exposure to hazardous chemicals. Now green chemistry attempts to eliminate the hazard altogether. If the industry can avoid using or producing hazardous chemicals, then the risk is avoided.

Green chemistry has the potential to bring benefits both to industry and to society. Innovative green chemistry can increase efficiency, cut costs, and help to avoid the dangers of hazardous chemicals.

Renewable feedstocks

One of the aims of green chemistry is to use renewable raw materials. At the moment, crude oil is the main source of chemical **feedstocks**. Less than 3% of crude oil is used to make chemicals. All the rest is burnt as fuel or used to make lubricants. Even so, reducing the reliance on petrochemicals would help to make the industry more sustainable.

Currently, most polymers are made from petrochemicals. This includes the polyester fibres that are widely used in clothing

DuPont has developed a way of making a new type of polyester by fermenting renewable plant materials. The company calls this polymer Sorona. Manufacturers convert Sorona into fibres for clothing and upholstery.

The chemical starting point for the synthesis of Sorona is malonic acid. DuPont has found that it can produce this acid by fermenting corn starch with bacteria.

Plants can be grown year after year. They are a **renewable resource**. However, growing plants for chemicals takes up land that could be used to grow food. Also energy is needed to make fertilizers and for harvesting crops.

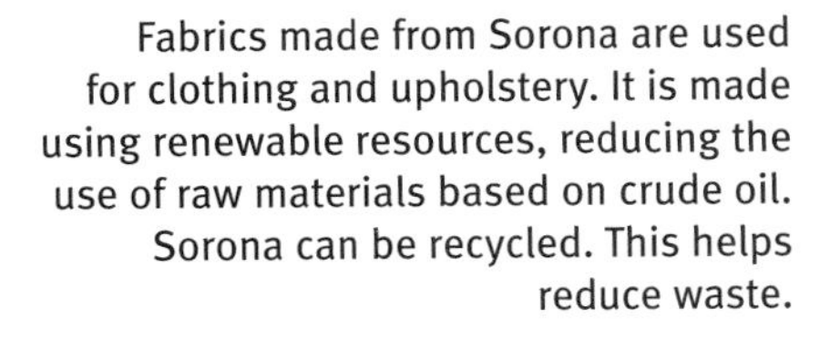

Fabrics made from Sorona are used for clothing and upholstery. It is made using renewable resources, reducing the use of raw materials based on crude oil. Sorona can be recycled. This helps reduce waste.

Key words

renewable resource

Questions

1 Classify these raw materials as 'renewable' or non-renewable':
   - **a** salt (sodium chloride)
   - **b** crude oil
   - **c** wood chippings
   - **d** limestone
   - **e** sugar beet

2 Which of the ways of making industry 'greener' are illustrated by the development of Sorona?

H

Converting more of the reactants to products

Yields

Chemists calculate the **percentage yield** from a production process to measure its efficiency. This compares the quantity of product with the amount predicted by the balanced chemical equation.

A high yield is a good thing, but it is not necessarily an indicator that the process is 'green'. This is illustrated on a small scale by the laboratory process to replace the —OH group in the alcohol butanol with a bromine atom. The product is bromobutane:

$$C_4H_9—OH + H_2SO_4 + NaBr \longrightarrow C_4H_9—Br + NaHSO_4 + H_2O$$

Carrying out this reaction in the laboratory uses 35 g sodium bromide and 50 g sulfuric acid to convert 20 g butanol into 30 g of pure bromobutane. The sodium bromide and sulfuric acid are in excess, so the yield is limited by the amount of the alcohol.

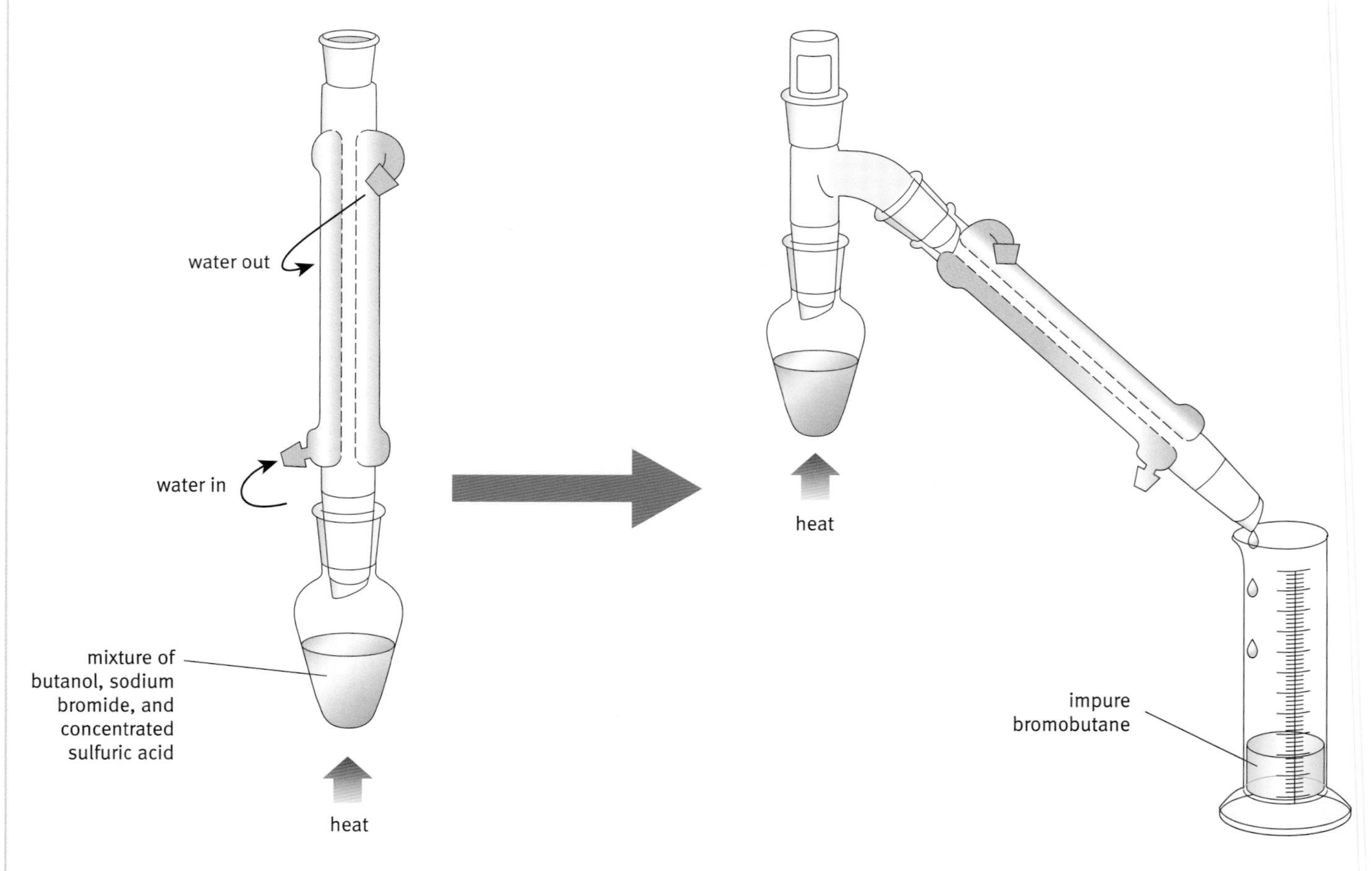

Converting butanol to impure bromobutane

H

According to the reacting masses in the equation, 74 g of butanol should produce 137 g bromobutane.

So the theoretical yield from 20 g butanol = $\frac{20}{74\text{ g}} \times 137\text{ g} = 37\text{ g}$

The percentage yield from the laboratory process = $\frac{30}{37\text{ g}} \times 100\% = 81\%$

This is a good yield for a laboratory preparation. However, the result is that only 30 g of product is made from 105 g of the starting chemicals.

Even with an impossible 100% yield the process could at best produce 37 g butanol from a total of 105 g of reactants. There is 68 g of waste.

Atom economy

In 1998, Barry Trost of Stanford University, USA, was awarded a prize for his work in green chemistry. He had introduced the term **atom economy** as a measure of the efficiency with which a reaction uses its reactant atoms.

$$\text{atom economy} = \frac{\text{mass of atoms in the product}}{\text{mass of atoms in reactants}} \times 100\%$$

H

$C_4H_9\text{—}OH + H_2SO_4 + NaBr \longrightarrow C_4H_9\text{—}Br + NaHSO_4 + H_2O$

totals of atoms in the reactants: 4C, 12H, 5O, Br, Na, S
(total relative atomic mass = 275)

totals of green atoms ending in the product: 4C, 9H, Br
(total relative atomic mass = 137)

totals of brown atoms ending as waste: 3H, 5O, Na, S
(total relative atomic mass = 138)

atom economy = $\frac{137\text{ g}}{275\text{ g}} \times 100\% = 50\%$

Calculating the atom economy for the laboratory preparation of bromobutane. The atoms in the reactants that end up in the product are shown in green. The atoms that end up as waste are brown.

So, at the very best, only half of the mass of starting materials can end up as product. This is not a green process.

This approach does not take yield into account and does not allow for the fact that many real-world processes use deliberate excess of reactants. It does, however, help in comparing different pathways to a desired product.

Key words

percentage yield
atom economy

Questions

3 Explain the purposes of the condensers used in the apparatus shown for making bromobutane.

4 Heating with a catalyst converts cyclohexanol, $C_6H_{11}OH$, to cyclohexene, C_6H_{10}.

a What is the percentage yield if 20 g hexanol gives 14.5 g cylohexene?

b What is the atom economy, assuming that the catalyst is recovered and reused?

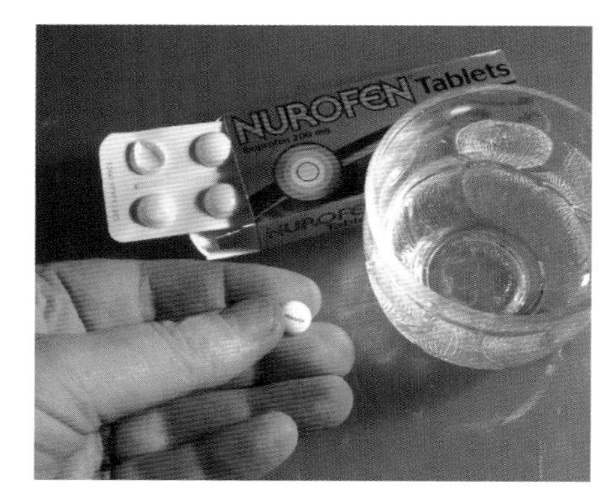

A greener pain reliever

Every year in the UK, people spend around £300 million buying pain killers. One of them is ibuprofen

Boots patented ibuprofen in the 1960s. At that time there were six stages in the complex processes to make the drug.

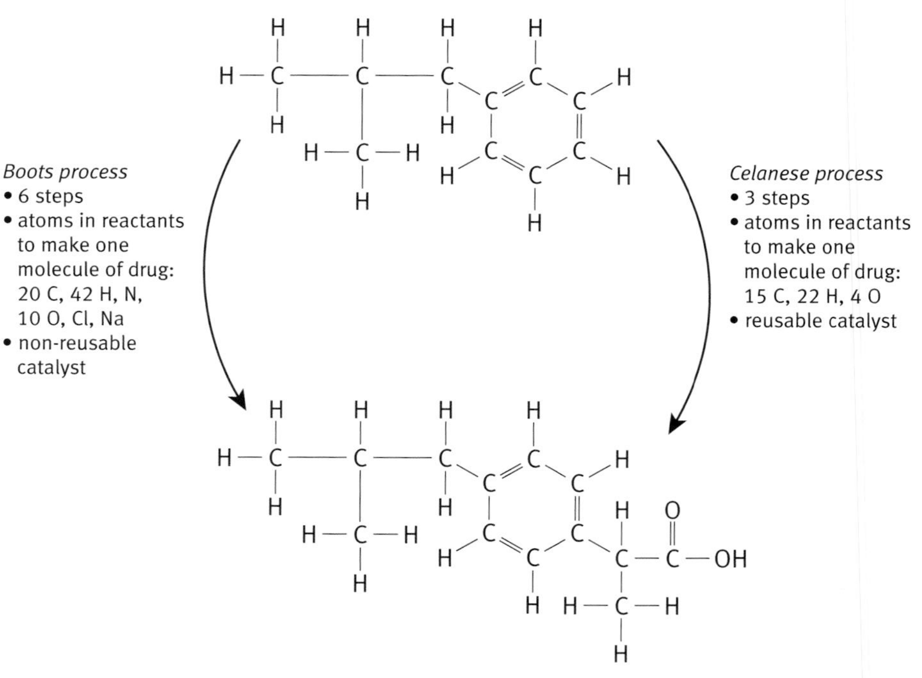

Two routes for making ibuprofen from the same starting material

There are 33 atoms of all kinds in one ibuprofen molecule. Adding up all the atoms in the chemicals used by Boots shows that the process needed 74 atoms for each ibuprofen molecule formed. So 41 atoms ended up making unwanted chemicals. This was wasteful.

Once the Boots patent for ibuprofen had run out after 20 years, it was possible for other companies to make and sell the drug. A new synthesis was developed by the Celanese Corporation which was more efficient. This new process has only three stages and needs only 41 atoms in the reacting chemicals to make each ibuprofen molecule. It has a much more efficient 'atom economy'. It gives fewer harmful by-products. It is a 'greener' synthesis because it creates less useless waste.

Another feature of the Boots process was that it used aluminium chloride as a catalyst in one key step of the process. Although a catalyst, the aluminium chloride could not be recovered and reused. This added to the waste.

The newer method uses other catalysts: hydrogen fluoride and an alloy of nickel and aluminium. Both of these catalysts can be recovered and reused many times.

Questions

5 What is the molecular formula for ibuprofen?

6 a What is the atom economy for the Boots process for making ibuprofen?

b What is the atom economy for the Celanese process for making ibuprofen?

c Calculate a new value of the atom efficency for the Celanese process, assuming that the ethanoic acid (CH_3COOH) formed as a by-product in one step is recycled in the process and does not go to waste.

Designing new catalysts

Matthew Davidson works at the University of Bath, where he designs new catalysts for the synthesis of polymers. His main strategy is to wrap an organic structure round a metal ion. For example, he has helped a manufacturer to devise a catalyst for making a type of polyester by wrapping citrate ions round titanium ions. This catalyst produces a clear rather than a yellowish polymer, and replaces a catalyst made of antimony, which is a toxic metal.

Matthew Davidson in his laboratory

'The main thing about a catalyst is that it has to be highly reactive towards the starting materials but not to the products.'

Matthew enjoys making new complex molecules and then analysing their structure to see what makes them active. 'I have to think about two main things – firstly the size and shape of the catalyst molecules and secondly how they interact with the electrons of other molecules.'

Matthew Davidson uses physical models and computer models in his work.

The need for new catalysts

But why do we need new catalysts? Matthew says there are several reasons: 'First, many older catalysts were toxic or contained harmful metals. Greater appreciation of the environment and health means that replacement catalysts need to be made. That is why we wanted to replace the antimony used to make polyesters. It is also the reason why we are researching for catalysts to replace the mercury and tin currently used to make the polyurethane used in training shoe soles.

'Second, old catalysts may not be as efficient as possible. Our new titanium-citrate catalyst is up to 15% more efficient than the traditional antimony catalyst. This allows more polymer to be made with less catalyst – good for both commercial and environmental reasons.

'Third, there are many useful chemical transformations for which there are no suitable catalysts yet. Catalysts are important for society, in that new ones can help our lives by making new medicines or new polymers.'

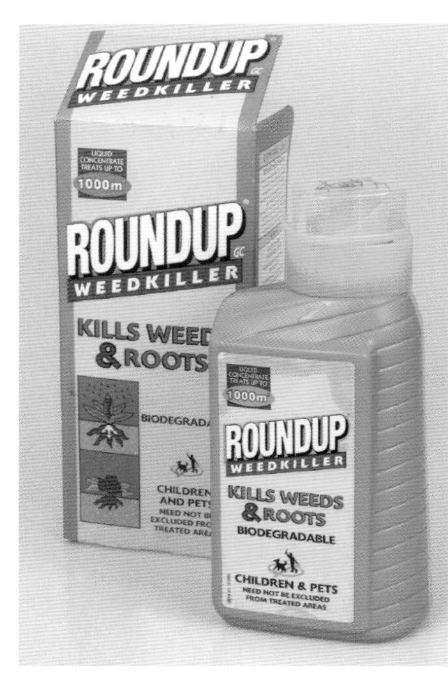

This weedkiller can now be made by a method that does not involve the use of toxic cyanide compounds.

Avoiding chemicals that are hazardous to health

The chemical industry produces a large number of synthetic chemicals. Some of these are reactive intermediates which are only used in manufacturing processes.

The aim of green chemistry is to replace reactants that are highly toxic with alternative chemicals that are not a threat to human health or the environment.

The aim is to protect the health of people working in the industry and also people who live near to industrial plants. It is important to avoid chemical accidents, including accidental releases of chemicals through explosions and fires.

Originally the company Monsanto used hydrogen cyanide in the process to produce the weedkiller that the company markets as 'Roundup'. Hydrogen cyanide is extremely toxic. The company has now developed a new route for making the herbicide. The new method has a different starting material and runs under milder conditions because of a copper catalyst.

Similarly, a new process for making polycarbonate plastics has replaced the gas phosgene with safer starting materials: methanol and carbon monoxide. Carbon monoxide is poisonous but it is not as dangerous as phosgene, which is so nasty that it has been used as a poison gas in warfare.

Energy efficiency

All manufacturing processes need energy to convert raw materials into useful products. In the chemical industry, energy is used in several ways, such as:

- to raise the temperature of reactants so a reaction begins or continues
- to heat mixtures of liquids to separate and purify products by distillation
- to dry product material
- to process waste treatment

The energy used in separation, drying, and waste management may be more than that used in the reaction stages.

Burning natural gas or other fossil fuels is the usual source of energy. Often the energy from burning is used to produce super-heated steam, which can then be used for heating around the chemical plant.

The most direct way of reducing the use of energy is to prevent losses of steam from leaking valves on steam-pipes and by installing efficient insulation on reaction vessels or pipes.

A large heat exchanger works like a laboratory condenser. One liquid or gas flows through pipes surrounded by another liquid flowing in the opposite direction. The hotter liquid or gas heats the cooler fluid.

Some of the reactions in the chemical process may be so exothermic that they provide the energy to raise steam and generate electricity which can be 'exported' to other processes nearby. The first step in the manufacture of sulfuric acid is to burn sulfur. This is so exothermic that a sulfuric acid plant has no fuel bills and can raise enough steam to generate sufficient electric power to contribution significantly to the income of the operation.

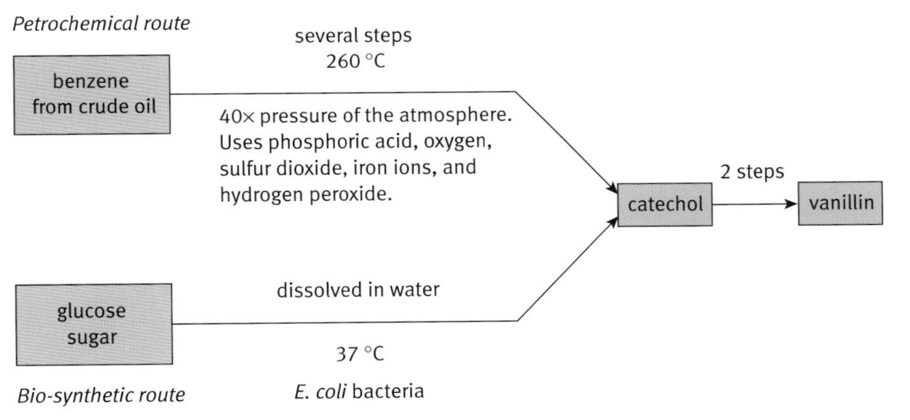

Comparing two routes to the flavouring agent vanillin

Chemical production in general has become much more energy efficient than in the past. The average energy required per tonne of chemical product is less than half that needed 50 years ago.

The aim of green chemistry is not only to make processes more energy efficient but also to lower their energy demand. One of the targets of current research is to develop new processes that run at much lower temperatures. One way of doing this is to use biocatalysts – the enzymes produced by microorganisms. Genetic modification of bacteria can lead to the development of safer processes that run at temperatures only a little above room temperature.

Questions

7 The reaction involving cyanide in the older process for making the active ingredient for Roundup was exothermic. The replacement reaction in the newer process is endothermic. Suggest why this difference contributes to safety.

8 Why can a laboratory condenser be described as a 'heat exchanger'.

9 Write a short paragraph to explain why the biosynthetic route to vanillin is 'greener' than the petrochemical route.

Reducing waste and recycling

One of the principles of green chemistry is that it is better to prevent waste than to treat or clean up waste after it is formed.

One way of cutting down on waste is to develop processes with higher atom economies. Another way is to increase recycling at every possible stage of the life cycle of a chemical product. A third way is to find uses for by-products that were previously dumped as waste.

Recycling

Industries have always tried to recycle waste produced during manufacturing processes. Recycling is easier when the composition of the scrap material is known.

The main UK manufacturer of polypropylene, Basell Polyolefins, used to burn unreacted propene. Much was burnt in an open flame so that not even the energy was recovered. Now the company has installed a distillation unit that makes it possible to separate chemicals from the waste gases. The recovery unit has cut the amount of waste. It collects over 3000 tonnes of propene per year.

A plant for recycling chemicals on a large scale

Open-loop recycling

In some cases, waste from one product is recovered and used in the manufacture of another, lower-quality product. This is open-loop recycling. It cuts down the amount of fresh feedstock needed, and the amount of waste going to landfill.

An example is the use of discarded soft-drinks bottles. The polymer in these bottles is a polyester (PET). Once collected, the PET bottles are fed through grinders that reduce them to flake form. The flake then proceeds through a separation and cleaning process that removes all foreign particles such as paper, metal, and other plastic materials.

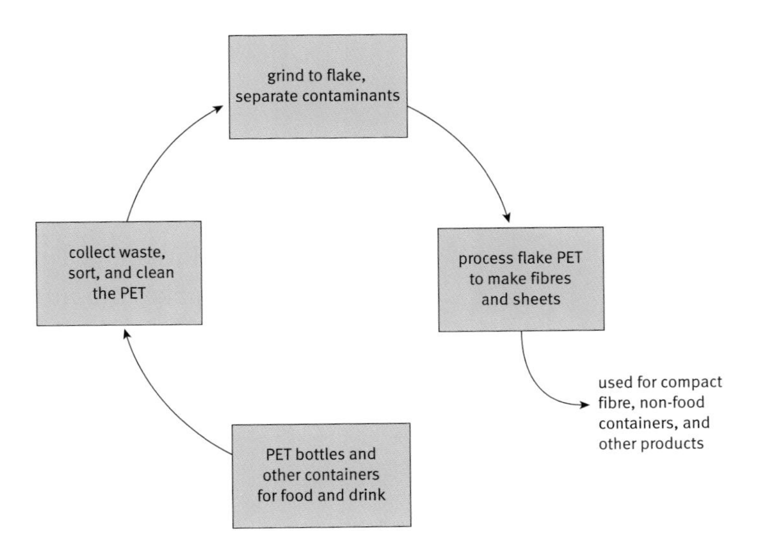

Open-loop recycling

The recovered PET is sold to manufacturers that convert it into a variety of useful products such as carpet fibre, moulding compounds, and non-food containers. Carpet companies can often use 100% recycled polymer to make polyester carpets. PET is also spun to make fibre filling for pillows, quilts, and jackets.

Closed-loop recycling

Recycling is better value if the waste material can be used to manufacture the same product with no loss in quality. With plastics this can be done by breaking down the waste into the monomers originally used to make the polymer. Several companies have developed processes for depolymerizing the polyester in soft-drinks bottles. The result is fresh feedstock for making new polymer.

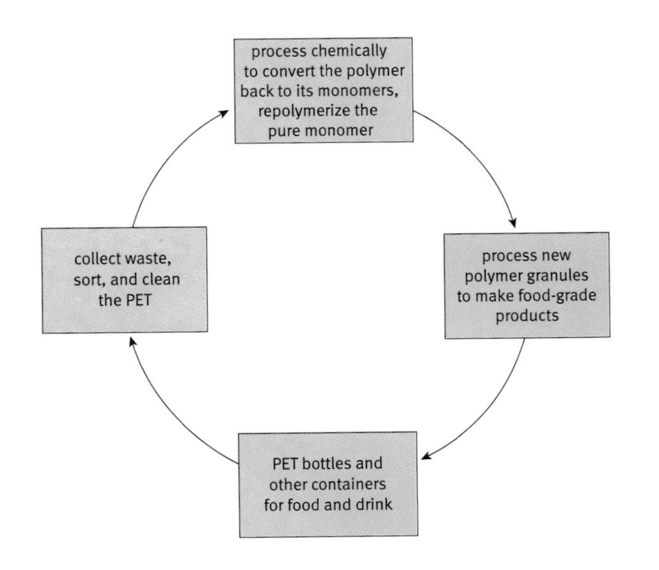

Closed-loop recycling

Questions

10 a Explain in a short paragraph the difference between open-loop and closed-loop recycling.

b Suggest one possible advantage and one possible disadvantage of each of these approaches to recycling.

Titanium dioxide is the white pigment in the paint protecting a railway bridge in Newcastle-upon-Tyne.

New uses for by-products

All chemical processes give a mixture of products: the one that the chemists want to make and others – the by-products. The process is more sustainable if the by-products can be used to make another product, so that less waste has to be dumped.

Huntsman Tioxide has found ways to cut down waste. The company makes titanium dioxide, the world's most important white pigment. Their research and development manager is Tony Jones: 'You'd be amazed at how much titanium dioxide you use in your life. It's mainly used in paint, but also in plastics, paper cosmetics, and toothpaste. It's so safe it's even used in food – when you suck the colour off a Smartie, the white sugar shell you find is coloured with TiO_2.

'The source of the TiO_2 is the mineral ilmenite, which is $FeTiO_3$. We extract the TiO_2 using sulfuric acid. After the process we neutralize the mixture using calcium carbonate to make useful by-products.

'First we get iron sulfate, which we sell to water companies. It is used in the water-purifying process. The iron causes muck in the water to coagulate and settle, leaving clear water for drinking. It can also be used as an additive for cement.

'Then the dirty sulfuric acid is passed over calcium carbonate to form calcium sulfate, also known as gypsum, which is used to make plasterboard. This not only cuts down the waste to landfill sites by 60%, but also reduces the need to mine natural gypsum.

Huntsman Tioxide's plant at Grimsby. The company is the world's third largest producer of titanium dioxide.

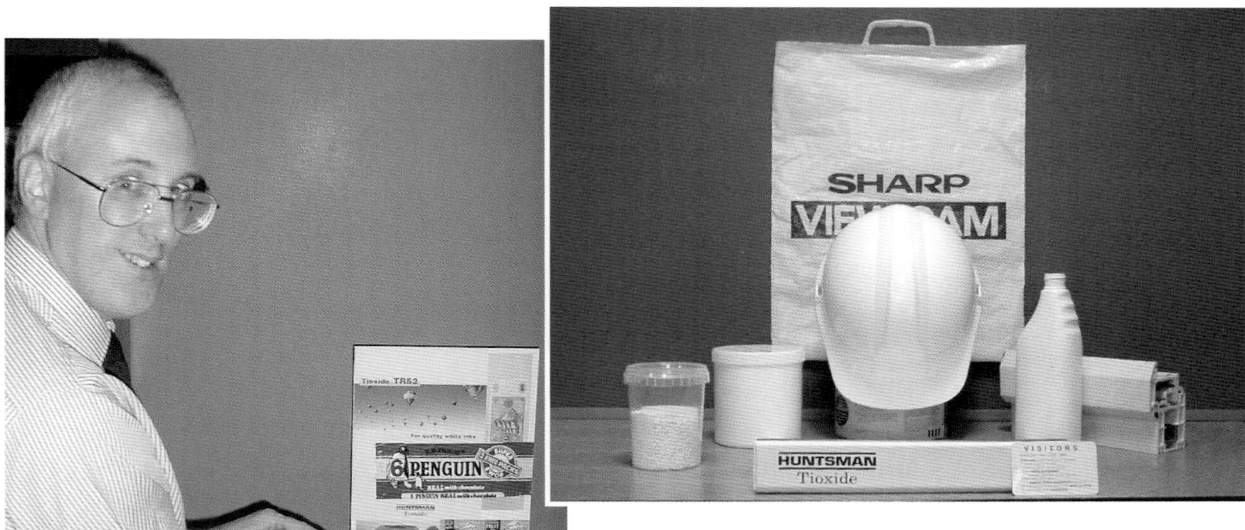

Tony Jones of Huntsman Tioxide with just a few of the many products containing titanium dioxide.

'Then we are left with red gypsum, which is just calcium sulfate stained with iron. That ends up as plaster too – but it's a pink colour. It can also be used by farmers as a conditioner on clay soil.

'The amazing thing is we now sell 500 000 tonnes of titanium dioxide every year, but 1 000 000 tonnes of the co-products. That's nearly twice as much, and remember that the ore has a straight 50/50 Ti/Fe composition!'

In 2001, Huntsman Tioxide installed a new power plant. This is a combined heat and power (CHP) plant that generates electricity and provides steam for heating in the production process. CHP plants are more efficient than conventional power stations.

Cutting pollution by wastes

It is usually impossible to eliminate waste completely. This means that it is important to remove or destroy any harmful chemicals before wastes are released into air, water, or landfills. On many sites, ground water must also be collected and processed, as it may contain traces of the chemicals made and used on the site.

Many manufacturing sites have a single processing plant for dealing with wastes. A wide range of separation techniques may be used, including filtering, centrifuging, and distillation.

Wastes may also be treated chemically to neutralize acids or alkalis, to precipitate toxic metal ions, or to convert chemicals to less harmful materials. Microorganisms or reed beds may be used to break down some chemicals.

Questions

11 What has Huntsman Tioxide done to make the manufacture of titanium dioxide more 'green'?

12 Draw an outline flow diagram to illustrate titanium dioxide production, showing the inputs to the process, the outputs and how they are used.

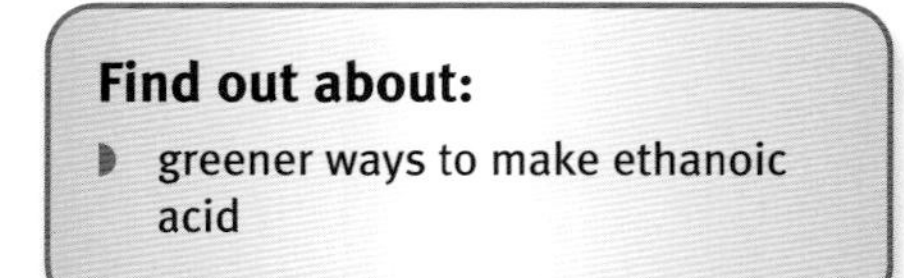

4C Manufacturing ethanoic acid

The processes that have been developed to manufacture ethanoic acid illustrate the application of the principles of green chemistry.

The industry produces over 8 million tonnes of the acid every year. Such a large amount of the acid is needed because it can be converted to many other useful products.

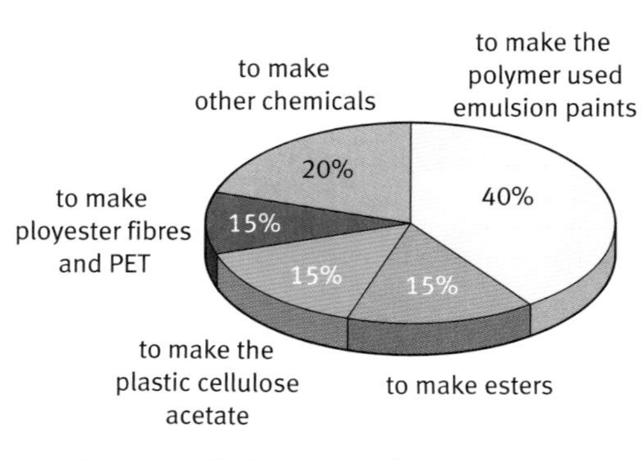

The uses of ethanoic acid

Oxidation of hydrocarbons

Up to the 1970s, ethanoic acid was produced industrially by the direct oxidation of hydrocarbons from crude oil. There are some countries where this process still operates.

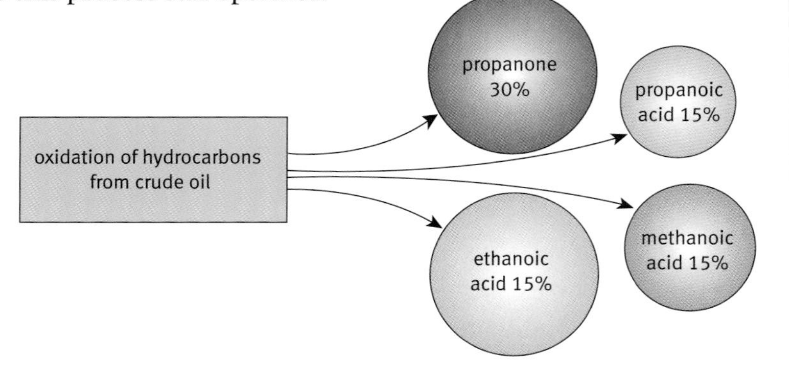

A summary showing the percentages of the products from manufacturing ethanoic acid from hydrocarbons

The process operates at 180–200 °C and at 40–50 times the pressure of the atmosphere. The catalyst is cobalt(II) ethanoate. The outcome is a wide range of products, including ethanoic acid, methanoic acid, propanoic acid, and propanone. The atom economy of this process is only about 35%.

This means that there is the cost of separating the products by fractional distillation. There are possible uses for all the products, but the economics of the operation depend on finding a market to sell all the chemicals produced.

Propanoic acid, and its calcium and sodium salts, can be used as mould inhibitors. Propanone is used as a solvent and in the manufacture of plastics. Methanoic acid is used in textile dyeing, in leather tanning, and in the manufacture of latex rubber.

The Monsanto process

From about 1970, ethanoic acid was mainly produced by a new process developed by Monsanto. In this process methanol and carbon monoxide combine to make ethanoic acid in the presence of a catalyst:

$$CH_3OH(l) + CO(g) \longrightarrow CH_3COOH(l)$$

The catalyst is a compound made from rhodium metal and iodide ions.

The Monsanto process is much greener than the earlier process. It has an atom economy of 100%, with all atoms in the reactants going into the product. So there is much less waste, and much less energy is needed to separate and purify the ethanoic acid. The reaction is extremely fast, and the catalyst has a long life.

The yield of ethanoic acid is high. About 98% of the methanol is converted to ethanoic acid.

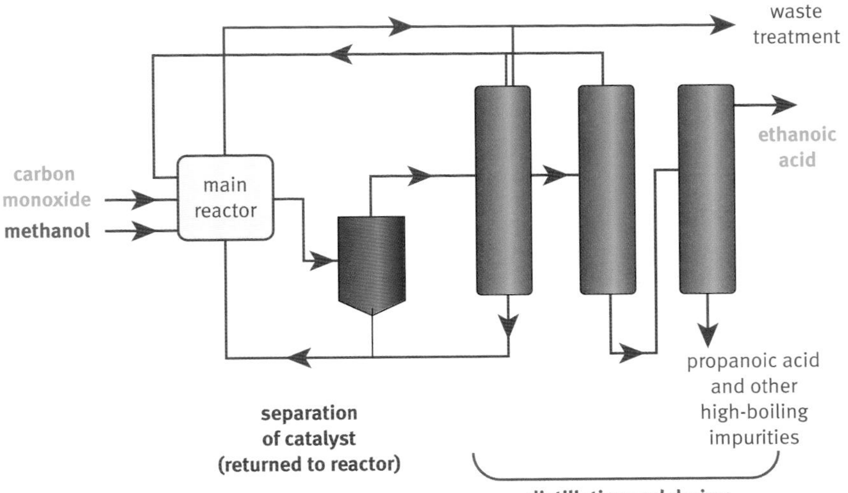

The Monsanto process for making ethanoic acid

Methanol is a cheaper feedstock than hydrocarbons from oil. The methanol is usually made from carbon monoxide and hydrogen. However, waste wood can be converted to methanol, so in time the process could be based on renewable biomass instead of on oil.

The Cativa process

In 1986, the oil company BP bought all the rights to this process from Monsanto. They now run a variant of the Monsanto process that uses a different catalyst. It is called the Cativa process. The metal iridium replaces rhodium.

The atom economy for the Cativa process is also 100%. The iridium-based catalyst is cheaper. Also there are also green benefits as BP technical manager, Mike Muskett, explains:

- 'The process is faster and so we need smaller reactors.
- 'The catalyst is even more selective which cuts the energy cost of purifying the product.
- 'We don't need as much water in the plant as with the older process, so we need less energy to dry the product.
- 'Existing plant can be converted to run the Cativa process: a converted plant can produce 75% more ethanoic acid than was previously possible using the original Monsanto process.'

Questions

1. The catalysts used to make ethanoic acid are based on three metals. Where are these metals in the periodic table? What is the relationship between the three metals?
2. What are the main reasons why the older process based on oxidation of hydrocarbons is so much less 'green' than the newer processes?
3. Why are there fewer distillation and drying steps in the Cativa process than in the Monsanto process?

C7 Chemistry for a sustainable world

Summary

Alcohols, carboxylic acids, and esters

- Chemists use molecular and structural formulae as well as models to represent organic molecules.
- Methane (CH_4), ethane (C_2H_6), propane (C_3H_8) and butane (C_4H_{10}) belong to the alkane series of hydrocarbons.
- Alkanes burn but are inert to common aqueous reagents.
- Methanol (CH_3OH) and ethanol (C_2H_5OH) are alcohols with important uses.
- The —OH functional group gives alcohols their characteristic properties.
- The physical and chemical properties of ethanol can be compared to the properties of water and ethane.
- Methanoic acid ($HCOOH$) and ethanoic acid (CH_3COOH) are carboxylic acids.
- Ethanoic acid is the acid in vinegar.
- The —COOH functional group gives organic acids their characteristic properties.
- Aqueous solutions of carboxylic acids show the characteristic reactions of acids with metals, alkalis, and carbonates.
- Alcohols react with carboxylic acids, in the presence of a strong acid catalyst, to make esters.
- Esters have fruity smells and are used in a variety of products for their odour, taste, inertness, and solvent properties.
- Chemists synthesize pure esters using techniques such as heating under reflux, distillation, and purification by treatment with reagents in a tap funnel, as well as drying.
- Fats and oils, which are esters of glycerol and fatty acids, act as an energy store in living things.
- The acids in fats are mostly saturated compounds (with single C—C bonds in the carbon chains), while the acids in vegetable oils are likely to be unsaturated (with one or more C=C bonds in the carbon chains).

Energy changes in chemistry

- Some reaction are exothermic and give out energy; others are endothermic and take in energy.
- Chemists use energy level diagrams to represent energy changes during reactions.
- Bond breaking is endothermic, while bond forming is exothermic.
- The activation energy for a reaction is the energy needed to break bonds to start a reaction.

Reversible reactions and equilibria

- Some chemical reactions are reversible.
- Reversible reactions can reach a state of dynamic equilibrium.
- In a dynamic equilibrium the forward and back reactions are going on at the same rate so that overall there is no change.
- In a solution of an acid there is a dynamic equilibrium between the ionized and un-ionized forms of the acid.
- Carboxylic acids are weak acids (only slightly ionized) while hydrochloric acid is a strong acid (fully ionized).

Analysis

- Qualitative analysis identifies the chemicals present in a sample

- Quantitative analysis measures how much of each chemical there is in a sample.
- Samples for analysis should be representative of the bulk of the material under test.
- Samples are often dissolved in a suitable solvent before analysis.
- Water is an aqueous solvent; organic solvents are non-aqueous.
- Standard procedures for collecting, storing and preparing samples help to ensure that analysis is valid.
- Methods of chromatography involve a mobile liquid or gas moving through a stationary phase which, may be a solid, or a liquid held by a solid.
- Types of chromatography include paper, thin-layer (TLC), and gas chromatography (GC).
- Reference materials can help to identify chemicals separated by chromatography.
- R_f values are a measure of the distance moved by a spot relative to the solvent front in paper or thin-layer chromatography.
- In GC the chemicals in a sample separate because their retention times differ.
- An acid–base titration, using a burette and a pipette, is a procedure for quantitative analysis.
- Concentrations of solutions can be measured in g/dm^3.
- A standard solution is one with a known concentration which can be used in a titration to find the concentration of an unknown solution.
- A standard solution can be prepared by weighing out a specimen of a suitable chemical, dissolving it in water, and then making the volume of solution up to a mark in a graduated flask.
- The mean and range of a set of repeat results for titrations with replicate samples can indicate the degree of uncertainty in the measured value.

Green chemistry

- The chemical industry makes bulk chemicals on a large scale (such as ammonia, sulfuric acid, sodium hydroxide, and phosphoric acid)
- The industry makes fine chemicals on a much smaller scale (such as drugs, food additives, and fragrances).
- Government agencies are responsible for regulating industry to protect the environment as well as the health and safety of workers in the industry, people who live nearby, and users of the products.
- Research by the industry aims to develop new products and processes that are based on renewable resources while being more efficient, in that they use less energy while producing less waste.
- The yield and atom economy of a process are measures of its efficiency.
- A chemical process involves a series of stages, including preparation of the feedstock, synthesis, separation of the products, handling of by-products and wastes, and the monitoring of purity.
- A catalyst provides an alternative route for a reaction with a lower activation energy.
- Ethanol is an example of a chemical which can be made in different ways, some of which are 'greener' than others.
- Ethanol is useful as a solvent, a fuel, and a feedstock for other processes.
- Three methods of making ethanol are: by fermentation of sugar with yeast; use of genetically modified bacteria to act on waste biomass; and by synthesis from petrochemicals.
- In the petrochemical route, ethane is first converted to ethene and then to ethanol.

Glossary

abundant Abundance measures how common an element is. Silicon is abundant in the lithosphere. Nitrogen is abundant in the atmosphere.

accuracy How close a quantitative result is to the true or 'actual' value.

acid A compound that dissolves in water to give a solution with a pH lower than 7. Acid solutions change the colour of indicators, form salts when they neutralize alkalis, react with carbonates to form carbon dioxide, and give off hydrogen when they react with a metal. An acid is a compound that contains hydrogen in its formula and produces hydrogen ions when it dissolves in water.

activation energy The minimum energy needed in a collision between molecules if they are to react. The activation energy is the height of the energy barrier between reactants and products in a chemical change.

actual yield The mass of the required chemical obtained after separating and purifying the product of a chemical reaction.

alcohols Alcohols are organic compounds containing the reactive group —OH. Ethanol is an alcohol. It has the formula C_2H_5OH

alkali A compound that dissolves in water to give a solution with a pH higher than 7. An alkali can be neutralized by an acid to form a salt. Solutions of alkalis contain hydroxide ions.

alkane Alkanes are hydrocarbons found in crude oil. All the C—C bonds in alkanes are single bonds. Ethane is an alkane. It has the formula C_2H_6.

allergy People with an allergy suffer symptoms when they eat some foods which most people find harmless. Symptoms can include itchy skin, shortness of breath or an upset stomach.

amino acid Small molecules made when proteins are digested.

antioxidant A chemical added to food to stop it going bad by reaction with oxygen in the air.

aqueous An aqueous solution is a solution in which water is the solvent.

atmosphere The layer of gases that surrounds the Earth.

atom The smallest particle of an element. The atoms of each element are the same and are different from the atoms of other elements.

atom economy A measure of the efficiency of a chemical process. The atom economy for a process shows the mass of product atoms as a percentage of the mass of reactant atoms.

attractive forces (between molecules) Forces that try to pull molecules together. Attractions between molecules are weak. Molecular chemicals have low melting points and boiling points because the molecules are easy to separate.

balanced equation An equation showing the formulae of the reactants and products. The equation is balanced when there is the same number of each kind of atom on both sides of the equation.

best estimate When you are measuring a variable, this is the value in which you have most confidence.

biosphere All the living organisms on Earth. This includes all the plants, animals, and microorganisms.

bleach A chemical that can destroy unwanted colours. Bleaches also kill bacteria. A common bleach is a solution of chlorine in sodium hydroxide.

branched chain Chain of carbon atoms with short side branches.

bulk chemicals Chemicals made by industry on a scale of thousands or millions of tonnes per year. Examples are sulfuric acid, nitric acid, sodium hydroxide, ethanol, and ethanoic acid.

burette A graduated tube with taps or valves used to measure the volume of liquids or solutions during quantitative investigations such as titrations.

by-products Unwanted products of chemical synthesis. By-products are formed by side-reactions that happen at the same time as the main reaction, thus reducing the yield of the product required.

carbohydrate A natural chemical made of carbon, hydrogen, and oxygen. The hydrogen and oxygen are present in the same proportions as in water. An example is glucose, C6H12O6. Carbohydrates includes sugars, starch, and cellulose.

carbon cycle The cycling of the element carbon in the environment between the atmosphere, biosphere, hydrosphere, and lithosphere. The element exists in different compounds in these spheres. In the atmosphere it is mainly present as carbon dioxide.

carbonate A compound which contains carbonate ions, CO_3^{2-}. An example is calcium carbonate, $CaCO_3$.

carboxylic acid Carboxylic acids are organic compounds containing the reactive group —COOH. Ethanoic acid (acetic acid) is an example. It has the formula CH_3COOH.

carrier gas The mobile phase in gas chromatography.

catalyst A chemical which speeds up a chemical reaction but is not used up in the process.

cause When there is evidence that changes in a factor produce a particular outcome, then the factor is said to cause the outcome. For example, increases in the pollen count cause increases in the incidence of hay fever.

cellulose The chemical which makes up most of the fibre in food. The human body cannot digest cellulose.

chemical change/reaction A change that forms a new chemical.

chemical equation A summary of a chemical reaction showing the reactants and products with their physical states (see balanced chemical equation).

chemical formula A way of describing a chemical that uses symbols for atoms. It gives information about the number of different types of atom in the chemical.

chemical properties A chemical property describes how an element or compound interacts with other chemicals, for example the reactivity of a metal with water.

chemical synthesis Making a new chemical by joining together simpler chemicals.

chromatography An analytical technique in which the components of a mixture are separated by the movement of a mobile phase through a stationary phase.

collision theory The theory that reactions happen when molecules collide. The theory helps to explain the factors that affect the rates of chemical change. Not all collisions between molecules lead to reaction.

combustion When a chemical reacts rapidly with oxygen, releasing energy.

compression A material is in compression when forces are trying to push it together and make it smaller.

concentration The quantity of a chemical dissolved in a stated volume of solution. Concentrations can be measured in grams per litre.

conservation of atoms All the atoms present at the beginning of a chemical reaction are still there at the end. No new atoms are created and no atoms are destroyed during a chemical reaction.

correlation When an outcome happens if a specific factor is present, but does not happen when it is absent, or if a measured outcome increases (or decreases steadily) as the value of a factor increases, we say there is a correlation between the two. For example, a matching pattern in the variation of pollen count and the incidence of hay fever is evidence of a correlation.

corrosive A corrosive chemical may destroy living tissue on contact.

covalent bonding Strong attractive forces that hold atoms together in molecules. Covalent bonds form between atoms of non-metallic elements.

cross-links Links between polymer chains.

crude oil A dark, oily liquid found in the Earth, which is a mixture of hydrocarbons.

crust (of the Earth) The outer layer of the lithosphere.

crystalline polymer A polymer with molecules lined up in a regular way as in a crystal.

density A dense material is heavy for its size. Density is mass divided by volume.

diabetes An early sign of diabetes is high levels of sugar in a person's blood. In type 1 diabetes the pancreas cannot make the insulin that helps to control sugar levels in blood. In type 2 diabetes the pancreas does not make enough insulin, or body cells do not respond normally to insulin.

digestion Breaking down large food molecules into smaller ones. This is needed so that they can pass into your blood.

dissolve Some chemicals dissolve in liquids (solvents). Salt and sugar, for example, dissolve in water.

DNA The chemical that makes up the chromosomes – deoxyribonucleic acid. DNA carries the genetic code, which controls how an organism develops.

drying agent A chemical used to remove water from moist liquids or gases. Anhydrous calcium chloride and anhydrous sodium sulfate are examples of drying agents.

durable A material is durable if it lasts a long time in use. It does not wear out.

dynamic equilibrium Chemical equilibria are dynamic. At equilibrium the forward and back reactions are still continuing but at equal rates so that there is no overall change.

E number Every food additive has an E number. E numbers show that the additive has passed safety tests and been approved for use throughout the European Union.

electrode A conductor made of a metal or graphite through which a current enters or leaves a chemical during electrolysis. Electrons flow into the negative electrode (cathode) and out of the positive electrode (anode).

electrolysis Splitting up a chemical into its elements by passing an electric current through it.

electrolyte A chemical which can be split up by an electric current when molten or in solution is the electrolyte. Ionic compounds are electrolytes.

electron configuration The number and arrangement of electrons in an atom of an element.

electrons Tiny particles in atoms. Electrons are found outside the nucleus. Electrons have negligible mass and are negatively charged, 1–.

electrostatic attraction The force of attraction between objects with opposite electric charges. A positive ion, for example, attracts a negative ion.

emissions Something given out by something else, for example, the emission of carbon dioxide from combustion engines.

emulsifier Emulsifiers are chemicals which help to mix together two liquids that would normally separate such as oil and water. In an emulsion one liquid is spread through the other in tiny droplets.

end point The point during a titration at which the reaction is just complete. For example, in an acid–alkali titration, the end point is reached when the indicator changes colour. This happens when exactly the right amount of acid has been added to react with all the alkali present at the start.

endothermic An endothermic process takes in energy from its surroundings.

energy level The electrons in an atom have different energies and are arranged at distinct energy levels.

energy level diagram A diagram to show the difference in energy between the reactants and the products of a reaction.

equilibrium A state of balance in a reversible reaction when neither the forward nor the backward reaction is complete. The reaction appears to have stopped. At equilibrium reactants and products are present and their concentrations are not changing.

esters An organic compound made from a carboxylic acid and an alcohol. Ethyl ethanoate is an ester. It has the formula $CH_3COOC_2H_5$.

exothermic An exothermic process gives out energy to its surroundings.

extraction (of metals) The process of obtaining a metal from a mineral by chemical reduction or electrolysis. It is often necessary to concentrate the ore before extracting the metal.

fat Fats are esters of glycerol with long-chain carboxylic acids (fatty acids). The fatty acids in animal fats are mainly saturated compounds.

feedstocks A chemical, or mixture of chemicals, fed into a process in the chemical industry.

fertile Soil that is fertile contains all the chemicals plants need to grow.

fertilizer A chemical or mixture of chemicals that is put on the soil to help plants grow better.

financial incentives Money which is received by (or not taken away from) a person or organization to encourage them to behave in a certain way. For example, higher car tax duty on large cars is aimed at encouraging people to buy small cars that use less petrol.

fine chemicals Chemicals made by industry in smaller quantities than bulk chemicals. Fine chemicals are used in products such as food additives, medicines, and pesticides.

flame colour A colour produced when a chemical is held in a flame. Some elements and their compounds give characteristic colours. Sodium and sodium compounds, for example, give bright yellow flames.

flexible A flexible material bends easily without breaking.

food additive A chemical that is not a nutrient but is added to food, for example to improve its appearance of make it keep longer.

food chain In the food industry this covers all stages from where the food grows, through harvesting, processing, preservation, and cooking to being eaten.

food labelling Food labelling on packages gives people information to help them decide what to buy. Labels list the ingredients. They may give a summary of the nutritional value of the food. Sometimes that includes advice about allergies.

Food Standards Agency The Food Standards Agency is an independent food safety watchdog set up by an Act of Parliament to protect the public's health and consumer interests in relation to food.

formulae (chemical) A way of describing a chemical that uses symbols for atoms. A formula gives information about the numbers of different types of atom in the chemical. The formula of sulfuric acid, for example, is H_2SO_4.

functional group A reactive group of atoms in an organic molecule. The hydrocarbon chain making up the rest of the molecule is generally unreactive with common reagents such as acids and alkalis. Examples of functional groups are —OH in alcohols and —COOH in carboxylic acids.

gemstone A crystalline rock or mineral that can be cut and polished for jewellery.

giant covalent structure A giant, three-dimensional arrangement of atoms that are held together by covalent bonds. Silicon dioxide and diamond have giant covalent structures.

giant ionic structure The structure of solid ionic compounds. There are no individual molecules, but millions of oppositely charged ions packed closely together in a regular, three-dimensional arrangement.

glycerol Glycerol is an alcohol with three —OH groups. Its chemical name is propan-1,2,3-triol. Its formula is $CH_2OH—CHOH—CH_2OH$.

granite A hard igneous rock with clearly visible crystals of various minerals.

group Each column in the periodic table is a group of similar elements.

halogens The family name of the group 7 elements.

hard A hard material is difficult to dent or scratch.

harvest Farmers harvest their ripe crops. What they gather in is their harvest.

heat under reflux Heating a reaction mixture in a flask fitted with a vertical condenser. Vapours escaping from the flask condense and flow back into the reaction mixture.

hormone A chemical messenger secreted by specialized cells in animals and plants. Hormones bring about changes in cells or tissues in different parts of the animal or plant.

hydrocarbon A compound of hydrogen and carbon only. Ethane, C_2H_4, is a hydrocarbon.

hydrogen ion A hydrogen atom that has lost one electron. The symbol for a hydrogen ion is H^+. Acids produce aqueous hydrogen ions, $H^+(aq)$, when dissolved in water.

hydrosphere All the water on Earth. This includes oceans, lakes, rivers, underground reservoirs, and rainwater.

hydroxide ion A negative ion, OH^-. Alkalis give aqueous hydroxide ions when they dissolve in water.

incinerator A factory for burning rubbish and generating electricity.

indicator A chemical that shows whether a solution is acidic or alkaline. For example, litmus turns blue in alkalis and red in acids. Universal indicator has a range of colours that show the pH of a solution.

insulin A hormone produced by the pancreas. It is a chemical which helps to control the level of sugar (glucose) in the blood.

intensive farming Modern farming methods that try to grow the maximum crop or maximum numbers of animals per area of land.

ionic bonding Very strong attractive forces that hold the ions together in an ionic compound. The forces come from the attraction between positively and negatively charged ions.

ionic compounds Compounds formed by the combination of a metal and a non-metal. They contain positively charged metal ions and negatively charged non-metal ions.

ions An electrically charged atom or group of atoms.

kidneys Organs that remove waste chemicals from your blood and excrete them in the urine.

landfill Dumping rubbish in holes in the ground.

life cycle assessment A way of analysing the production, use, and disposal of a material or product to add up the total energy and water used and the effects on the environment.

line spectrum A spectrum made up of a series of lines. Each element has its own characteristic line spectrum.

lithosphere The rigid outer layer of the Earth, made up of the crust and the part of mantle just below it.

locating agent A chemical used to show up colourless spots on a chromatogram.

long-chain molecule Polymers are long-chain molecules. They consist of long chains of atoms.

macroscopic Large enough to be seen without the help of a microscope.

mantle The layer of rock between the crust and the outer core of the Earth. It is approximately 2900 kilometers thick.

material The polymers, metals, glasses, and ceramics that we use to make all sorts of objects and structures.

mean value A type of average, found by adding up a set of measurements and then dividing by the number of measurements. You can have more confidence in the mean of a set of measurements than in a single measurement.

measurement uncertainty Variations in analytical results owing to factors that the analyst cannot control. Measurement uncertainty arises from both systematic and random errors.

metal Elements on the left side of the periodic table. Metals have characteristic properties: they are shiny when polished and they conduct electricity. Some metals react with acids to give salts and hydrogen. Metals are present as positive ions in salts.

metal hydroxide A compound consisting of metal positive ions and hydroxide ions. Examples are sodium hydroxide, NaOH, and magnesium hydroxide, $Mg(OH)_2$.

metal oxide A compound of a metal with oxygen.

metallic bonding Very strong attractive forces that hold metal atoms together in a solid metal. The metal atoms lose their outer electrons and form positive ions. The electrons drift freely around the lattice of positive metal ions and hold the ions together.

mineral A naturally occurring element or compound in the Earth's lithosphere.

mobile phase The solvent that carries chemicals from a sample through a chromatographic column or sheet.

molecular models Models to show the arrangement of atoms in molecules, and the bonds between the atoms.

molecule A group of atoms joined together. Most non-metals consist of molecules. Most compounds of non-metals with other non-metals are also molecular.

molten A chemical in the liquid state. A chemical is molten when the temperature is above is melting point but below its boiling point.

nanometre A unit of length 1 000 000 000 times smaller than a metre.

natural A material that occurs naturally but may need processing to make it useful, such as silk, cotton, leather, and asbestos.

natural cycle (in the environment) The cycling of an element between the atmosphere, hydrosphere, lithosphere, and biosphere as a result of natural processes.

natural resource A resource which exists naturally. It is not artificial. Examples include air, water, wood, crude oil, and metal ores.

negative ion An ion that has a negative charge (an anion).

neutralization A reaction in which an acid reacts with an alkali to form a salt. During neutralization reactions, the hydrogen ions in the acid solution react with hydroxide ions in the alkaline solution to make water molecules.

neutrons An uncharged particle found in the nucleus of atoms. The relative mass of a neutron is 1.

nitrogen cycle The continual cycling of nitrogen, which is one of the elements that is essential for life. By being converted to different chemical forms, nitrogen is able to cycle between the atmosphere, lithosphere, hydrosphere, and biosphere.

non-aqueous A solution in which a liquid other than water is the solvent.

nucleus The tiny central part of an atom (made up of protons and neutrons). Most of the mass of an atom is concentrated in its nucleus.

obesity People are obese if they have put on so much weight that their health is in danger.

ore A natural mineral that contains enough valuable minerals to make it profitable to mine.

organic chemistry The study of carbon compounds. This includes all the natural carbon compounds from living things and synthetic carbon compounds.

organic farm Farming using natural fertilizers and limited use of pesticides and herbicides.

outlier A measured result that seems very different from other repeat measurements, or from the value you would expect, which you therefore strongly suspect is wrong.

oxidation A reaction that adds oxygen to a chemical.

oxide A compound of an element with oxygen.

pancreas An organ in the body which produces some hormones and digestive enzymes.

percentage yield A measure of the efficiency of a chemical synthesis.

$$\text{percentage yield} = \frac{\text{actual yield}}{\text{theoretical yield}}$$

period In the context of chemistry, a row in the periodic table.

periodic In chemistry, a repeating pattern in the properties of elements. In the periodic table one pattern is that each period starts with metals on the left and ends with non-metals on the right.

pest Any living thing that damages crops or animals that are grown for food or other human needs.

pesticide Any chemical used to kill or control pests.

petrochemical plant A factory for making chemicals from crude oil (petroleum) or natural gas. The products are petrochemicals.

photosynthesis A chemical reaction that happens in green plants using the energy in sunlight. The plant takes in water and carbon dioxide, and uses sunlight to convert them to glucose (a nutrient) and oxygen.

pH scale A number scale that shows the acidity or alkalinity of a solution in water.

physical properties Properties of elements and compounds such as melting point, density, and electrical conductivity. These are properties that do not involve one chemical turning into another.

pipette A pipette is used to measure small volumes of liquids or solutions accurately. A pipette can be used to deliver the same fixed volume of solution again and again during a series of titrations.

plasticizer A chemical added to a polymer to make it more flexible.

polymer A material made up of very long molecules. The molecules are long chains of smaller molecules.

polymerize The joining of lots of small molecules into a long chain forms a polymer.

positive ions Ions that have a positive charge (cations).

precautionary principle Take steps to minimize the risks associated with specific human actions when no one knows how serious they are.

precision A measure of the spread of quantitative results. If the measurements are precise all the results are very close in value.

preservative A chemical added to food to stop it going bad.

primary pollutants A harmful chemical that human activity adds directly to the atmosphere.

products The new chemicals formed during a chemical reaction.

properties Physical or chemical characteristics of a chemical. The properties of a chemical are what make it different from other chemicals.

protein Nutrients that your body needs to make new cells.

proton number The number of protons in the nucleus of an atom (also called the atomic number). In an uncharged atom this also gives the number of electrons.

protons Tiny particles that are present in the nuclei of atoms. Protons are positively charged, 1+.

qualitative Qualitative analysis is any method for identifying the chemicals in a sample. Thin-layer chromatography is usually a qualitative method of analysis.

quantitative Quantitative analysis is any method for determining the amount of a chemical in a sample. An acid–base titration is an example of quantitative analysis.

quartz A crystalline form of silicon dioxide, SiO_2.

range The difference between the highest and the lowest of a set of measurements.

rate of reaction A measure of how quickly a reaction happens. Rates can be measured by following the disappearance of a reactant or the formation of a product.

reactants The chemicals on the left-hand side of an equation. These chemicals react to form the products.

reactants The chemicals that react together in a chemical reaction.

reacting mass The masses of chemicals that react together, and the masses of products that are formed. Reacting masses are calculated from the balanced symbol equation using relative atomic masses and relative formula masses.

reactive metal A metal with a strong tendency to react with chemicals such as oxygen, water and acids. The more reactive a metal, the more strongly it joins with other elements such as oxygen. So reactive metals are hard to extract from their ores.

real difference You can be sure that the difference between two mean values is real if their ranges do not overlap.

recycling A range of methods for making new materials from materials that have already been used.

reducing agent A chemical that removes oxygen from another chemical. For example, carbon acts as a reducing agent when it removes oxygen from a metal oxide. The carbon is oxidized to carbon monoxide during this process.

reduction A reaction that removes oxygen from a chemical. For example, some metal oxides can be reduced to metals by heating them with carbon.

reference materials Known chemicals used in analysis for comparison with unknown chemicals.

regulations Rules that can be enforced by an authority, for example government. The law says that all vehicles that are three years or more old must have an annual exhaust emission test is a regulation that helps to reduce atmospheric pollution.

relative atomic mass The mass of an atom of an element compared to the mass of an atom of carbon. The relative atomic mass of carbon has been defined as 12.

relative formula mass The combined relative atomic masses of all mass the atoms in a formula. To find the relative formula mass of a chemical, you just add up the relative atomic masses of the atoms in the formula.

renewable resource Resources that can be replaced as quickly as they are used. An example is wood from the growth of trees.

replicate sample Two or more samples taken from the same material. Replicate samples should be as similar as possible and analysed by the same procedure to help judge the precision of the analysis.

representative sample A sample of a material that is as nearly identical as possible in its chemical composition to that of the larger bulk of material sampled.

retardation factor A retardation factor, R_f, is a ratio used in paper or thin-layer chromatography. If the conditions are kept the same, each chemical in a mixture will move a fixed fraction of the distance moved by the solvent front. The R_f value is a measure of this fraction.

retention time In chromatography, the time it takes for a component in a mixture to pass through the stationary phase.

reversible processes A change which can go forwards or backwards depending on the conditions. Many reversible processes can reach a state of equilibrium.

risk A possible hazard that might result from something that happens.

risk assessment A check on the hazards involved in a scientific procedure. A full assessment include the steps to be taken to avoid or reduce the risks from the hazards identified.

risk factor A variable linked to an increased risk of disease. Risk factors are linked to disease but may not be the cause of the disease.

salt An ionic compound formed when a metal reacts with a non-metal or when an acid neutralizes an alkali.

sample A small portion collected from a larger bulk of material for laboratory analysis (such as a water sample or a soil sample).

sandstone A rock made of sand grains stuck together.

saturated In the molecules of a saturated compound, all of the bonds are single bonds. The fatty acids in animal fats are all saturated compounds.

scale up To redesign a synthesis to produce a chemical in larger amounts. A process might be scaled up first from a laboratory method to a pilot plant; then from a pilot plant to a full-scale industrial process.

secondary pollutants A harmful chemical formed in a atmosphere by reactions involving other pollutants.

shell A region in space (around the nucleus of an atom) where there can be electrons.

small molecules Particles of chemicals that consist of small numbers of atoms bonded together. Chemicals made up of one or more non-metallic element and which have low boiling and melting points consist of small molecules.

soft A soft material is easy to dent or scratch.

solvent front The furthest position reached by the solvent during paper or thin-layer chromatography

spectroscopy The use of instruments to produce and analyse spectra. Chemists use spectroscopy to study the composition, structure and bonding of elements and compounds.

stabilizer A food additive which helps to keep ingredients evenly and smoothly mixed.

starch A type of carbohydrate found in bread, potatoes, and rice. Plants produce starch to store the food they make by photosynthesis. Starch molecules are long chains of glucose molecules.

stationary phase The medium through which the mobile phase passes in chromatography.

stiff A stiff material is difficult to bend or stretch.

strong A strong material is hard to pull apart or crush.

strong acid A strong acid is fully ionized to produce hydrogen ions when it dissolves in water.

sugar A carbohydrate that tastes sweet and is soluble in water. Common sugars are table sugar (sucrose), milk sugar (lactose), and the sugar made by photosynthesis (glucose).

surface area (of a solid chemical) The area of a solid in contact with other reactants that are liquids or gases.

sustainable Using the Earth's resources in a way that can continue in future, rather than destroying them.

sustainable development A plan for meeting people's present needs without spoiling the environment for the future.

synthetic A material made by a chemical process, not naturally occurring.

technological development An advance in tools and devices, for example the changes to modern car engines that make them more efficient.

tension A material is in tension when forces are trying to stretch it or pull it apart.

theoretical yield The amount of product that would be obtained in a reaction if all the reactants were converted to products exactly as described by the balanced chemical equation.

titration An analytical technique used to find the exact volumes of solutions that react with each other.

toxic A chemical which may lead to serious health risks, or even death, if breathed in, swallowed or taken in through the skin.

toxin A poisonous chemical produced by a microorganism, plant, or animal.

trends A description of the way a property increases or decreases along a series of elements or compounds which is often applied to the elements (or their compounds) in a group or period.

unsaturated There are double bonds in the molecules of unsaturated compounds. There is no spare bonding. The fatty acids in vegetable oils include a high proportion of unsaturated compounds.

urea A chemical made in the liver when amino acids are broken down. Urea is excreted in the kidneys.

vegetable oil Vegetable oils are esters of glycerol with fatty acids (long-chain carboxylic acids). More of the fatty acids in vegetable oils are unsaturated when compared with the fatty acids in animal fats.

vinegar A sour-tasting liquid used as a flavouring and to preserve foods. It is a dilute acetic (ethanoic) acid made by fermenting beer, wine, or cider.

vulcanization A process for hardening natural rubber by making cross-links between the polymer molecules.

weak acids Weak acids are only slightly ionized to produce hydrogen ions when they dissolve in water.

weather Atmospheric conditions, including temperature, wind (air movements), and rain. Air quality is affected by changes in these conditions.

weathering Chemical changes of the minerals in rocks caused by reactions with air and water.

word equation A summary in words of a chemical reaction.

yield The crop yield is the amount of crop that can be grown per area of land.

Index

Publisher's acknowledgements

Oxford University Press wishes to thank the following for their kind permission to reproduce copyright material:

p10 ESA/ PLI/Corbis UK Ltd.; **p12r** John Wilkinson/Ecoscene/Corbis UK Ltd.; **p12l** Harvey Pincis/Science Photo Library; **p15** NETCEN; p20tl Tek Image/ Science Photo Library; **p20tr** Raoux John/Orlando Sentinel/Sygma/Corbis UK Ltd.; **p22** Nick Hawkes; Ecoscene/Corbis UK Ltd.; **p23** Charles D. Winters/ Science Photo Library; **p26b** David Scharf/Science Photo Library; **p26tl** Dr Jeremy Burgess/Science Photo Library; **p26tr** Burkard Manufacturing Co. Limited; **p27** Wellcome Trust; **p28** Sipa Press (SIPA)/Rex Features; **p29** Medical-on-Line; **p30** Caroline Penn/Corbis UK Ltd.; **p31b** Jim Winkley/ Corbis UK Ltd.; **p31t** Martin Bond/Science Photo Library; **p32b** NASA/Zooid Pictures; **p32t** Hulton-Deutsch Collection/Corbis UK Ltd.; **p33** David Townend/Photofusion Picture Library/Alamy; **p38b** Neil Rabinowitz/Corbis UK Ltd.; **p38t** Oxford University Press; **p39** David Muscroft/Superstock Ltd.; **p40bl** Taryn Cass/Zooid Pictures; **p40br** Yves Forestier/SYGMA/Corbis UK Ltd.; **p40tc** David Constantine/Science Photo Library; **p40tl** PhotoCuisine/ Corbis UK Ltd.; **p40tr** Empics; **p41bc** Alexis Rosenfeld/Science Photo Library; **p41bl** K.M. Westermann/Corbis UK Ltd.; **p41br** Bernardo Bucci/ Corbis UK Ltd.; **p41tc** Dennis Gilbert/VIEW Pictures Ltd/Alamy; **p41tl** David Keith Jones/Images of Africa Photobank/Alamy; **p41tr** Tom Tracy Photography/ Alamy; **p42b** John Cleare Mountain Camera; **p42t** Janine Wiedel Photolibrary/Alamy; **p43b** J & P Coats Ltd; **p43t** Instron® Corporation; **p44cl** Andrew Syred/Science Photo Library; **p44cr** Eye Of Science/Science Photo Library; **p44l** Steve Prezant/Corbis UK Ltd.; **p45** Dr Tim Evans/Science Photo Library; **p47** Science & Society Picture Library; **p48l** Dan Sinclair/Zooid Pictures; p48r Taryn Cass/Zooid Pictures; p50 Tim Pannell/Corbis UK Ltd.; **p51l** Zooid Pictures; **p51r** ABACA/Empics; **p52** W. L. Gore & Associates, Ltd.; **p53b** Eye Of Science/Science Photo Library; **p53t** Du Pont (UK) Ltd; **p54b** Corbis UK Ltd.; **p54t** James L. Amos/Corbis UK Ltd.; **p57** David Hoffman Photo Library/Alamy; **p58c** G P Bowater/Alamy; **p58b** Geoff Tompkinson/ Science Photo Library; **p58t** Pictor International/ImageState/ Alamy; **p59** Ken Hawkins/Focus Group, LLC/Alamy; **p60b** Avecia Ltd; **p60t** Avecia Ltd; **p66t** John James/Alamy; **p66c** Richard Morrell/Corbis UK Ltd.; **p66b** Jason Ingram/ Alamy; **p67tl** Gideon Mendel/Corbis UK Ltd.; **p67tr** Corbis UK Ltd.; **p67b** gkphotography/Alamy; **p69bl** Nigel Cattlin/Holt Studios International Ltd/Alamy; **p69br** Peter Dean/Agripicture Images/ Alamy; **p69t** Peter Dean/ Agripicture Images/Alamy; **p70** Wolfgang Kaehler/ Corbis UK Ltd.; **p72l** Ed Bock/Corbis UK Ltd.; **p72t** Nigel Cattlin/Holt Studios International Ltd/Alamy; **p73** geogphotos/Alamy; **p74c** John Garrett/Corbis UK Ltd.; **p74t** Paul Glendell/Alamy; **p74** Nic Hamilton/Alamy; **p75** Soil Association; **p76b** foodfolio/Alamy; **p76t** Zooid Pictures; **p77t** Adrienne Hart-Davis/Science Photo Library; **p77b** Zooid Pictures; **p80t** George McCarthy/Corbis UK Ltd.; **p80bc** Nigel Cattlin/Holt Studios International Ltd/Alamy; **p80br** Taryn Cass/Zooid Pictures; **p80bl** F. Waliyar/International Crops Research Institute for the Semi-Arid Tropics; **p81b** Lavendelfoto/ INTERFOTO Pressebildagentur/ Alamy; **p81t** Sheila Terry/Science Photo Library; **p82** Niehoff/imagebroker/ Alamy; **p83** Mark Harmel/Alamy; **p84** Richard Eaton/Photofusion Picture Library/Alamy; p86 Vincent Kessler/Reuters/Corbis UK Ltd.; **p88** Nigel Cattlin/Holt Studios International Ltd/Alamy; **p94b** Brijesh Singh/Reuters/ Corbis UK Ltd.; **p94t** Andrew Lambert Photography; **p96b** Andrew Lambert Photography; **p96t** Andrew Lambert Photography/Science Photo Library; **p97** Charles D. Winters/Science Photo Library; **p100** Herve Berthoule/Jacana/ Science Photo Library; **p101** Andrew Lambert Photography; **p102tl** William B. Jensen/Oesper Collections: University of Cincinnati.; **p102r** Dept. of Physics, Imperial College/Science Photo Library; **p102bl** Martyn F. Chillmaid; **p103** Roger Ressmeyer/Corbis UK Ltd.; **p105** David Parker/Science Photo Library; **p106** Dept. Of Physics/Imperial College/Science Photo Library; **p110c** Josè Manuel Sanchis Calvete/Corbis UK Ltd.; **p110b** Dirk Wiersma/ Science Photo Library; **p110tcl** Andrew Lambert Photography; **p110tcr** Charles D. Winters/Science Photo Library; **p110tl** Arnold Fisher/Science Photo Library; **p110tr** Andrew Lambert Photography; **p111b** Charles D. Winters/Science Photo Library; **p111t** Andrew Lambert Photography; **p112** Science Photo Library; **p116** Charles D. Winters/Science Photo Library; **p117b** Martyn F. Chillmaid; **p117c** Arnold Fisher/Science Photo Library; **p117t** Andrew Lambert Photography; **p125b** Adam Hart-Davis/Science Photo Library; **p125bc** Nina Towndrow/Nuffield Curriculum Centre; **p125t** Adam Hart-Davis/Science Photo Library; **p125tc** Nina Towndrow/Nuffield Curriculum Centre; **p126** Galen Rowell/Corbis UK Ltd.; **p128tl** George Bernard/Science Photo Library; **p128bl** Roberto de Gugliemo/Science Photo Library; **p128c** Josè Manuel Sanchis Calvete/Corbis UK Ltd.; **p128r** Josè Manuel Sanchis Calvete/Corbis UK Ltd.; **p129r** Peter Falkner/Science Photo Library; **p129l** Arnold Fisher/ Science Photo Library; **p130bl** Left Lane Productions/Corbis UK Ltd.; **p130r** John Mead/Science Photo Library; **p130tl** Sinclair Stammers/Science Photo Library; **p131** Mike Widdowson; **p132b** Nina Towndrow/Nuffield Curriculum Centre; **p132t** Nina Towndrow/Nuffield Curriculum Centre; **p133b** Sciencephotos/Alamy; **p133t** Nina Towndrow/ Nuffield Curriculum Centre; **p136b** James L. Amos/Corbis UK Ltd.; **p136t** Layne Kennedy/Corbis UK Ltd.; **p137** Kevin Fleming/Corbis UK Ltd.; **p142tl** Alexis Rosenfeld/Science Photo Library; **p142tr** John Van Hasselt SYGMA/Corbis UK Ltd.; **p142bl** H. David Seawell/Corbis UK Ltd.; **p142br** Charles E. Rotkin/Corbis UK Ltd.; **p145** Nik Wheeler/Corbis UK Ltd.; **p150** Maximilian Stock Ltd/Science Photo Library; **p151l** Geoff Tompkinson/ Science Photo Library; **p151r** William Taufic/Corbis UK Ltd.; **p152bl** Martyn F. Chillmaid; **p152br** Martyn F. Chillmaid; **p152tl** Dave Bartruff/Corbis UK Ltd.; **p152tr** Andrew Lambert Photography; **p153r** Lurgi Metallurgie/ Outokumpu; **p153c** Martyn F. Chillmaid; p153l Martyn F. Chillmaid; **p154** Zooid Pictures; **p155** Andrew Lambert Photography/Science Photo Library; **p158bl** Zooid Pictures; **p158br** Zooid Pictures; **p158tl** Zooid Pictures; **p158tr** Richard Megna/ Fundamental Photos/Science Photo Library; **p159** BSIP/Beranger/Science Photo Library; p160 Martyn F. Chillmaid; **p161tl** Martyn F. Chillmaid; **p161tr** Martyn F. Chillmaid; **p161bl** Martyn F. Chillmaid; **p161tr** Martyn F. Chillmaid; **p162** Peter Bowater/Alamy; **p166** Gary Banks/BP Saltend; **p170** Holt Studios International; **p173** Sidney Moulds/Science Photo Library; **p178b** Corbis UK Ltd, **p178t** Christine Osborne/CORBIS; **p181** Andrew Lambert Photography/Science Photo Library; **p184** Imagebroker/Alamy; **p194t&b** Zooid Pictures; **p196** Corbis UK Lrd; **p197b** Getty Images; **p198** Corbis UK Ltd; **p200** Martin Bond/Science Photo Library; **p204b** Crown Copyright Health & Safety Laboratory/Science Photo Library; **p205** Corbis UK Ltd; **p207** Science Photo Library; **p211** National Portrait Gallery; **p212l** Zooid Pictures, **p212r** Lou Chardonnay/Corbis UK Ltd.; **p213** Zooid Pictures; **p214t** Maximilian Stock Ltd/Science Photo Library, **p214b** Photo courtesy of LGC; **p215t** Pullman, **p215b** Nick Laham/Allsport/ Getty Images; **p216** Bayer AG; **p217** www.ars.usda.gov; **p218t&b** Zooid Pictures; **p219l** Adrian Arbib/Corbis UK Ltd., **p219r** Zooid Pictures; **p220** Environmental Health Department, London Borough of Camden/ Oxford University Press; **p221l** BBC Photograph Library, **p221r** David Stoecklein/ Corbis UK Ltd.; **p224** Analtech Inc.; **p225** Analtech Inc.; **p226l&r** CAMAG; **p228** Wellcome Trust; **p229** James Holmes/Thomson Laboratories/Science Photo Library; **p234t&b** Anna Grayson; **p235t&b** Anna Grayson; **p236t&b** Anna Grayson; **p240** Maximilian Stock Ltd/Science Photo Library; **p241** R. Estall/Robert Harding Picture Library Ltd/Alamy; **p243t** Steve Bicknell/The Steve Bicknell Style Library/Alamy, **p243b** William Taufic/Corbis UK Ltd.; **p244b** Laurance B. Aiuppy/Stock Connection/Alamy; **p245** Du Pont (UK) Ltd; **p248** Zooid Pictures; **p250** Nigel Cattlin/Holt Studios International; **p251** Alamy; **p254t** Leslie Garland/Leslie Garland Picture Library/Alamy, **p254b** Huntsman Tioxide, **p255** Tony Jones/Huntsman Tioxide.

Illustrations by IFA Design, Plymouth, UK and Clive Goodyer.

OXFORD
UNIVERSITY PRESS

Great Clarendon Street, Oxford OX2 6DP

Oxford University Press is a department of the University of Oxford.
It furthers the University's objective of excellence in research, scholarship,
and education by publishing worldwide in

Oxford New York

Auckland Cape Town Dar es Salaam Hong Kong Karachi
Kuala Lumpur Madrid Melbourne Mexico City Nairobi
New Delhi Shanghai Taipei Toronto

With offices in

Argentina Austria Brazil Chile Czech Republic France Greece
Guatemala Hungary Italy Japan Poland Portugal Singapore
South Korea Switzerland Thailand Turkey Ukraine Vietnam

First published 2006

British Library Cataloguing in Publication Data

Data available

ISBN-13: 978-0-19-915050-2
ISBN-10: 0-19-915050-8

10 9 8 7 6 5 4

Typeset by IFA Design Ltd, Plymouth, UK; Q2A Design, Delhi, India; and Oxford University Press

Printed in Thailand by Imago

Project Team acknowledgements

These resources have been developed to support teachers and students undertaking a new OCR suite of GCSE science specifications, *Twenty First Century Science*.

Many people from schools, colleges, universities, industry, and the professions have contributed to the production of these resources. The feedback from over 75 Pilot Centres was invaluable. It led to significant changes to the course specifications, and to the supporting resources for teaching and learning.

The University of York Science Education Group (UYSEG) and Nuffield Curriculum Centre worked in partnership with an OCR team led by Mary Whitehouse, Elizabeth Herbert, and Emily Clare to create the specifications, which have their origins in the *Beyond 2000* report (Millar & Osborne, 1998) and subsequent Key Stage 4 development work undertaken by UYSEG and the Nuffield Curriculum Centre for QCA. Bryan Milner and Michael Reiss also contributed to this work, which is reported in: *21st Century Science GCSE Pilot Development: Final Report* (UYSEG, March 2002).

Sponsors

The development of *Twenty First Century Science* was made possible by generous support from:

- The Nuffield Foundation
- The Salters' Institute
- The Wellcome Trust